Advances in Cement Composite Materials

Advances in Cement Composite Materials

Editor

Tomáš Dvorský

MDPI

Basel • Beijing • Wuhan • Barcelona • Belgrade • Novi Sad • Cluj • Manchester

Editor
Tomáš Dvorský
Department of
Environmental Engineering
VSB–Technical University of Ostrava
Ostrava
Czech Republic

Editorial Office
MDPI
St. Alban-Anlage 66
4052 Basel, Switzerland

This is a reprint of articles from the Special Issue published online in the open access journal *Buildings* (ISSN 2075-5309) (available at: www.mdpi.com/journal/buildings/special_issues/Cement_Composite_Materials).

For citation purposes, cite each article independently as indicated on the article page online and as indicated below:

Lastname, A.A.; Lastname, B.B. Article Title. *Journal Name* **Year**, *Volume Number*, Page Range.

ISBN 978-3-7258-0714-7 (Hbk)
ISBN 978-3-7258-0713-0 (PDF)
doi.org/10.3390/books978-3-7258-0713-0

Contents

About the Editor

Tomáš Dvorský

Tomas Dvorsky is an Associate Professor at the Department of Environmental Engineering, Faculty of Mining and Geology, VSB-Technical University of Ostrava. His area of research interest is the use of industrial waste as a secondary raw material in the segment of cement composites. His scientific and research activities are very closely linked with industrial partners, with whom he tries to apply new solutions in practice. In the past, he has been involved in solving challenges in the use of waste hard PUR foam or cellulose fibers in cement composites.

He is currently focusing on the use of other emerging industrial wastes in the construction segment. His work has been recognized several times at international invention shows, where he has been awarded gold, silver, and bronze medals.

Preface

Concrete, in the sense of a cement composite consisting of filler, binder, and water, is one of the main building materials of the past as well as the modern history of mankind. At present, the area of cement composites, or concrete, is going through a period of relatively rapid development. The efforts of researchers around the world are focused on the development of new additives and admixtures that affect the final properties of fresh and hardened concrete. Another relatively strong group are scientists dealing with the replacement of natural aggregates with various secondary materials or by-products (i.e., waste materials) from industrial production to protect the natural resources of aggregates. The resulting cement composites use the historically proven properties of concrete, but, at the same time, this new material has added value in the form of reduced waste production. When suitable components are used, so-called eco-friendly cement composites with minimal negative impact on the environment can be produced.

The field of cement composites is very broad, and this Special Issue provided a venue for researchers to present their latest findings.

Tomáš Dvorský
Editor

buildings

Article

Influence of Superabsorbent Polymer in Self-Compacting Mortar

Michel Henry Bacelar de Souza [1], Lucas Ramon Roque Silva [1], Vander Alkmin dos Santos Ribeiro [2,*], Paulo César Gonçalves [2], Mirian de Lourdes Noronha Motta Melo [1], Carlos Eduardo Marmorato Gomes [3] and Valquíria Claret dos Santos [2]

1 Institute of Mechanical Engineering, Unifei-Federal University of Itajubá, Av. BPS, 1303, Itajubá 37500-903, MG, Brazil; michelhenry@unifei.edu.br (M.H.B.d.S.); lucasramonroque@gmail.com (L.R.R.S.); mirianmottamelo@unifei.edu.br (M.d.L.N.M.M.)
2 Institute of Natural Resources, Federal University of Itajubá (UNIFEI), Itajubá 37500-903, MG, Brazil; paulocg9@unifei.edu.br (P.C.G.); valquiria@unifei.edu.br (V.C.d.S.)
3 Department of Construction, State University of Campinas (UNICAMP), Campinas 13083-970, SP, Brazil; cemgomes@unicamp.br
* Correspondence: vanderalkmin@gmail.com

Abstract: Self-compacting concrete (SCC) is an innovative type of concrete that does not require vibration for compaction; however, it needs attention in relation to the control of thermally generated cracks, due to the hydration heat generated mainly during the curing process in pieces with large concrete volume. In this study we investigated the addition of Superabsorbent Polymers (SAP) as internal curing elements in self-compacting mortar (SCM), as well as its thermal and mechanical characteristics, looking to obtain the optimal proportion of materials in a way that is useful as the basis of self-compacting concrete use, focusing on large volume structures. This work stands out for studying an alternative for shrinkage control, in an unconventional cementitious composite, highlighting the thermal analysis of the mixture. In the experimental program, laboratory trials were conducted using self-compacting mortar with the addition of 0%, 0.1%, and 0.2% of SAP either for CPII-E-32 and CPV-ARI and with samples undergoing both dry and submerged curing. Among the results obtained, the reduction of variation in temperature in samples with added SAP stand out, and it is possible to presume it as being a viable way of mitigating the temperature spikes in large volume concrete parts. By contrast, the compression and tensile tests indicated a reduced strength, except in the tensile trial on the CPII-E-32, which the addition of SAP resulted in an increase in strength.

Keywords: self-compacting mortar; self-compacting concrete; superabsorbent polymer; SAP; internal curing; heat of hydration

check for updates

Citation: de Souza, M.H.B.; Silva, L.R.R.; Ribeiro, V.A.d.S.; Gonçalves, P.C.; Melo, M.d.L.N.M.; Gomes, C.E.M.; dos Santos, V.C. Influence of Superabsorbent Polymer in Self-Compacting Mortar. *Buildings* **2023**, *13*, 1640. https://doi.org/10.3390/buildings13071640

Academic Editor: Tomáš Dvorský

Received: 29 May 2023
Revised: 22 June 2023
Accepted: 26 June 2023
Published: 28 June 2023

1. Introduction

The construction industry went through adaptations and updates in the last decades, presenting a growing technological advance aimed at the sector, either as the introduction of new material or even new building techniques. These factors allow for a better workability in a shorter executing time, which in many cases results in saving in labor and costs [1].

The self-compacting concrete (SCC) is an example of that, being an innovative type of concrete that does not require vibration to compact. With concrete being the most used material in building around the world, the evolution of this cement composite significantly impacted the development of new engineering projects [2].

Several research studies present promising materials to be utilized to control such harmful events; some of these materials, which are currently being studied, are known as superabsorbent polymers (SAP) [3]. It is a material that is becoming a potential point of technological innovation in cement internal curing and hydration. These are materials with a high capacity of water absorption, acting as an internal reservoir to not only improve the workability, but also reduce the water/cement ratio [4].

In the last two decades, there was an increase in the number of studies proposing the utilization of SAP residues in distinct types of cement blends, analyzing the way that they can positively influence the limitation of the effects of retraction, help the internal curing process, and improve the workability [5]. Highlights of such studies led to the belief that the pathological manifestations commonly found in these elements will be better countered. Still, there are few studies that refer to the influence of adding SAP in other cement-based materials, such as self-compacting concrete [6].

Studies that relate the development of concrete as a technology alongside with environmental solutions, are of extreme relevance for the ecological and responsible development of the building industry. These materials that are among the most used in the sector are essential in most ventures.

To ensure the development of the physical characteristics of concrete or hardened mortar, such as mechanical strength and durability, it is necessary to provide adequate conditions of temperature and humidity to material, especially in the fresh and early-age stages. This curing process aims to optimize the hydration of the cement and reduce its retraction [7].

Big concrete structures, such as dams and foundation blocks, can be susceptible to cracks, particularly in the early-age of the Portland cement due to the thermic tensions and the induction of autogenous contraction [8].

The making of high-volume concrete parts is a challenge to technologists, as these present significant increases in temperature that can result in damaging effects to the structure [9]. In these structures, the heat is generated by the hydration reaction, particularly by the cement reaction, while the temperature increase happens because the heat generation rate is superior to the dissipation rate to the environment due to the large dimensions of the parts and the thermic characteristics of the concrete [10].

According to [11], the main hydration reaction, which generates the largest amount of heat, usually happens in the initial hours of the mixing of the cement material with water, usually in the first 72 h. During the hydration phase and the concrete hardening, the structural element also exchanges heat with the environment, to the point which its internal temperature tends to equalize with the ambient temperature beginning by the structure external surface [9]. In this stage, the risk of cracking is increased due to the element of the concrete that was compressed and expanded in the early-age now contracting, often unevenly, creating tension to traction gradients, at the same time that there is little tensile strength to endure the demand [12]. The overdone thermic evolution can lead to a temperature range that causes destructive fissures in dams.

The use of SAP with self-compacting cement and mortar is underexplored; some studies have proven that it has advantages in the utilization of conventional cement composites, being of significant help to counter retractions in the curing process [13]. In [14], SAP in two diameters (600 μm and 800 μm) was added to self-compacting cement pastes, where increases in compressive strength of 35.2% and 34.3%, respectively, were obtained. The works of [15–18] are some examples of tests of the efficiency of SAPs as an internal curing agent to mitigate autogenous shrinkage in mortars. In parallel, studies such as [19,20] showed the addition of residues of other types of polymers in cementitious composites that also obtained gains in compressive strength.

The aim of this study is to analyze the addition of superabsorbent polymers (SAP) as the element of internal curing in self-compacting mortar (SCM), as well as its mechanical and thermic characteristics, and its usability in large volume structure, bearing in mind the production of self-compacting concrete. The SCM mix was made using type II and type V cement, with one reference mix and two other mixes with different percentages of SAP for each type of cement. The influence of SAP is checked in the submerged and dry curing process, the thermic behavior, and the mechanical and physical characteristics of the SCM with SAP in the fresh and hardened states.

This work stands out for studying an alternative to the control of retraction, analyzing the heat of hydration in SCM with the addition of SAP as internal curing agent, which may ensure the longer lifespan and reduce the probability of fissures. To facilitate satisfactory results in self-compacting concretes, the dosing of mortar is performed. In the mortar study, it is possible to determine if the material presents excessive fluidity, low viscosity, separation, and exudation, and by dosing the mortar, the material ratio is designed in a way that does not reduce the SCC workability.

2. Materials and Methods

The methodological procedure of this article consists in the creation of reference self-compacting mortar mixes, using CPII and CPV without the addition of SAP, adjusted mixes of SCM with added SAP, performing the tests in the fresh state, waiting for the curing time (dry and wet), and performing the test in the hardened state. The two types of cement belong to a class with high initial resistance; therefore, they need greater attention to prevent the appearance of shrinkage of thermal origin. According to [21,22], the proportion of SAP additions that have shown favorable results over the years ranges from 0.1% to 1%. However, as this study proposes the use of SAP in conjunction with a non-conventional cementitious compound, lower percentages of the polymer (0.1% and 0.2%) were chosen. These steps are shown in Figure 1, through a flowchart.

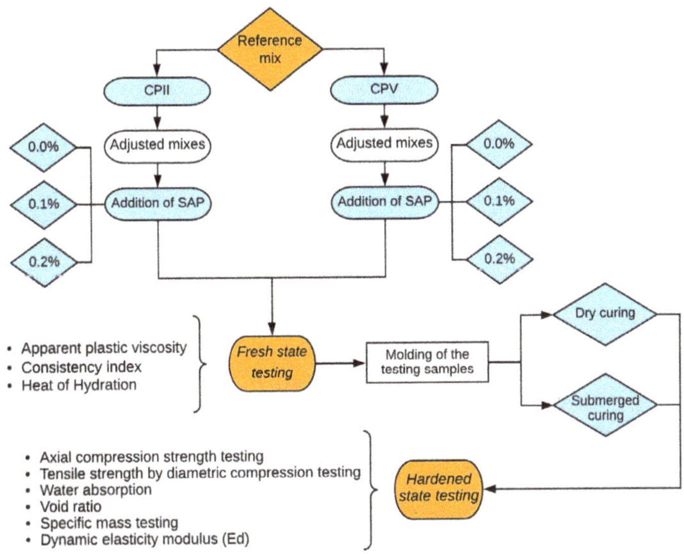

Figure 1. Study's methodological procedure flowchart.

The type II Portland cement used, type CPII-F-32, was made by Grupo Votorantim Cimentos, an ITAU company. The properties and characteristics of the cement used in this article are presented in Table 1, and were provided by the manufacturer's website.

Table 1. Properties of CPII-F-32.

Standard ABNT NBR		Value	Unit
Specific Mass (ABNT NBR 23:2001)		2.99	Kg/dm^3
Setting Time	Start	170	min
(ABNT NBR 65:2003)	Finish	241	min
	1 day	-	MPa
Compression strength	3 days	25	MPa
(ABNT NBR 7215:1997)	7 days	31	MPa
	28 days	39	MPa

Source: Adapted from VOTORANTIMCIMENTOS, 2022 [23].

The high strength type V Portland cement used, type CPV-ARI Structure, was made by Group Intercement, a CAUÊ company. The properties and characteristics of the cement used in this article are shown in Table 2, and were provided by the manufacturer's website.

Table 2. Properties of CPV-ARI.

Standard ABNT NBR		Value	Unit
Specific Mass (ABNT NBR 23:2001)		3.04	Kg/dm^3
Setting Time (ABNT NBR 65:2003)	Start	142	min
	Finish	220	min
Compression strength (ABNT NBR 7215:1997)	1 day	27	MPa
	3 days	37	MPa
	7 days	42	MPa
	28 days	48	MPa

Source: Adapted from BRASILINTERCEMENT, 2022 [24].

The activated silica used in the mix as the aggregate was provided by the company Tecnosil. It is a fine pulverized powder originated from the fabrication of metallic silicon or ferrosilicon. The high content of SiO_2 in the amorphous form (non-crystalline), combined with the high grit, provides a high reactivity with the products resulting from the cement hydration, ensuring the best performance in concrete and mortar. The activated silica has spheric, vitreous particles, with the average diameter smaller than one µm and a specific surface area between 15 and 25 m^2/kg, providing a high specific surface area and low apparent specific mass.

The sand used is from the city of Itajubá-MG, and it was split in an exceptionally fine portion (passing through the 600 µm sieve) and another portion (trapped in the 600 µm sieve and of maximum diameter of 4.8 mm). This separation was made to help the final dosing of the mortar mixes. As mixing water, the water from the public utilities of the city of Itajubá-MG was used.

The superplasticizer additive (SPA) used in the mix was the SILICON NS HIGH 210, manufactured and provided by the company Tecnsoil. According to the company, this SPA is a latest generation polycarboxylate-based organic additive that provides a high dispersion power, high workability, reduces water, and increases mechanical strength. The potassium-based superabsorbent polymer is a polymer developed by the Instituto Granado de Tecnologia da Poliacrilonitrila (IGTPAN) and its main characteristic is retaining water. The SAP-k reticulate has the capacity to absorb 300 times its weight in water. (Figure 2).

Figure 2. SAP samples.

When in contact with water, it transforms into a gelatinous substance known as HIDROGEL. After releasing all its water, the SAP-k reverts to its original state (granulate), with the average active capacity of three to five years, able to absorb new irrigation water or rainwater without changing its structure [25]. Table 3 shows the characteristics of the SAP-k used in the tests.

Table 3. Characteristics of the used Superabsorbent Polymer.

Appearance	Yellow Granulates
Particle size (mm)	0.6–2.5
Relative density (g/cm^3)	1.1
pH concentration 1 g/L	7–9
Water solubility	Insoluble
Absorption (g water/g SAP)	200–400
Time (min) to reach 60% absorption	10
UV rays' stability	High resistance to UV rays
Product average stability in ground (years)	3–5
Acrylamide residual (ppm)	<1.0
Acrylic acid residual (ppm)	<1.0
Potassium concentration (%)	>20%
Nitrogen concentration (%)	>10%

Source: Adapted from IGTPAN, 2023 [25].

The SAP-k provided by the IGTPAN is obtained through the chemical recycling of polyacrylonitrile, commonly used in textile applications due to characteristics similar to wool. It has special characteristics, such as low density, thermic stability, high strength and elasticity modulus, and UV degradation stability.

To achieve the desired SCM reference mix (CPII-F32 and CPV-ARI) and perform the addition of Superabsorbent Polymer, this research used as a base the mix developed by Silva, 2019, which used the dosing method proposed by Tutikian, 2007 [26], and part of the practical and experimental method of Helene and Terzian, 1992 [27].

The experimental program is divided into two parts, the first one being the tuning of the mix proposed by Silva, 2019 [28], to obtain the refence mix, making the proper adjustments to be used in the study. Starting with an a/c ratio of 0.5 and with 1.5% of superplasticizer by cement weight, until achieving the desired consistency.

The second part, with the possession of the reference mix, the mix with added SAP was developed, observing the behavior of the mortar, and making the necessary correction so that they continue with the self-compacting characteristics.

In the designing of the mix, the dry aggregate (cement, activated silica, fine sand, coarse sand, and SAP) was added first and mixed. Next, the water was added followed by the addition of SPA. Following that, the proper additions were made to achieve the properties. In this manner, the mixes presented in Table 4 were achieved.

Table 4. Unitary mix by mass used in SCM.

Mix	Cement	Silica	Fine Sand	Coarse Sand	SPA	a/c	SAP
CPII-F-32 REF	1.00	0.10	1.5	1.3	2.3%	0.6	0.00
CPII-F-32 0.1%	1.00	0.10	1.6	1.4	2.7%	0.7	0.01
CPII-F-32 0.2%	1.00	0.10	1.6	1.4	2.4%	0.6	0.02
CPV-ARI REF	1.00	0.06	1.5	1.3	1.7%	0.5	0.00
CPV-ARI 0.1%	1.00	0.06	1.6	1.4	2.0%	0.6	0.01
CPV-ARI 0.2%	1.00	0.06	1..6	1.4	2.1%	0.6	0.02

After this stage, the fresh state tests were made, followed by the molding of the testing samples. The demolding was made 24 h later and from that, half the test samples were submerged in potable water for curing, and the other half were separated for dry curing, both for a period of 28 days. Therefore, there was a total of 12 (twelve) mixes, differentiated by the percentage of SAP (0; 0.1; 0.2), curing type (dry or submerged), and cement type (CPII-F-32 and CPV-ARI).

After this stage, the fresh state tests were made, followed by the molding of the testing samples. The demolding was made 24 h later and from that, half the test samples were submerged in potable water for curing and the other half were separated for dry curing, both for a period of 28 days. Therefore, there was a total of 12 (twelve) mixes, differentiated by the percentage of SAP (0; 0.1; 0.2), curing type (dry or submerged), and cement type (CPII-F-32 and CPV-ARI).

2.1. Fresh State Testing

The fresh state test was the first to be made; it was used to ensure and prove the self-compacting performance of the mortar, show the values of specific mass of the recently fill testing bodies, and obtain the thermic data in the beginning of the curing of the mortar. Table 5 shows the quantity of testing samples used in each test, the percentage of SAP and cement type, as well as the method used in making.

Table 5. Number of test samples used in each fresh state test.

Fresh State Testing	Standard/Method	N° Test Samples Ø 5 × 10 cm					
		CPII-F-32			CPV-ARI		
		0%	0.10%	0.20%	0%	0.10%	0.20%
Apparent specific mass	NBR 9833:2009	3	3	3	3	3	3
Consistency index	Okamura e Ouchi, 1999	Did not use test samples					
Apparent plastic viscosity							
Heat of hydration	NBR 12006:1990	3	3	3	3	3	3

To obtain the flow diameter, the consistency index test was performed. This procedure enables the analysis of the mortar's flow capacity under the influence of its own weight and its filling capability. The test consists of filling a cone shape (Figure 3) without compacting, that is vertically suspended, and after the mortar ceases movement, two orthogonal diameters are measured.

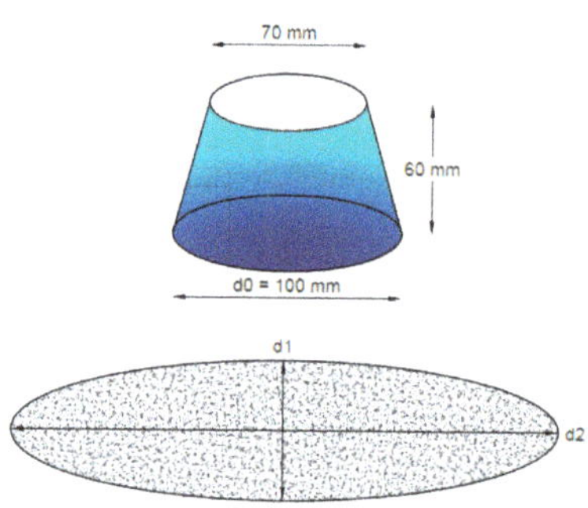

Figure 3. Mortar slump test. Source: Adapted from Okamura and Ouchi, 2003 [2].

Applying the dosing method proposed by Okamura and Ouchi, 2003 [2], it is possible to calculate the relative spread index for mortar (Gm) using the two perpendicular measurements obtained in the mortar slump test.

It was used as optimal value Gm = 5, using as limits the values from 3 to 7 that equate to a slump diameter of 200 to 280.

To determine the mortar dosing according to the Okamura and Ouchi, 2003 [2] method, V-funnel testing of the rectangular section was performed. The test consists in measuring the time necessary for the mortar to flow through the funnel resisting separation. The funnel was filled without compacting the mortar, and the inferior hatch was open. With the help of a stopwatch, the time mortar flow time was measured until little was visible through the top part of the equipment. Figure 4 shows the dimensions of the V-funnel used to measure the relative flow time of mortar.

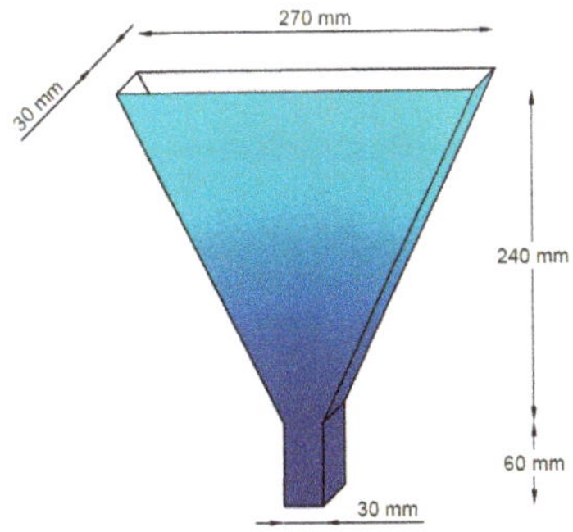

Figure 4. Mortar Funnel testing. Source: Adapted from Okamura and Ouchi, 2003 [2].

After timing the flow time in the V-funnel, it is possible to determine the relative flow time of mortar (Rm) using the dosing method proposed by Okamura and Ouchi, 1999 [29].

According to Takada et al., 1998 [30], mortar with Gm values of 5 and Rm values of 1 are classified as acceptable to obtain concrete with self-compacting characteristics. Edamatsu et al., 1999 [31] obtained Gm values between 3 and 7, corresponding to the mortar slump diameter from 280 mm to 283 mm, and the Rm value between 1 and 2, corresponding to the flow time of 5 to 10 s. Domone and Jin, 1999 [32] suggest that Gm higher than 8, equivalent to slump diameters bigger than 300 mm and Rm values of 1 to 5 equate to flow time (t) of 2 to 10 s. A high Gm value indicates a higher deformability, and a lower Rm value indicates a higher viscosity.

To measure the heat of hydration in the mortar, mixtures used a semi-adiabatic calorimeter, according to the NBR 120.006 1990 Cement—Determining the Heat of Hydration using the Langavant Bottle Method [33].

To perform this test, a thermal data acquisition board made by National Instruments, model cDAQ-9171 was used. A Type-K Thermocouple was used in conjunction to receive the signal emitted by the board, using the software LABTRIX version XP1502.

The analysis time for each of the mixes was two hours, starting from the time the sample finished confection.

2.2. Hardened State Testing

The tests made were axial compression strength, tensile strength by diametric compression, void ratio, specific mass, and dynamic elasticity modulus. The number of test samples used and the standard or method used are shown in Table 6.

Table 6. Number of test samples used in each hardened state test.

Hardened State Testing	Standard/Method	N° Test Samples Ø 5 × 10 cm					
		CPII-F-32			CPV-ARI		
		0%	0.10%	0.20%	0%	0.10%	0.20%
Axial compression strength	NBR 5739:2007	6	6	6	6	6	6
Tensile strength	NBR 7222:2011	6	6	6	6	6	6
Dynamic elasticity modulus	ASTM 1876-01	6	6	6	6	6	6
Water absorption, void ratio, and specific mass	NBR 9778:2005	6	6	6	6	6	6

Note: the amount shown is regarding the sum of dry and submerged samples (3 each).

The measurement of the mortar axial compression strength was made using a INSTRON 8001 machine, with the cylindrical test samples molded to 50 mm in diameter and 100 mm in height. The test samples must be kept in the wet curing process until the time of testing. Before the execution of the test, the base of the test samples must be prepared.

Next, the load of (0.45 ± 0.15) MPa/s is applied until the rupture of the test sample (ABNT NBR 5739:2007) [34].

According to NBR 7222:2011 [35], through tensile testing by axial compression, the results of tensile strength are indirectly obtained. In the test, the test samples are also molded in a cylinder of 50 mm in diameter and 100 mm in height. The test consists in positioning the test samples in the compression machine in the horizontal position, supported by two wooden planks. The load is continuously applied, constantly increasing the tensile tension, at a rate of 0.05 MPa/s until the rupture of the test sample.

According to the NBR 9778:2005 [36] standard, the Water absorption, void ratio, and specific mass tests were carried out at the Universidade Federal de Itajubá (UNIFEI). The mass values were collected using a SHIMASZU scale, model UX6200H, with a precision of 0.01 g. Three test samples were made for each of the self-compacting mixes that were evaluated at the 28th day. For the testing, the test sample of 150 cm^3 are placed in the heating chamber at a temperature of (105 ± 5) °C for a period of 72 h, later recording the mass in the dry state. To measure the submerged mass, the samples must remain underwater at a temperature of (23 ± 5) for 72 h, weighing hydrostatically. The samples are removed from the water next and dried with a damp cloth to measure the samples saturated mass.

The dynamic elasticity modulus test was performed at the Materials Lab for Civil Engineering of the Universidade Federal de Itajubá (UNIFEI). The equipment Sonelastic was used to perform the impulse excitation technique (IET) in the following setting:

- Sonelastic software version 3.0;
- Adjustable bar support;
- SA-BC cylinders;
- Directional acoustic receiver CA-DP;
- Manual pulsator.

Based in the ASTM E1876 2015 [37], the test samples were subjected to the Impulse Excitation Technique (IET). Initially the length, diameter, and mass of the test samples were recorded, then they were supported by an adjustable support. Next, using a manual pulsator, each test sample was subjected to light knocks causing acoustic responses that were detected by an acoustic receiver. These acoustic responses were processed by the Sonelastic software version 3.0 to calculate the elasticity modulus and the dampening of each test sample.

Three test samples were made for each of the self-compacting mixes that were evaluated at the 28th day. Figure 5 respectively shows the equipment during the execution of the test and the Software Sonelastic version 3.0 interface.

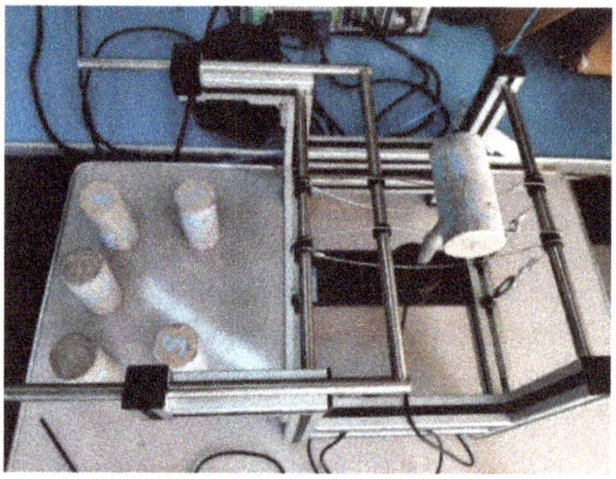

Figure 5. Support used to perform the dynamic elasticity modulus testing.

3. Results

This section may be divided by subheadings. It should provide a concise and precise description of the experimental results, their interpretation, as well as the experimental conclusions that can be drawn.

3.1. Fresh State Properties

The results of the fresh state SCM are presented in this section. For a better organization of the analysis, this section is subdivided.

3.1.1. Slump Flow Test

The results of the slump flow test of the reference SCM and the samples with 0.1% and 0.2% SAP are shown in Tables 7 and 8, performed according to the Okamura and Ouchi, 2003 [2] method, and using the parameters mentioned in this article methodology. Based in the parameters presented by the author previously mentioned in the Materials and Methods chapter, it was used as optimal value Gm = 5, and limiting values of 3 and 7 that amount to slump diameters of 200 to 280.

Table 7. Slump flow test—CPII.

Mix	% SAP	Cure	Cement Type	d1	d2	d0	Gm (3 < Gm < 7)
1	0	Sub	CPII	270	265	100	6.2
3	0	Dry	CPII	270	265	100	6.2
5	0.1	Sub	CPII	265	255	100	5.8
7	0.1	Dry	CPII	265	255	100	5.8
9	0.2	Sub	CPII	230	225	100	4.2
11	0.2	Dry	CPII	230	225	100	4.2

Table 8. Slump flow test—CPV.

Mix	% SAP	Cure	Cement Type	d1	d2	d0	Gm (3 < Gm < 7)
2	0	Sub	CPV	275	275	100	6.6
4	0	Dry	CPV	275	275	100	6.6
6	0.1	Sub	CPV	260	270	100	6.0
8	0.1	Dry	CPV	260	270	100	6.0
10	0.2	Sub	CPV	275	280	100	6.7
12	0.2	Dry	CPV	275	280	100	6.7

Using the results presented in Tables 7 and 8, it is possible to assert that the addition of the superabsorbent polymer alters the slump behavior of the SCM in the fresh state. It can be observed that all the mortar mixes are within the limits to be considered self-compacting, however this was only possible due to the adjustments made to the mixes so that they had the desired consistency. During the testing, it was observed that the SAP, which has as a characteristic to absorb massive quantities of liquid, "sucked" part of the available water volume in the mix, therefore reducing the flow of the test samples, and stopped them from reaching the minimal slump diameter (200 mm). It is important to emphasize that the mortar using CPV-ARI presented a larger slump diameter than the CPII-F-32 ones, as expected due to it being a type of cement more commonly used in self-compacting cement composites, and which allows for a better flow in the mix in the fresh state.

These results converge with the ones of Schröfl et al., 2012 [15], in which the authors investigated the absorption capacity of SAP by adding in the mixes different amounts and types of SAP. Once the SAP absorbs the water in the mixture, it is important to make up for this (lost) water with the addition of extra water to ensure the workability required by the mixture in the fresh state.

In Figures 6 and 7, the slump flow tests of the mixes using CPII-F-32 and CPV-ARI are shown, respectively. It is possible to notice by the images that all the mixes, after the already mentioned adjustments, present a good flow and uniformity, meeting the requirements to be considered as self-compacting, and without presenting exudation or separation.

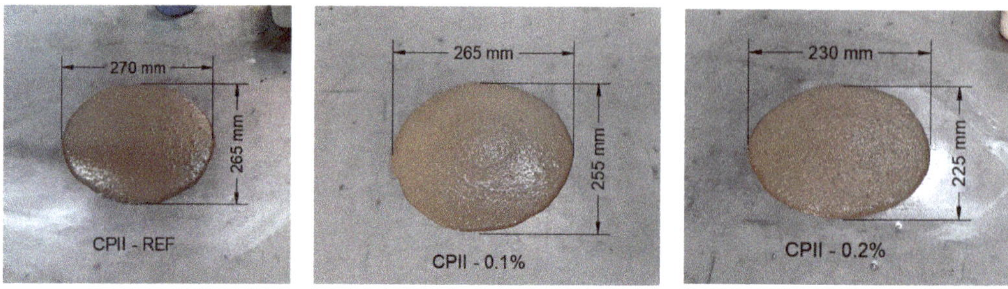

Figure 6. Slump flow test of the CPII-F-32 mix, 0% (**left**), 0.1% (**middle**), and 0.2% (**right**).

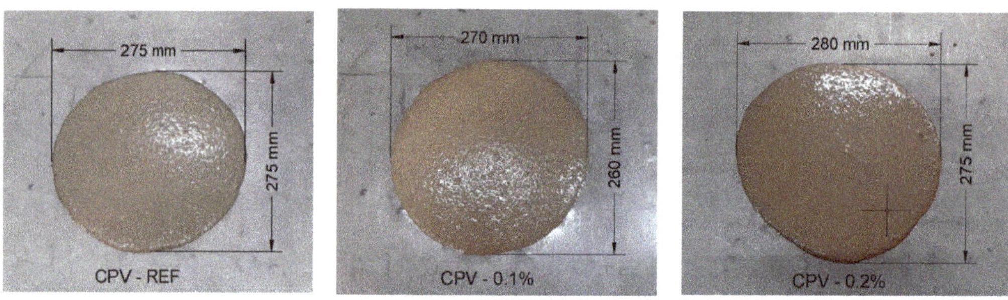

Figure 7. Slump flow test of the CPV-ARI mix, 0% (**left**), 0.1% (**middle**), and 0.2% (**right**).

3.1.2. Apparent Plastic Viscosity—V-Funnel

The results of the plastic apparent plastic viscosity of mortar made using CPII-F-32 and CPV-ARI are, respectively, shown in Tables 9 and 10, and were performed according to the Okamura and Ouchi, 2003 [2] method, using a minimal Rm value of 1 and maximum of 2, that corresponds to a flow time inferior to 10 s.

According to Tables 9 and 10, it is possible to observe that all the mixes are within the limitations of the proposed methodology (1 < Rm < 2), with a flow time lower than 10 s. This was also possible due to the adjustments made in the slump flow test phase, which allowed the mixes to achieve adequate apparent plastic viscosity. As expected, the mortar made using CPV-ARI presented a faster flow time than the ones using CPII-F, due to the cement type being more adequate to use in self-compacting composites, as it offers a better flow in the fresh state.

Table 9. Apparent plastic viscosity of CPII.

Mix	% SAP	Cure	Cement Type	Time (s)	Rm (1 < Rm < 2)
1	0	Sub	CPII	9	1.1
3	0	Dry	CPII	9	1.1
5	0.1	Sub	CPII	9	1.1
7	0.1	Dry	CPII	9	1.1
9	0.2	Sub	CPII	10	1.0
11	0.2	Dry	CPII	10	1.0

Table 10. Apparent plastic viscosity of CPV.

Mix	% SAP	Cure	Cement Type	Time (s)	Rm (1 < Rm < 2)
2	0	Sub	CPV	8	1.3
4	0	Dry	CPV	8	1.3
6	0.1	Sub	CPV	7	1.4
8	0.1	Dry	CPV	7	1.4
10	0.2	Sub	CPV	9	1.1
12	0.2	Dry	CPV	9	1.1

3.1.3. Heat of Hydration

The heat of hydration testing was performed according to the NBR 12.006 1990 [33], and an adaptation was made to simulate the conditions inside a semi-adiabatic chamber.

The results are described in Tables 11 and 12, explaining the values detected by each of the thermocouples, for the CPII-F-32 mixes (Odd mixes: 1; 3; 5; 7; 8 and 11), as well for the CPV-ARI mixes (Even mixes: 2; 4; 6; 8; 10 and 12). At the end of the tables, a comparison is made to find out the variation of temperature during the experiment, showing the delta between the initial and final temperature.

Table 11. Individual gathering of thermocouples in CPII-F-32.

	Thermocouple 1			Thermocouple 2			Thermocouple 3		
Sample Number	1 and 3	5 and 7	9 and 11	1 and 3	5 and 7	9 and 11	1 and 3	5 and 7	9 and 11
% of SAP	0.00%	0.10%	0.20%	0.00%	0.10%	0.20%	0.00%	0.10%	0.20%
00:00:00	23.1	23.9	21.7	23.1	23.4	21.5	23.6	23.9	22
00:15:00	24.2	25.2	22.3	23	23.6	20.2	24.2	24.9	22
00:30:00	25.8	25.8	22.8	24.5	24.1	20.6	26	25.6	23.8
00:45:00	26.7	26.5	23.7	25.5	24.6	21.2	27	26.4	24.7
01:00:00	27.4	26.6	23.6	26.1	24.4	20.9	27.8	26.4	25
01:15:00	27.6	27.2	23.6	26.2	25.1	21.1	28.2	26.8	24.8
01:30:00	28.2	27.4	24.4	26.9	25.4	21.7	28.9	27.1	26
01:45:00	28.5	27.1	24.4	27.2	24.9	21.9	29.1	26.9	25.2
02:00:00	28.6	27.3	24.9	27.3	25.4	22	29.3	26.8	26.5
Tf-Ti	5.5	3.4	3.2	4.2	2	0.5	5.7	2.9	4.5

Source: Author, 2021.

Table 12. Individual gathering of thermocouples in CPV-ARI.

	Thermocouple 1			Thermocouple 2			Thermocouple 3		
Sample Number	2 and 4	6 and 8	10 and 12	2 and 4	6 and 8	10 and 12	2 and 4	6 and 8	10 and 12
% of SAP	0.00%	0.10%	0.20%	0.00%	0.10%	0.20%	0.00%	0.10%	0.20%
00:00:00	23.3	21.4	22.5	21.9	20.2	21.5	23.6	19.7	20.7
00:15:00	23.5	18.9	18.7	20.8	16.2	16.9	24.1	16.4	16.4
00:30:00	24.4	20.2	19.6	21.5	18.1	18.1	25.7	18.6	17.6
00:45:00	25.3	22	20	22.4	19.8	18.8	26.8	20.3	18.6
01:00:00	25.7	22.9	21.1	22.7	20.5	20	27.2	20.7	19.9
01:15:00	26.8	23.2	21.5	23.7	22	20.2	28.3	22.3	20.2
01:30:00	26.8	23.7	21.3	23.5	22.5	20.1	28.6	22.8	19.8
01:45:00	27	23.8	21.7	23.8	22.5	20.3	28.9	22.7	20.2
02:00:00	27.5	23.8	22.3	24.6	22.4	20.9	29.4	22.7	20.9
Tf-Ti	4.2	2.4	−0.2	2.7	2.2	−0.6	5.8	3	0.2

Tables 13 and 14 show the average temperature detected by the Thermocouples, in which it is possible to observe a reduction in its variation as the SAP addition increases in the CPII-F-32 and the CPV-ARI.

Table 13. Thermocouple average in CPII-F-32.

Thermocouple Average			
Sample Number	**1 and 3**	**5 and 7**	**9 and 11**
% of SAP	**0.00%**	**0.10%**	**0.20%**
00:00:00	23.27	23.73	21.73
00:15:00	23.80	24.57	21.50
00:30:00	25.43	25.17	22.40
00:45:00	26.40	25.83	23.20
01:00:00	27.10	25.80	23.17
01:15:00	27.33	26.37	23.17
01:30:00	28.00	26.63	24.03
01:45:00	28.27	26.30	23.83
02:00:00	28.40	26.50	24.47
Tf-Ti	5.13	2.77	2.73

Table 14. Thermocouple average in CPV-ARI.

Thermocouple Average			
Sample Number	**2 and 4**	**6 and 8**	**10 and 12**
% of SAP	**0.00%**	**0.10%**	**0.20%**
00:00:00	22.93	20.43	21.57
00:15:00	22.80	17.17	17.33
00:30:00	23.87	18.97	18.43
00:45:00	24.83	20.70	19.13
01:00:00	25.20	21.37	20.33
01:15:00	26.27	22.50	20.63
01:30:00	26.30	23.00	20.40
01:45:00	26.57	23.00	20.73
02:00:00	27.17	22.97	21.37
Tf-Ti	4.23	2.53	−0.2

This reduction can be better seen in Figure 8, noting that the mixes with the SAP addition had a reduction in the heat peak. In the CPV, there was also a reduction in temperature in the first minutes. The controlling of the heat of hydration in the beginning of the curing process is especially important because it avoids the self-drying phenomenon in which the water is consumed by this heat, reducing the available quantity for the hydration of the cement base, which may cause retraction fissures [38].

These results overlap with the ones presented by Kumm, 2009 [39], in which the temperature variation, and the relative humidity of the cement paste were measured. The mixes with added SAP showed lower temperature peaks than the reference mixes, which favor the internal curing process of the pastes by controlling the heat of hydration.

3.2. Hardened State Properties

In this stage of the experimental program, the analysis of the hardened concrete properties was made. To this end, the following tests were made: axial compression strength, tensile strength by diametric compression, void ratio, specific mass, and dynamic elasticity modulus.

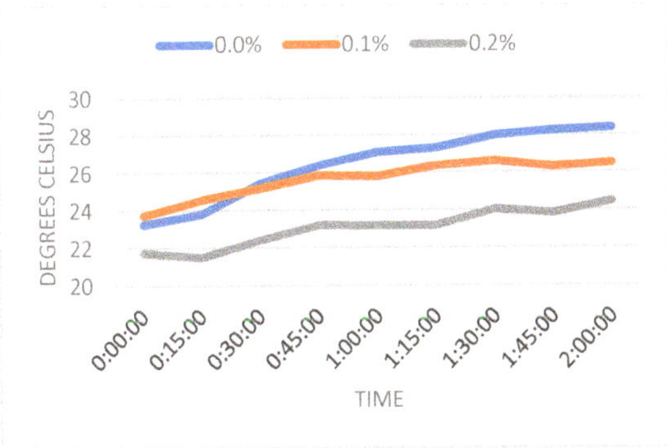

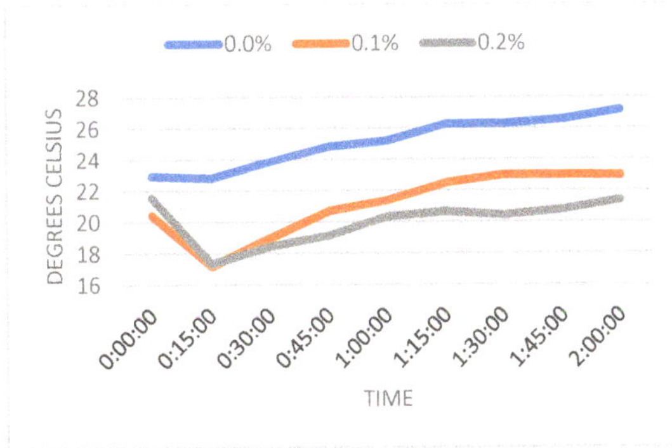

Figure 8. Temperature fluctuation in time (CPII-F-32 and CPV-ARI).

3.2.1. Axial Compression Strength Testing

Table 15 presents the results from the axial compression strength testing using the type II cement mixes. For these tests, three test samples were used for each of the mixes, identified in the Tables as samples 1, 2, and 3.

Table 15. Compression strength test results (CPII-F-32).

Mix	% SAP	Cure	Cement Type	Compression (MPa)			Average
				Sample 1	Sample 2	Sample 3	
1	0	Sub	CPII	27.54	25.52	29.02	27.36
3	0	Dry	CPII	17.71	16.68	17.28	17.23
5	0.1	Sub	CPII	31.89	22.67	24.34	26.30
7	0.1	Dry	CPII	13.86	14.31	12.44	13.54
9	0.2	Sub	CPII	25.37	27.38	25.09	25.95
11	0.2	Dry	CPII	19.81	14.74	22.80	19.12

Observing Table 15 and Figure 9, it is noticeable that the addition of SAP influenced the compression strength of the mixes using CPII-F-32 cement, although with distinct behavior in the dry and submerged curing. For the samples that were cured underwater, the SAP did not significantly compromise, reducing 3.8% to 0.1% of SAP, and 5.15% to 0.2% of SAP in relation to the reference mix. In the dry cured samples, the SAP reduced its strength by 21.42% for the samples with 0.1% of SAP.

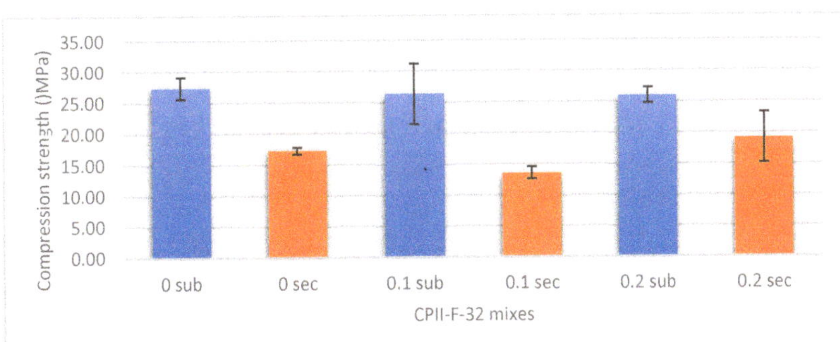

Figure 9. Average results of compression strength of CPII-F-32.

For the 0.2% SAP samples, there was an 10.97% increase in strength. It is known that the increase in porosity results in strength losses and, as seen, that is the result of adding

SAP. However, the properties and advantages of SAP allow for a better hydration of the cement base, and justify the low reduction of strength in the submerged samples and the strength increase in the dry cured 0.2% SAP samples.

Table 16 presents the results of the axial compression strength testing in the mixes produced with type V cement. For these tests, three test samples were used for each of the mixes, identified in the Tables as samples 1, 2, and 3.

Table 16. Compression strength test results (CPV-ARI).

Mix	% SAP	Cure	Cement Type	Compression (MPa)			Average
				Sample 1	Sample 2	Sample 3	
2	0	Sub	CPV	50.91	50.45	50.59	50.65
4	0	Dry	CPV	28.67	41.28	40.29	36.75
6	0.1	Sub	CPV	29.50	31.05	31.52	30.69
8	0.1	Dry	CPV	20.63	21.76	23.48	21.96
10	0.2	Sub	CPV	27.36	27.80	38.03	31.06
12	0.2	Dry	CPV	20.96	25.97	23.28	23.40

Observing Table 16 and Figure 10, it is noticeable that the addition of SAP influenced the compression strength of the mixes using CPV-ARI cement. Different from the samples using CPII-F-32 cement, in the type V cement, there was a substantial reduction in strength in the mixes with added SAP. The SAP samples cured underwater had a substantial reduction in strength, lowering by 39.41% for the 0.1% of SAP, and 36.68% for the 0.2% of SAP in comparison to the reference mix. In the dry cured samples, the 0.1% SAP mix reduced the strength by 40.25%, and the 0.2% SAP mix reduced the strength by 36.33%.

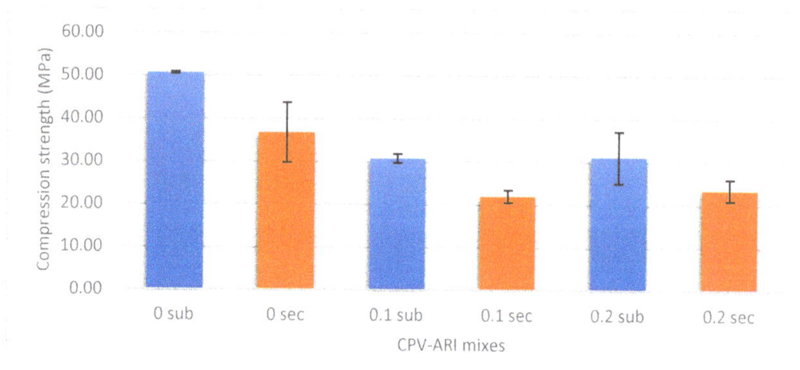

Figure 10. Average results of compression strength of CPV-ARI.

The general overview indicates that the obtained results point to a reduction in compression strength in the samples with added SAP (0.1% and 0.2%); however, the test samples with 0.2% added SAP present themselves as better than the ones with 0.1% addition, suggesting a potential ideal dosage for the mix.

These results support what is presented in the work of Muthalvan et al., 2021 [40], in which the cement mortar samples with a higher dosage of SAP present better compression strength results than the ones with a lower content. To Suarez, 2015 [41], the content that presented the best results in compression strength was the 0.2% by mass of cement. In its work, it was also stated that the reduction in compression strength with the use of SAP can be attributed to effects caused by the SAP in the hydration of the cement. The polymer expands when in contact with water. In the hydration phase, the water needed for the internal cure is pulled from the polymer, which returns to its original size and leaves voids in the structure.

According to Schröfl et al., 2012 [42], when SAP particulates are used in the mix in cement-based materials, several voids are formed, which can affect the material strength in a negative way. In articles by Jensen and Hensen, 2002 [43] and Pereira and Matos, 2011 [44], reduced compression strength of the samples containing superabsorbent polymers was found.

3.2.2. Tensile Strength by Diametric Compression Testing

Table 17 presents the results of the tensile strength by diametric compression test. For these tests, three test samples were used for each of the mixes, identified in the Tables as samples 1, 2, and 3.

Table 17. Tensile strength test results—CPII.

Mix	% SAP	Cure	Cement Type	Traction (MPa)			Average
				Sample 1	Sample 2	Sample 3	
1	0	Sub	CPII	2.34	3.84	3.10	3.09
3	0	Dry	CPII	1.48	1.56	1.99	1.68
5	0.1	Sub	CPII	3.84	3.79	3.61	3.75
7	0.1	Dry	CPII	2.13	2.00	2.00	2.04
9	0.2	Sub	CPII	3.86	3.56	3.86	3.76
11	0.2	Dry	CPII	1.98	3.72	2.92	2.87

Observing Table 17 and Figure 11, it is possible to notice that the addition of SAP influenced the tensile strength of the CPII-F-32 cement mix. For the samples cured underwater, the SAP provided a 21.36% increase in 0.1% mixes, and 21.69% in 0.2% mixes, compared to the reference mix. In the dry cured samples, the increment was significant, increasing the tensile strength by 21.43% in 0.1% mixes, and in the 0.2% mix, a 70.83% increase in strength.

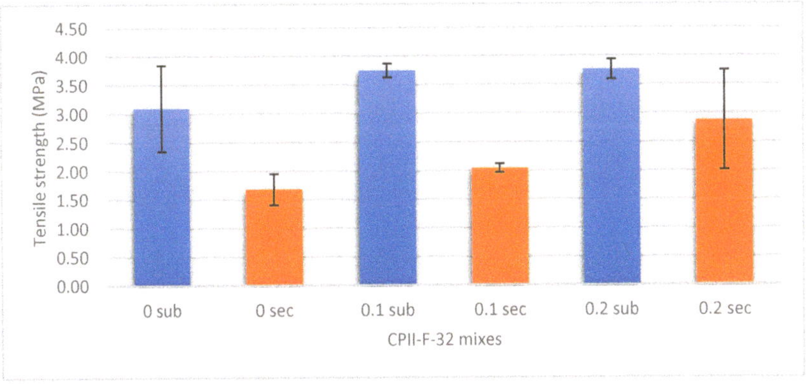

Figure 11. Average results of tensile strength of CPII-F-32.

Both the submerged samples and dry cured samples, with a 0.2 percent of SAP, show increased values, proving that this polymer content in the mix is adequate to this end.

Table 18 presents the results of the tensile strength by diametric compression test of the mixes using type V cement. For these tests, three test samples were used for each of the mixes, identified in the Tables as samples 1, 2, and 3.

Table 18. Tensile strength test results—CPV.

Mix	% SAP	Cure	Cement Type	Traction (MPa)			Average
				Sample 1	Sample 2	Sample 3	
2	0	Sub	CPV	3.62	4.58	3.90	4.03
4	0	Dry	CPV	4.06	4.11	3.69	3.96
6	0.1	Sub	CPV	3.62	3.15	3.20	3.32
8	0.1	Dry	CPV	2.35	2.61	2.45	2.47
10	0.2	Sub	CPV	2.91	3.88	3.25	3.35
12	0.2	Dry	CPV	2.54	2.78	2.82	2.71

Observing Table 18 and Figure 12, it is possible to observe that the addition of SAP influenced the tensile strength of the CPV-ARI cement mix, but differently from the CPII-F-32. For the samples cured underwater, the samples with SAP had the strength reduced by 17.62% for 0.1% SAP, and 16.87% for 0.2% SAP, compared to the reference mix. In the dry cured samples, the SAP reduced the strength by 27.63% in the 0.2% SAP samples, and for the 0.2% SAP samples, there was a 31.57% reduction in strength.

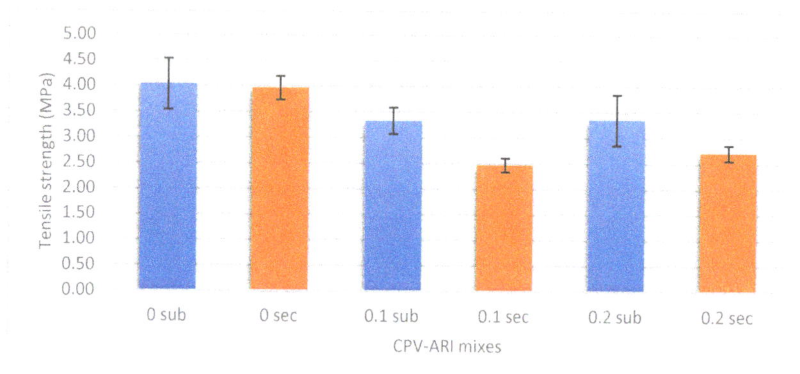

Figure 12. Average results of tensile strength of CPV-ARI.

In conclusion, the results from the test point to a reduction in tensile strength in CPV-ARI testing samples with added SAP (0.1 and 0.2%) compared to the reference mix. However, the CPII-F-32 mix with added SAP presented an improvement in tensile strength, both in the submerged samples and dry cured ones.

Still, it is important to highlight that samples with a higher content of SAP, namely the 0.2%, presented better results in comparison to the lower content ones, as is the case for CPV-ARI mixes or when compared to the reference mix, as is the case for the CPII-F-32.

The results using CPV-ARI support the ones presented by Suarez, 2015 [41], in which a reduction in tensile strength was observed with the addition of the superabsorbent polymer in micro concretes if compared to the refence mix; however, even increasing the SAP content, large reductions in tensile strength were not observed.

3.2.3. Water Absorption, Void Ratio, and Specific Mass Testing

Table 19 shows the results of the real specific mass tests, Table 20 shows the void ratio, and Table 21 shows the water absorption of the self-condensing mortar using CPII-F-32 cement, performed according to ABNT NBR 9778:2009. For these tests, three test samples were used for each of the mixes, identified in the Tables as samples 1, 2, and 3.

Table 19. Specific mass of SCM in the hardened state—CPII-F-32.

Mix	% SAP	Cure	Cement Type	Specific Mass (g/cm³)			Average
				Sample 1	Sample 2	Sample 3	
1	0	Sub	CPII	2.29	2.27	2.26	2.27
3	0	Dry	CPII	2.33	2.35	2.31	2.33
5	0.1	Sub	CPII	2.14	2.15	2.20	2.16
7	0.1	Dry	CPII	2.35	2.32	2.34	2.34
9	0.2	Sub	CPII	2.20	2.17	2.18	2.18
11	0.2	Dry	CPII	2.30	2.16	2.18	2.21

Table 20. Void ratio of SCM in the hardened state—CPII-F-32.

Mix	% SAP	Cure	Cement Type	Void ratio			Average
				Sample 1	Sample 2	Sample	
1	0	Sub	CPII	5.07	5.21	5.11	5.13
3	0	Dry	CPII	7.38	7.46	6.78	7.21
5	0.1	Sub	CPII	4.24	4.16	4.60	4.33
7	0.1	Dry	CPII	8.94	8.14	8.87	8.65
9	0.2	Sub	CPII	4.66	4.43	4.12	4.40
11	0.2	Dry	CPII	6.91	5.06	5.02	5.66

Table 21. Water absorption of SCM in the hardened state—CPII-F-32.

Mix	% SAP	Cure	Cement Type	Absorption %			Average
				Sample 1	Sample 2	Sample 3	
1	0	Sub	CPII	5.34	5.50	5.39	5.41
3	0	Dry	CPII	7.97	8.06	7.28	7.77
5	0.1	Sub	CPII	4.43	4.34	4.82	4.53
7	0.1	Dry	CPII	9.82	8.87	9.73	9.47
9	0.2	Sub	CPII	4.88	4.63	4.30	4.61
11	0.2	dry	CPII	7.42	5.33	5.28	6.01

For the mortar specific mass test in the hardened state, it is possible to notice a decrease in the samples with added SAP. For the 0.1% and 0.2% SAP samples that were cured underwater, the average reductions were 4.84% and 3.96%, respectively. The dry cured samples increased by 0.43% in the 0.1% samples, and reduced by 5.15% in the 0.2% samples.

Mechtcherine et al., 2009 [45] performed tests to determine the porosity and pores size distribution in high strength concrete. Different tests were made, and for all the samples considered for the test, the authors observed an increase in porosity of the concrete samples with SAP.

Regarding the void ratio, it is possible to observe in Table 20 that the addition of SAP in the mix results in variance. For the 0.1% SAP samples, there was a 5.85% decrease for the underwater samples and an 19.97% increase for the dry cured samples. For the 0.2% SAP samples, there was a 14.23% reduction for the underwater samples, and a 21.50% reduction for the dry cured samples.

Regarding water absorption, it can be observed in Table 21 that the addition of SAP to the mix causes variance. The samples cured underwater with 0.1% added SAP decreased by 16.27%, and the dry cured samples increased by 21.88%. The samples cured underwater with 0.2% added SAP increased by 14.79%, and the dry cured samples reduced by 22.65%, when compared to the reference mix.

Table 22 shows the results of the real specific mass tests, Table 23 shows the void ratio, and Table 24 shows the water absorption of the self-compacting mortar using CPV-ARI, performed according to ABNT NBR 9778:2009 [36]. For these tests, three test samples were used for each of the mixes, identified in the Tables as samples 1, 2, and 3.

Table 22. Specific mass of SCM in the hardened state—CPV-ARI.

Mix	% SAP	Cure	Cement Type	Specific Mass (g/cm^3)			Average
				Sample 1	Sample 2	Sample 3	
2	0	Sub	CPV	2.30	2.31	2.35	2.32
4	0	Dry	CPV	2.34	2.33	2.33	2.33
6	0.1	Sub	CPV	2.26	2.16	2.18	2.20
8	0.1	Dry	CPV	2.22	2.25	2.24	2.24
10	0.2	Sub	CPV	2.25	2.26	2.25	2.25
12	0.2	Dry	CPV	2.30	2.31	2.30	2.30

Table 23. Void ratio of SCM in the hardened state—CPV-ARI.

Mix	% SAP	Cure	Cement Type	Void Ratio			Average
				Sample 1	Sample 2	Sample 3	
2	0	Sub	CPV	5.39	5.47	4.83	5.23
4	0	Dry	CPV	6.90	5.18	6.33	6.14
6	0.1	Sub	CPV	6.38	4.17	3.63	4.73
8	0.1	Dry	CPV	5.50	6.41	5.96	5.96
10	0.2	Sub	CPV	5.39	5.72	5.04	5.39
12	0.2	Dry	CPV	6.67	6.99	6.88	6.85

Table 24. Water absorption of SCM in the hardened state—CPV-ARI.

Mix	% SAP	Cure	Cement Type	Absorption %			Average
				Sample 1	Sample 2	Sample 3	
2	0	Sub	CPV	5.69	5.79	5.07	5.52
4	0	Dry	CPV	7.42	5.46	6.76	6.54
6	0.1	Sub	CPV	6.81	4.35	3.77	4.98
8	0.1	Dry	CPV	5.82	6.84	6.34	6.34
10	0.2	Sub	CPV	5.70	6.07	5.31	5.69
12	0.2	Dry	CPV	7.15	7.52	7.39	7.35

For the mortar specific mass test in the hardened state, it is possible to observe a reduction in the samples with added SAP. The 0.1% and 0.2% samples cured underwater showed an average reduction of 5.17% and 3.02%, respectively. The 0.1% dry cured samples reduced by 3.86%, and the 0.2% samples reduced by 1.29%.

Lura et al., 2012 [6] explains that once the SAP reaches its final size, it forms stable inclusion filled with water. This water is then aspired to the interior of smaller capillary pores and consumed in the hydration of the cement. The SAP ends up as empty pores in the cement slurry, corroborating the specific mass test results, as the SAP samples have lower specific mass, both for the CPII-F-32 and the CPV-ARI cement.

Regarding the void ratio, it is possible to observe in Table 23 that the addition of SAP to the mix results in variance. The samples cured underwater with 0.1% added SAP decreased by 9.56%, and the dry cured samples reduced by 2.93%. The samples cured underwater with 0.2% added SAP increased by 3.06%, and the dry cured samples increased by 11.56%.

Regarding water absorption, it can be observed in Table 24 that the addition of SAP to the mix results in variance. The samples cured underwater with 0.1% added SAP decreased by 9.78%, and the dry cured samples reduced by 3.06%. The samples cured underwater with 0.2% added SAP increased by 3.08%, and the dry cured increased by 12.39%, when compared to the reference mix.

3.2.4. Dynamic Elasticity Modulus (Ed)

Tables 25 and 26, and Figures 13 and 14 show the results of the mortar dynamic elasticity modulus tests, according to ASTM E1876, 2015 [37].

Table 25. Dynamic elasticity modulus of CPII-F-32.

Mix	% SAP	Cure	Cement Type	Ed (GPa)			Average
				Sample 1	Sample 2	Sample 3	
1	0	Sub	CPII	26.84	29.33	24.33	26.83
3	0	Dry	CPII	19.86	18.18	18.59	18.88
5	0.1	Sub	CPII	24.55	20.08	22.61	22.41
7	0.1	Dry	CPII	16.24	14.55	14.19	14.99
9	0.2	Sub	CPII	25.12	23.00	23.89	24.00
11	0.2	Dry	CPII	17.72	15.61	19.55	17.63

Table 26. Dynamic elasticity modulus of CPV-ARI.

Mix	% SAP	Cure	Cement Type	Ed (GPa)			Average
				Sample 1	Sample 2	Sample 3	
2	0	Sub	CPV	30.75	32.45	30.66	31.29
4	0	Dry	CPV	21.95	27.96	28.67	26.19
6	0.1	Sub	CPV	23.07	24.01	24.06	23.71
8	0.1	Dry	CPV	21.16	20.34	21.74	21.08
10	0.2	Sub	CPV	26.12	26.90	25.24	26.09
12	0.2	Dry	CPV	20.12	20.66	20.66	20.48

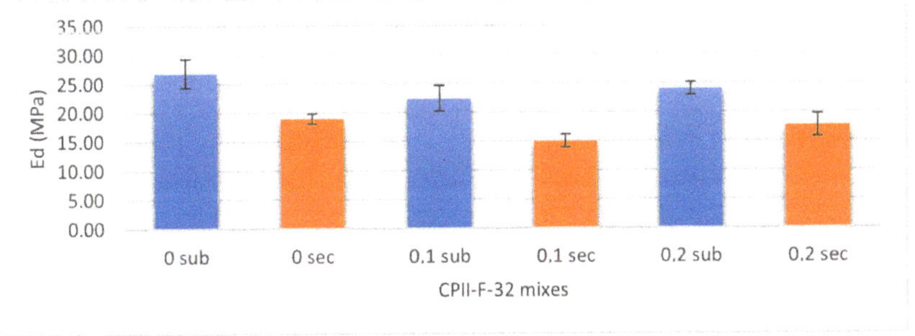

Figure 13. Average dynamic elasticity modulus of CPII-F-32.

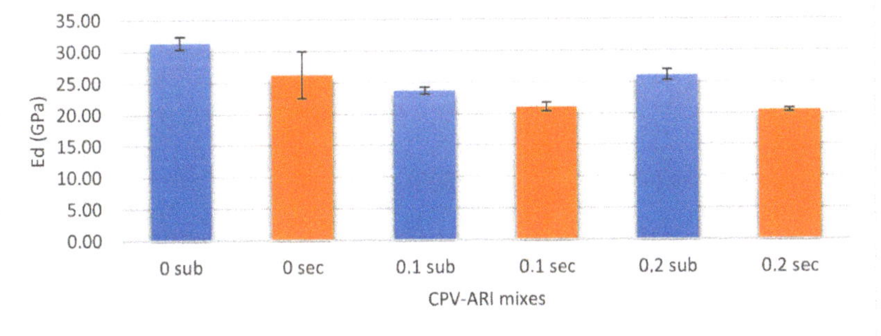

Figure 14. Average dynamic elasticity modulus of CPV-ARI.

It is possible to state that, according to the results presented in Table 25 and Figure 13, the elasticity modulus of self-compacting mortar with CPII-F-32 cement is affected by the addition of SAP. This variation is not linear, but follows a pattern that presents itself both in the underwater curing process and in the dry curing. The mortar cured underwater with 0.1% added SAP decreased the elasticity modulus by 15.47%, and the dry cured mortar reduced by 20.60%, when compared to the reference mix. Mortar with 0.2% added SAP had smaller decreases, with a 10.55% reduction for samples cured underwater and 6.62% for dry cured samples. These results reassure the results of previous tests, with the 0.2% SAP addition being more advantageous to the mixes.

The elasticity modulus in the cement mixes with the inclusion of less rigid particles, such as superabsorbent polymers, is generally smaller if compared to conventional concrete, for which the aggregate used is more rigid, such as quartz, basalt, and limestone. The addition of less rigid particles to the concrete makes the stress-deformation curve less inclined in the elastic region, making the final product have a lower elasticity modulus [46].

Table 26 and Figure 14 present the results, making it possible to state that the elasticity modulus of self-compacting mortar with CPV-ARI cement is affected by the addition of SAP. The mortar cured underwater with 0.2% added SAP presented lower reductions (16.62%) than the 0.1% samples (24.24%), when compared to the reference mix. The dry cured mortar presented similar reductions, with a 19.51% reduction in 0.1% SAP samples, and a 21.80% reduction in 0.2% SAP. Differently from the type II cement, the two types of curing process of the CPV cement did not behave in the same manner, improving in 0.2 percent samples cured underwater, but not repeating the results in the dry cured samples.

According to the obtained results that are presented in Tables 25 and 26, and by analyzing Figure 15, which presents the variation in the dynamic elasticity modulus for the two types of cures, it is possible to observe that CPII-F-32 cement presented a reduction in the values with the addition of the superabsorbent polymer, indicating that it may have helped in the dry curing process, providing extra hydration due to the swelling of the SAP particles in the samples. However, in the CPV-ARI samples, there was a higher reduction in mixes with 0.1% SAP, and a significant increase in the 0.2% SAP mix, demonstrating that the SAP had a better behavior with 0.1% SAP for this type of cement, in the dry curing process.

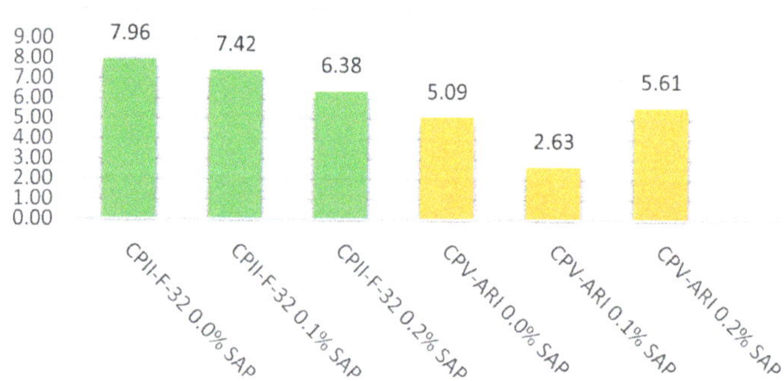

Figure 15. Dynamic elasticity modulus variance.

Figure 16 shows the Pareto graphs for the standardized effects of the results of the compression (left) and traction (right) tests performed on the mortars with CPII-F-32 cement with SAP additions. Pareto charts show the absolute values of the standardized effects from highest to lowest effect, and have a reference line to indicate which effects are statistically significant. Thus, the bars that cross the reference line have significance. As expected, the different types of curing have a great influence on the behavior of mortars, both for compression and traction. Regarding compression, it is noted that the inclusion of PAS did not affect the resistance of the mortars significantly, maintaining values close to those of the reference trait. For traction, the addition promoted significant changes in the mortars, obtaining higher values than the reference trait.

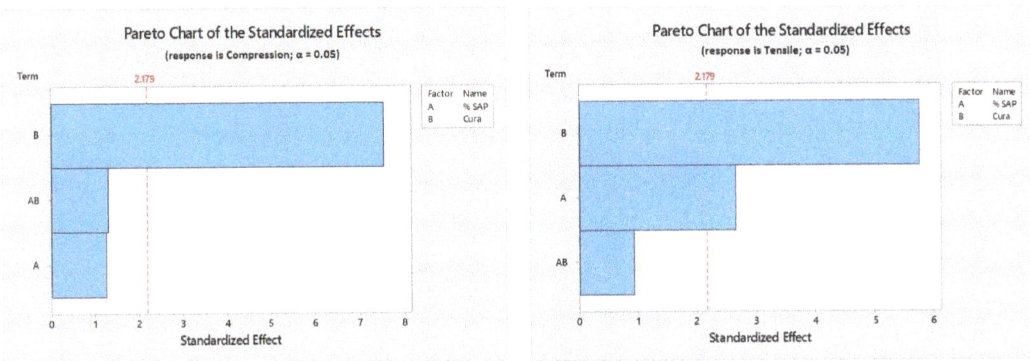

Figure 16. Pareto plot of standardized effects for CPII-E-32 (Compression and Tensile).

Figure 17 shows the Pareto charts for the standardized effects of the results of the compression (left) and traction (right) tests performed on the mortars with CPV-ARI cement with SAP additions. In them, it is possible to notice the factors that presented greater influence on the traits. As in type II cement, the different types of curing have a great influence on the behavior of the mortars, both for compression and traction. It is noted that the inclusion of PAS significantly affected the resistance of the mortars, both for compression and for traction, with a significant reduction in relation to the reference trait.

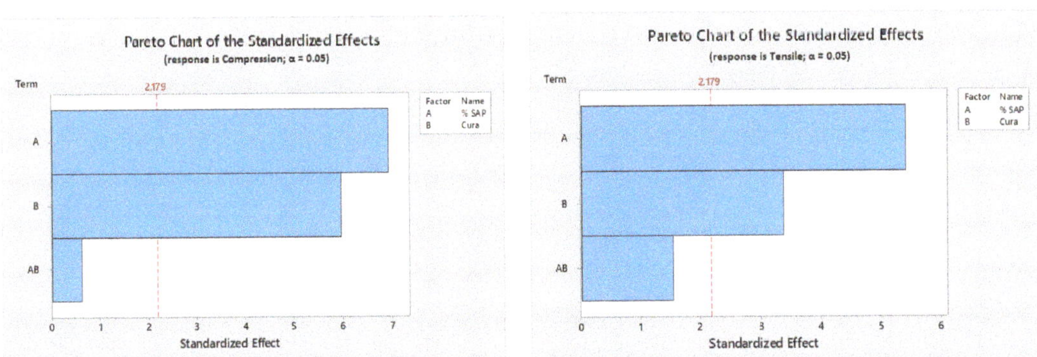

Figure 17. Pareto plot of standardized effects for CPV-ARI (Compression and Tensile).

Comparing the results of the Pareto charts between the two types of cement samples studied, it is noted that the PAS percentage parameter had higher significance for the CPV-ARI than the CPII-F-32. In Schröfl et al., 2012 [15], it was found that the SAP particles in the cement mixtures can create voids in the matrix, which can negatively affect the strength of the material, which justifies the loss of resistance of the CPV-ARI samples.

However, in their work, Agostinho et al., 2012 [47] found that CPV-ARI pastes with addition of 0.15% and 0.30% SAP showed lower rates of reaction acceleration due to a change in cement hydration kinetics, with peaks of lower C3S hydration due to internal curing or the presence of extra water that altered the w/c ratio. According to Baloch et al., 2019 [48], there are two possible behaviors for SAP. The first is the potential to intensify hydration processes, promoting internal healing and increasing endurance. The second is the possibility of creating voids in the matrix, weakening the resistance. In the studies of [49,50], it was found that a more uniform distribution of SAP throughout the cement matrix results in an increase in the effectiveness of internal curing in concrete. These works have shown that SAP has the ability to fill pores in a concrete mixture, which enables better control of autogenous shrinkage in cementitious composites.

4. Conclusions

In relation to the objectives initially set, it is possible to state that it was possible to analyze the addition of superabsorbent polymers (SAP) as internal curing elements in self-compacting mortar (SCM), as well as its mechanical and thermic characteristics, using the contents found as a base for the making of SCC applied to large volume structures.

In the fresh state testing, the addition of SAP resulted in a reduction in specific mass, a reduction in fluidity that had to be corrected with additional water, and an increase in slump time, which was corrected in the same manner, to both types of cement.

In relation to heat of hydration testing, both mortar using CPII-F-32 and CPV-ARI reached satisfactory results, allowing control of the temperature in the beginning of the curing process, something of importance to structures with a large volume of concrete, as is the case of hydraulic projects, in which temperature control is essential to avoid fissures caused by retraction. This can be an indication of a way of mitigating the temperature spikes that can occur in large volume concrete parts.

In the hardened state testing, the addition of SAP resulted in not significantly compromising the compression strength in mortar using CPII-F-32 cement, and in mortar using CPV-ARI cement, there was a substantial reduction in strength, reaching over 40%. The tensile strength of the submerged CPII-F-32 mortar presented similar increases, (around 21%); however, for the dry cured mortar, the increase in strength approached 70%. The mortar using CPV-ARI cement did not increase in strength when compared to the reference mix. The dynamic elasticity modulus presented a reduction with the addition of SAP; however, the reduction was lesser in samples with 0.2% of added SAP, both for type II as for type V cement.

Author Contributions: M.H.B.d.S.: conceptualization, methodological, data curation and writing—original draft preparation; L.R.R.S.: methodological, data curation and writing—original draft preparation; V.A.d.S.R.: writing—review and editing, visualization; P.C.G.: writing—review and editing, visualization; M.d.L.N.M.M.: writing—review and editing, visualization; C.E.M.G.: writing—review and editing, visualization; V.C.d.S.: conceptualization, methodological, writing—review and editing, supervision, project administration. All authors have read and agreed to the published version of the manuscript.

Funding: The authors thank CNPq for the financial support PIBIC CNPq.

Institutional Review Board Statement: Not applicable.

Data Availability Statement: Not applicable.

Acknowledgments: The authors thank company IGTPAN for the donation of SAP.

Conflicts of Interest: The authors declare no conflict of interest.

References

1. EFNARC. *The European Guidelines for Self-Compacting Concrete*; EFNARC: Flums Hochwiese, Switzerland, 2005; Volume 63.
2. Okamura, H.; Ouchi, M. Self-Compacting Concrete. *J. Adv. Concr. Technol.* **2003**, *1*, 5–15. [CrossRef]
3. Justs, J.; Wyrzykowski, M.; Bajare, D.; Lura, P. Internal Curing by Superabsorbent Polymers in Ultra-High Performance Concrete. *Cem. Concr. Res.* **2015**, *76*, 82–90. [CrossRef]
4. Hasholt, M.T.; Jensen, O.M.; Kovler, K.; Zhutovsky, S. Can Superabsorent Polymers Mitigate Autogenous Shrinkage of Internally Cured Concrete without Compromising the Strength? *Constr. Build. Mater.* **2012**, *31*, 226–230. [CrossRef]
5. Snoeck, D.; Jensen, O.M.; De Belie, N. The Influence of Superabsorbent Polymers on the Autogenous Shrinkage Properties of Cement Pastes with Supplementary Cementitious Materials. *Cem. Concr. Res.* **2015**, *74*, 59–67. [CrossRef]

6. Lura, P.; Friedemann, K.; Stallmach, F.; Mönnig, S.; Wyrzykowski, M.; Esteves, L.P. Kinetics of Water Migration in Cement-Based Systems Containing Superabsobent Polymers. In *Application of Super Absorbent Polymers (SAP) in Concrete Construction: State-of-the-Art Report Prepared by Technical Committee 225-SAP*; Springer: Empa, The Netherlands; Swiss Federal Laboratories for Materials Science and Technology: Dübendorf, Switzerland, 2012; pp. 21–37. ISBN 9789400727335/9789400727328.
7. Bentz, B.Y.D.P.; Lura, P.; Roberts, J.W. Mixture Proportioning for Internal Curing. *Concr. Int.* **2005**, *27*, 35–40.
8. Schroefl, C.; Mechtcherine, V.; Vontobel, P.; Hovind, J.; Lehmann, E. Sorption Kinetics of Superabsorbent Polymers (SAPs) in Fresh Portland Cement-Based Pastes Visualized and Quantified by Neutron Radiography and Correlated to the Progress of Cement Hydration. *Cem. Concr. Res.* **2015**, *75*, 1–13. [CrossRef]
9. Mehta, P.K.; Monteiro, P.J.M. *Concrete: Microstructure, Properties, and Materials*, 4th ed.; McGraw-Hill Education: New York, NY, USA, 2014; ISBN 9780071797870.
10. Domone, P. Self-Compacting Concrete: An Analysis of 11 Years of Case Studies. *Cem. Concr. Compos.* **2006**, *28*, 197–208. [CrossRef]
11. Jensen, O.; Hansen, P. Water-Entrained Cement-Based Materials: I. Principles and Theoretical Background. *Cem. Concr. Res.* **2001**, *31*, 647–654. [CrossRef]
12. Shen, D.; Liu, X.; Zeng, X.; Zhao, X.; Jiang, G. Effect of Polypropylene Plastic Fibers Length on Cracking Resistance of High Performance Concrete at Early Age. *Constr. Build. Mater.* **2020**, *244*, 117874. [CrossRef]
13. de Souza, M.H.B.; Gonçalves, P.C.; Silva, L.R.R.; de Melo, M.L.N.M.; dos Santos, V.C. Use of Superabsorbent Polymers in Cement-Based Compounds: A Bibliometric Analysis. *Res. Soc. Dev.* **2021**, *10*, e21818. [CrossRef]
14. de Souza, M.H.B.; Teixeira, B.A.; Gonçalves, P.C.; da Silva, L.R.R.; de Melo, M.L.N.M.; dos Ribeiro, V.A.S.; Sachs, D.; Capellato, P.; dos Santos, V.C. Effects on the Properties of Self-Compacting Cement Paste (PAA) with the Addition of Superabsorbent Polymer. *Materials* **2022**, *15*, 8478. [CrossRef] [PubMed]
15. Schröfl, C.; Mechtcherine, V.; Gorges, M. Relation between the Molecular Structure and the Efficiency of Superabsorbent Polymers (SAP) as Concrete Admixture to Mitigate Autogenous Shrinkage. *Cem. Concr. Res.* **2012**, *42*, 865–873. [CrossRef]
16. Rostami, R.; Klemm, A.J.; Almeida, F.C.R. Reduction of Shrinkage by Superabsorbent Polymers (SAP) in Fibre Reinforced Mortars. *Constr. Build. Mater.* **2021**, *288*, 123109. [CrossRef]
17. Shen, D.; Li, C.; Li, M.; Liu, C.; Kang, J. Experimental Investigation on Correlation between Autogenous Shrinkage and Internal Relative Humidity of Superabsorbent Polymer-Modified Concrete. *J. Mater. Civ. Eng.* **2022**, *34*, 4099. [CrossRef]
18. Wyrzykowski, M.; Assmann, A.; Hesse, C.; Lura, P. Microstructure Development and Autogenous Shrinkage of Mortars with C-S-H Seeding and Internal Curing. *Cem. Concr. Res.* **2020**, *129*, 105967. [CrossRef]
19. El-Sayed, A.M.; Faheim, A.A.; Salman, A.A.; Saleh, H.M. Sustainable Lightweight Concrete Made of Cement Kiln Dust and Liquefied Polystyrene Foam Improved with Other Waste Additives. *Sustainability* **2022**, *14*, 15313. [CrossRef]
20. Eskander, S.B.; Saleh, H.M.; Tawfik, M.E.; Bayoumi, T.A. Towards Potential Applications of Cement-Polymer Composites Based on Recycled Polystyrene Foam Wastes on Construction Fields: Impact of Exposure to Water Ecologies. *Case Stud. Constr. Mater.* **2021**, *15*, e00664. [CrossRef]
21. Hamzah, N.; Saman, H.M.; Baghban, M.H.; Sam, A.R.M.; Faridmehr, I.; Sidek, M.N.M.; Benjeddou, O.; Huseien, G.F. A Review on the Use of Self-Curing Agents and Its Mechanism in High-Performance Cementitious Materials. *Buildings* **2022**, *12*, 152. [CrossRef]
22. Zhang, B.; Zhu, H.; Feng, P.; Zhang, P. A Review on Shrinkage-Reducing Methods and Mechanisms of Alkali-Activated/Geopolymer Systems: Effects of Chemical Additives. *J. Build. Eng.* **2022**, *49*, 104056. [CrossRef]
23. Votorantim Cimentos. Available online: https://www.votorantimcimentos.com.br/produtos/cimentos-votoran/obras-especiais-industrial-meios-agressivos/ (accessed on 19 June 2023).
24. Brasil Intercement. InterCement. Available online: https://brasil.intercement.com/ (accessed on 19 June 2023).
25. Instituto Granado de Tecnologia da Poliacrilonitrila IGTPAN. Available online: https://www.igtpan.com/ (accessed on 19 June 2023).
26. Tutikian, B.F. Proposição de um Método de Dosagem Experimental Para Concretos Autoadensáveis. Ph.D. Thesis, Universidade Federal Do Rio Grande Do Sul, Rio Grande Do Sul, Brazil, 2007.
27. Helene, P.; Terzian, P. *Manual de Dosagem e Controle Do Concreto; Método Para Concreto Convencional IPT/EPSUP*; PINI: São Paulo, Brazil, 1992.
28. Da Silva, L.R.R. Análise de Propriedades do Concreto Autoadensável Com Resíduo Polimérico. Master's Thesis, Universidade Federal de Itajubá, Itajubá, Brazil, 2020.
29. Okamura, H.; Ouchi, M. Self-Compacting Concrete. Development, Present Use and Future. In Proceedings of the First International Rilem Symposiüm on Self-Compacting Concrete, Stockholm, Sweden, 13–14 September 1999.
30. Takada, K.; Pelova, G.; Walraven, J. The First Trial of Self-Compacting Concrete in The Netherlands According to the Japanese Design Method. In *Proceedings of FIP Congress*; FIP: Amsterdam, The Netherlands, 1998; pp. 113–115.
31. Edamatsu, H.; Da Nishi, N.; Oüchi, M. A Rational Mix-Design Method for Self-Compacting Concrete Considering Interaction between Coarse Aggregate and Mortar Particies. In Proceedings of the First International Rilem Symposiüm on Self-Compacting Concrete, Stockholm, Sweden, 13–14 September 1999; pp. 309–320.
32. Domone, P.L.; Jin, J. Properties of Mortar for Self-Compacting Concrete. In Proceedings of the First International Rilem Symposiüm on Self-Compacting Concrete, Stockholm, Sweden, 13–14 September 1999; pp. 109–120.
33. *NBR 120006*; Cement—Determining the Heat of Hydration Using the Langavant Bottle Method. NBR: Dhaka, Bangladesh, 1990.

34. *ABNT NBR 5739/2018*; Concreto—Ensaio de Compressão de Corpos-de-Prova Cilíndricos. Associação Brasileira de Normas Tecnicas: Rio de Janeiro, Brazil, 2018.
35. *ABNT NBR 7222:2011*; Concreto e Argamassa—Determinação da Resistência à Tração Por Compressão Diametral de Corpos de Prova Cilíndricos. ABNT—Associação Brasileira de Normas Técnicas: Rio de Janeiro, Brazil, 2011.
36. *ABNT NBR 9778:2009*; Argamassa e Concreto Endurecidos—Determinação da Absorção de Água, Índice de Vazios e Massa Específica. ABNT—Associação Brasileira de Normas Técnicas: Rio de Janeiro, Brazil, 2009.
37. *ASTM E1876*; International Standard Test Method for Dynamic Young's Modulus, Shear Modulus, and Poisson's Ratio by Impulse Excitation of Vibration. ASTM International: West Conshohocken, PA, USA, 2015; p. 17.
38. Monteiro, P.; Mehta, P. *Concrete: Microstructure, Properties and Materials*; McGraw-Hill: New York, NY, USA, 2006; ISBN 0071462899/9780071462891.
39. Kumm, T.C. *Influência do Emprego de Polímeros Superabsorventes Nas Propriedades de Materiais à Base de Cimento Portland*; Universidade Federal de Santa Catarina: Florianopolis, Brazil, 2009.
40. Muthalvan, R.S.; Ravikumar, S.; Avudaiappan, S.; Amran, M.; Aepuru, R.; Vatin, N.; Fediuk, R. The Effect of Superabsorbent Polymer and Nano-Silica on the Properties of Blended Cement. *Crystals* **2021**, *11*, 1394. [CrossRef]
41. Suarez, M.L.G. *Polímeros Super Absorventes (PSA) Como Agente de Cura Interna Para Previnir Fissuração em Concretos de Alta Resistência*; Universidade de Brasília: Brasília, Brazil, 2015; p. 82.
42. Schröfl, C.; Mechtcherine, V. Superabsorbent Polymers to Mitigate Autogenous Shrinkage of Cement Mortar. In Proceedings of the 10th International Conference on Superplasticizers and Other Chemical Admixtures, Prague, Czech Republic, 28–31 October 2012; Research Group for Microstructural Characterization, Institut für Baustoffe (Institute of Construction Materials), Technische Universität: Dresden, Germany, 2012; pp. 265–278.
43. Jensen, O.M.; Hansen, P.F. Water-Entrained Cement-Based Materials: II. Experimental Observations. *Cem. Concr. Res.* **2002**, *32*, 973–978. [CrossRef]
44. Pereira, D.F.; e Matos, V.N. *Combate à Retração Autógena Utilizando Polímeros Superabsorventes*; Monografia de Projeto Final; Departamento de Engenharia Civil e Ambiental, Universidade de Brasília: Rio de Janeiro, Brazil, 2011; 72p.
45. Mechtcherine, V.; Dudziak, L.; Hempel, S. Mitigating Early Age Shrinkage of Ultra-High Performance Concrete by Using Super Absorbent Polymers (SAP). In Proceedings of the 8th International Conference on Creep, Shrinkage and Durability Mechanics of Concrete and Concrete Structures, Ise-Shima, Japan, 30 September–2 October 2008; Institute of Construction Materials, TU: Dresden, Germany, 2009; Volume 2, pp. 847–853.
46. Herrera-Sosa, E.S.E.A. Recovery and Modification of Waste Tire Particles and Their Use as Reinforcements of Concrete. *Int. J. Polym. Sci.* **2015**, *2015*, 1. [CrossRef]
47. Agostinho, L.B.; Borges, J.G.; da Silva, E.F.; Cupertino, D.V.M.R. A calorimetry analysis of portland cement pastes containing superabsorbent polymer (Sap) and nanosilica (ns). *Rev. Mater.* **2020**, *25*, 1160. [CrossRef]
48. Baloch, H.; Usman, M.; Rizwan, S.A.; Hanif, A. Properties Enhancement of Super Absorbent Polymer (SAP) Incorporated Self-Compacting Cement Pastes Modified by Nano Silica (NS) Addition. *Constr. Build. Mater.* **2019**, *203*, 18–26. [CrossRef]
49. Kalinowski, M.; Woyciechowski, P.; Sokołowska, J. Effect of Mechanically-Induced Fragmentation of Polyacrylic Superabsorbent Polymer (SAP) Hydrogel on the Properties of Cement Composites. *Constr. Build. Mater.* **2020**, *263*, 120135. [CrossRef]
50. Mechtcherine, V.; Gorges, M.; Schroefl, C.; Assmann, A.; Brameshuber, W.; Ribeiro, A.B.; Cusson, D.; Custódio, J.; Da Silva, E.F.; Ichimiya, K.; et al. Effect of Internal Curing by Using Superabsorbent Polymers (SAP) on Autogenous Shrinkage and Other Properties of a High-Performance Fine-Grained Concrete: Results of a RILEM Round-Robin Test. *Mater. Struct. Constr.* **2014**, *47*, 541–562. [CrossRef]

buildings

MDPI

Article

Study on Mechanical Properties of Heap Deposited Fly Ash Based Geopolymers with Different Alkaline Activator Properties

Martin Sisol [1], Michal Marcin [1,*], Karel Dvořák [2], Michaela Suďová [1] and Viera Ivanková [1]

[1] Institute of Earth Resources, Faculty of Mining, Ecology, Process Control and Geotechnologies, Technical University of Košice, Letná 9, 040 01 Košice, Slovakia
[2] Institute of Technology of Building Materials and Components, Faculty of Civil Engineering, Brno University of Technolgy, Veveří 512/95, 602 00 Brno, Czech Republic
* Correspondence: michal.marcin@tuke.sk

Abstract: Geopolymers can improve environmental protection and are therefore considered a 21st-century material that requires special attention. The main objective of this study was to examine the utilization of fly ash deposited on a heap as a potential raw material for the geopolymerization process. In this work, flexural and compressive strengths, as well as water absorption of geopolymers, were evaluated based on alkaline activator properties. Na_2O content, SiO_2/Na_2O molar ratio and fly ash-to-water ratio were key observed factors. Mechanical strengths tests were conducted after 7, 28, 90, 180, and 360 days. The research findings showed that geopolymer cured at appropriate conditions and with a suitable alkaline activator can reach a compressive strength of more than 55 MPa. The presented study highlighted the possibility of the efficient use of fly ash to produce useful materials, i.e., geopolymers, which contribute to environmentally friendly solutions. This approach is fully in line with the principles of sustainable development.

Keywords: geopolymer; fly ash; mechanical properties; compressive strength; flexural strength; environmentally friendly materials; sustainable development; green alternative; environmental solutions

check for updates

Citation: Sisol, M.; Marcin, M.; Dvořák, K.; Suďová, M.; Ivanková, V. Study on Mechanical Properties of Heap Deposited Fly Ash Based Geopolymers with Different Alkaline Activator Properties. *Buildings* **2022**, *12*, 1780. https://doi.org/10.3390/buildings12111780

Academic Editor: Tomáš Dvorský

Received: 21 September 2022
Accepted: 20 October 2022
Published: 24 October 2022

1. Introduction

In line with the objectives set by the European Union, which also include achieving an economy with net-zero greenhouse gas emissions, it is desirable to reduce the amount of carbon dioxide released into the atmosphere and to avoid loss of the protection of the ozone layer [1,2]. In addition, climate action, responsible production, and sustainable industry are among the 17 Sustainable Development Goals (SDGs) [3]. However, the huge increase in carbon dioxide and waste production in recent decades is a crucial obstacle to sustainable development and sustainability [4]. For these reasons, solutions must be sought in various areas, and manufacturing and processing are no exception. In this context, a significant amount of carbon dioxide is emitted during the calcination process in the process of obtaining ordinary Portland cement (OPC) in which limestone is transformed to lime. Simultaneously, burning fossil fuels to deliver thermal energy for calcination leads to global warming.

In recent years, this issue has become increasingly prevalent. As a result, even a modest decrease in Portland cement usage will significantly reduce carbon dioxide emissions and lead to major environmental benefits. On this basis, it is possible to emphasize the huge potential of geopolymers with respect to environment protection. Alternative binders are required to minimize carbon dioxide emissions and to reduce greenhouse gas emissions [5–7].

This has prompted studies into finding new environmentally friendly, cementitious products that have high strength and longevity while using a reasonable amount of energy resources in the manufacturing process. Many studies have investigated geopolymer and alkali-activated materials (AAM), which are synthesized using fly ash, metakaolin, or slag, as an alternative to cement with alkaline activator over the last two decades [8–11].

In more detail, geopolymers are semi-crystalline alkali-activated aluminosilicates formed when aluminosilicates react in alkaline media [9]. Geopolymers have a similar chemical composition to zeolites, but their microstructure is amorphous [12]. The development of geopolymer also encourages the use of aluminosilicate materials as the key binding material in geopolymer, such as fly ash, a natural industrial by-product from coal combustion in power plants [13]. The structure of fly ash also entitles it to be included in the manufacture of construction materials [14]. A geopolymerization process is an effective approach for converting a wide variety of aluminosilicate waste materials into usable binders that can be used for building materials. A significant number of naturally occurring and industrially generated aluminosilicates are used for geopolymer synthesis [15,16].

Fly ash is an undesirable byproduct formed by the burning of pulverized coal. Industrialization has contributed to the production and deposition of fly ash in vast amounts over the years [5]. According to statistics, about 780 million tons of fly ash are generated in the world each year, but only about 17–20% are used [17–19]. The primary use of fly ash is in the manufacture of OPC. Furthermore, a recent study shows that fly ash can be used to make carbon nanotubes or as a source of rare elements [20]. The use of fly ash in geopolymers helps to alleviate the growing environmental impact of aluminosilicate minerals including fly ash and slag [21].

The structure of the fly ash along with the alkaline activator have a major impact on the final properties of the geopolymer, according to References [13,22]. There are several types of alkaline activators that can be used to activate fly ash. Sodium hydroxide (NaOH) and sodium silicate (Na_2SiO_3), as well as potassium hydroxide (KOH) and potassium silicate, are widely used alkaline activators in the geopolymerization procedure (K_2SO_3) [23]. The molarity of NaOH and Na_2SiO_3 in the alkaline activator solution is one of the parameters that affect the chemical reactions of the geopolymer paste. Numerous scientific studies have observed the impact of the molarity ratios of NaOH and Na_2SiO_3/NaOH on the compressive and flexural strengths of geopolymers [24–28]. Oxide composition of precursors has a direct influence on the compressive and flexural strength [29] as well as the structural and rheological characteristics of raw material [30]. The alkaline solution represents an essential role in the production of geopolymers with high mechanical strength. Its choice is primarily based on its reactivity and the cost of the materials used.

The Si/Al ratio, hydroxide concentration, efficiency of the geopolymerization reaction, and curing conditions all have a significant impact on the mechanical properties of geopolymers. Based on the materials used or the molar ratio, geopolymers may have a variety of mechanical, chemical, and thermal properties [31–34]. Some of these properties are high initial strength, resistance to chemical attacks from substances like acids or sulfates, thermal and fire stability at high temperature, as well as resistance to freezing [35]. As a result of all of these outstanding qualities, they are interesting renewable structural materials for the future. Since they are a non-flammable material with reasonably good insulation, geopolymers are an ideal alternative to materials such as polystyrene and glass [36].

At this point, it is also possible to speak of a circular economy with great opportunity for sustainable development. The circular economy is an economic model that bears in its name the policy of managing resources in one round, in circulation, continuously round and round, until the maximum possible use and closure of material flows. Accordingly, fly ash and other waste materials generated by the industrial and manufacturing sectors can be included in this circle and thus be used to make new products, which is in line with the circular economy. In fact, construction is a highly energy-intensive sector and one of the main sources of environmental degradation. On this basis, there is a great need for the research and development of new and sustainable approaches [37]. Evidence shows that new building materials based on alkaline activation technology are very promising and provide environmental and ecological benefits, especially in relation to products that need significant technological properties and durability [38]. In this context, geopolymers are a greener substitute for cement to significantly reduce environmental degradation [37,39]. Their use leads to large savings in energy and primary materials, which corresponds to sustainable production. Also, it reduces the need for OPC consumption and the consequent production of carbon dioxide [4].

Theoretically, this study contributes to the knowledge of environmentally friendly materials, i.e., geopolymers, in the conditions of Slovakia. It also helps policy makers to better understand the issue and develop more effective strategies and mechanisms to improve environmental protection, remove obstacles, and achieve sustainable development. Practical use is obvious in the construction industry. The experimental evidence of this study may encourage civil engineers in Slovakia to use green alternatives, which is currently cautious. This issue has long been neglected and underestimated in practice in Slovakia, and therefore this study can be a valuable platform for proactive and evidence-based solutions.

In Slovakia, almost all of the heating plant waste, especially fly ash, is not utilized. It can be concluded that little attention is paid to this issue in Slovakia, which is reflected in the lack of evidence-based practices [40]. In this paper, the utilization of fly ash from Slovakia, mainly deposited in a heap, which creates an environmental problem, was used as a potential aluminosilicate material for the geopolymerization process. In this study, the created alkali-activated materials are regarded the same as geopolymers. Examined mechanical properties, such as compressive strength, flexural strength, and water absorption, were observed over the course of a year. The influence of different parameters of alkaline activators, such as Na_2O amount, SiO_2-to-Na_2O molar ratio, and water-to-fly ash ratio, was the main observation point in terms of their effect on the above-mentioned mechanical properties of geopolymers.

2. Materials and Methods

The main objective of this study was to examine the utilization of fly ash deposited in a heap as a potential raw material for the geopolymerization process. The raw material for this experiment was a heap of deposited fly ash (Class F) (HDFA) from the District Heating Plant in Košice, Slovakia using melting boilers with a temperature of 1400–1550 °C. HDFA was collected from an upper layer sludge bed. Chemical and mineralogical analyses are shown in Tables 1 and 2. Based on its chemical composition provided by analysis, the SiO_2/Al_2O_3 ratio is 2.99. The particle distribution of HDFA is shown in Figure 1.

Table 1. Chemical analysis of a raw material.

wt.%	SiO_2	Al_2O_3	Fe_2O_3	CaO	MgO	Other	LOI
HDFA	45.63	15.28	8.74	3.62	1.68	2.68	23.87

Table 2. Mineralogical analysis of a raw material.

wt.%	Amorphous	Hematite	Quartz	Mullite
HDFA	83.52	1.57	6.42	8.46

Chemical analysis was performed by atomic absorption spectroscopy (AAS) (Perkin Elmer instrument, model 1100B). For mineralogical analysis, an X-ray diffractometer URD-6/ID 3003 (Rich. Seifert-FPM, Germany) was used, under the following conditions: X-ray radiation Co Kα, high voltage of 40 kV, current 35 mA, step scan mode with the step of 0.05° 2θ, time per step 3 s.

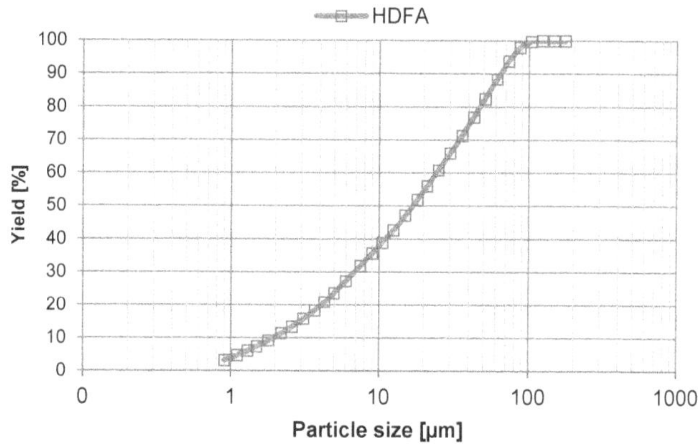

Figure 1. Particle size distribution of HDFA.

The particle size distribution of raw material is one of the most important characteristics that affect its reactivity [41,42]. The size distribution of HDFA was obtained using a Diffraction Spectrometer Helos 12 LA (Sympatec GmbH, Clausthal-Zellerfeld, Germany) with a dispersing Feeder Rodos 12 SR for dry, pulverulent solids. Particle size analysis was in the range of 0.9–170 μm. HDFA 50% comprehends particles smaller in size than 17 microns.

Scanning electron microscopy (SEM) was used to investigate the morphology of the HDFA (Figure 2) using a Jeol JSM 7000Fdevice.

Figure 2. SEM picture of HDFA.

The mechanical properties of geopolymers were examined depending on the alkaline activator used. The effect of Na_2O, the effect of SiO_2/Na_2O molar ratio, and the water-to-HDFA ratio were observed. Samples were prepared by mixing HDFA with an alkaline activator. The SiO_2/Na_2O molar ratio (Ms modulus) in alkaline activator was adjusted by NaOH addition to Na–water glass. The alkaline activator was made by combining NaOH solid pellets with Na–water glass and distilled water. Sodium–water glass (Kittfort Praha Co.) with a density of 1.328–1.378 g/cm^3 was used. The sodium–water glass contains 36–38% Na_2SiO_3 and the molar ratio of SiO_2/Na_2O is 3.2–3.5. Solid NaOH with a density of 2.13 g/cm^3 (Kittfort Praha Co., Neratovice, Czech Republic) containing at least 97–99.5% of NaOH.

The design of the experimental work was divided into 3 parts. Each part of the experiment was conducted with the same conditions, but the alkaline activator varied. Table 3 describes the composition of each tested sample with highlighted varied parameter.

Table 3. Design of experiment.

Material	Na_2O	SiO_2/Na_2O	W
HDFA	**6**	1.25	30
HDFA	**7**	1.25	30
HDFA	**8**	1.25	30
HDFA	**9**	1.25	30
HDFA	8	**0.75**	30
HDFA	8	**1**	30
HDFA	8	**1.25**	30
HDFA	8	**1.5**	30
HDFA	8	**1.65**	30
HDFA	8	1.25	**30**
HDFA	8	1.25	**32**
HDFA	8	1.25	**35**

Altogether, 12 samples were prepared with a different alkaline activator. The amount of Na_2O ranged from 6 to 9 (%), the SiO_2/Na_2O molar ratio (mol/mol) ranged from 0.75 to 1.65, and water to HDFA ratio (W) varied from 30 to 35 (%). HDFA mixture with alkaline activator was stirred for 10 min before the homogenous mixture was formed. The mixture was poured into molds with dimensions of 40 × 40 × 160 mm and compacted on a vibration table. Samples were then cured in a hot air-drying chamber for 6 h at 80 °C. Following that, the samples were taken out of the forms, labeled, and deposited in laboratory conditions with room temperature (20 °C) before the mechanical properties' tests. Mechanical strength measurements were made following the Slovak standard STN EN 12390-3, which indicates that compressive and flexural strengths were tested after 7, 28, 90, 180, and 360 days. The same three samples were split in half to assess flexural strength, and six halves were tested for compressive strength. The mechanical strengths of the cured samples were analyzed using the Form + Test MEGA 100-200-10D hydraulic device. A part of the samples was stored at laboratory temperature for 28 days before undergoing a water absorption test following Slovak Standard STN 73 1316.

3. Results and Discussion

3.1. Effect of Na_2O Amount

The first series of strengths and water absorption tests were focused on Na_2O concentration changing in alkaline activator solution. This concentration was in the range of 6–9%. Figure 3 shows all flexural and compressive test results.

The geopolymer with the lowest strength was obtained with 6% Na_2O in the mix. After 360 days, the flexural strength was 5.0 MPa, and the compressive strength was 27.7 MPa. This is a factor that noticeably affects the alkaline solution's content is the concentration of an alkaline solution. Higher strength is acquired by applying an alkaline solution of a higher concentration, which indicates that with lower concentrations, the strengths were lower [43,44]. A sample with 8% Na_2O in the mixture had the highest strength, with a flexural strength of 11.0 MPa after 360 days and a compression strength of 57.1 MPa. The resulting strength difference was approximately twice as high. Intense increases in mechanical strength occurred during the first 90 days when the samples reached their maximum strength. Their strengths did not change significantly, except for a sample containing 7% Na_2O, for which strength increased throughout the testing.

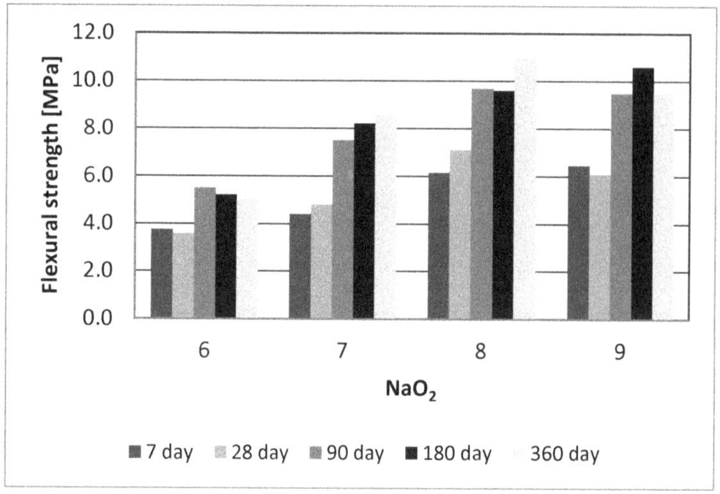

(a)

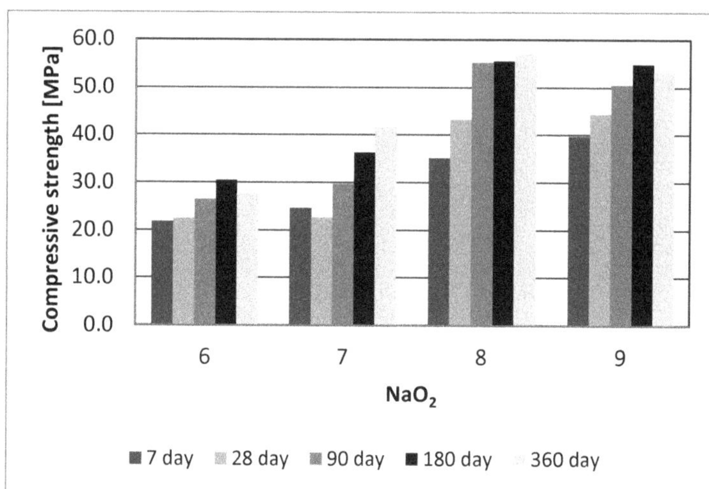

(b)

Figure 3. Flexural strengths (**a**) and compressive strengths (**b**) results with different Na_2O amounts in the alkaline activator.

Water absorption results with different Na_2O amounts in alkaline activator are shown in Figure 4. The increased amount of Na_2O in the mixture had a positive effect on the water absorption of the cured geopolymer samples. Samples with 8 and 9% Na_2O had a very similar trend of water absorption over time and reached the lowest values. The water absorption of the sample was 6.68% with 8% Na_2O in the mixture and 6.71% with 9% Na_2O. The sample with 6% Na_2O in the mixture (12.98%) had the highest water absorption. Again, samples with higher strength were shown to have lower water absorption.

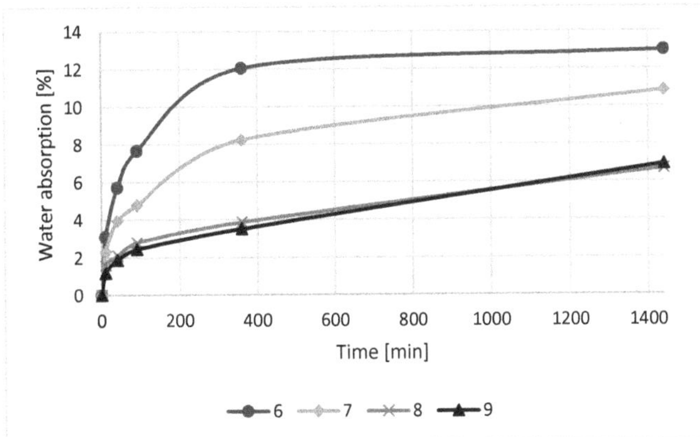

Figure 4. Water absorption results with different Na_2O amounts in the alkaline activator.

3.2. *Effect of SiO_2/Na_2O Molar Ratio*

The effects of different alkaline activators on the compressive and flexural strengths of samples with the different SiO_2/Na_2O molar ratios are shown in Figure 5. The values of flexural strength of HDFA geopolymers increase with SiO_2/Na_2O molar ratio from 0.75 to 1.25. However, increasing the SiO_2/Na_2O molar ratio further reduces the flexural strengths achieved. Higher amounts of SiO_2 in the alkaline activator impacted the flexural strengths, hence the strengths were decreased. The highest value was observed after 360 days and is equal to 11.0 MPa (1.25 sample).

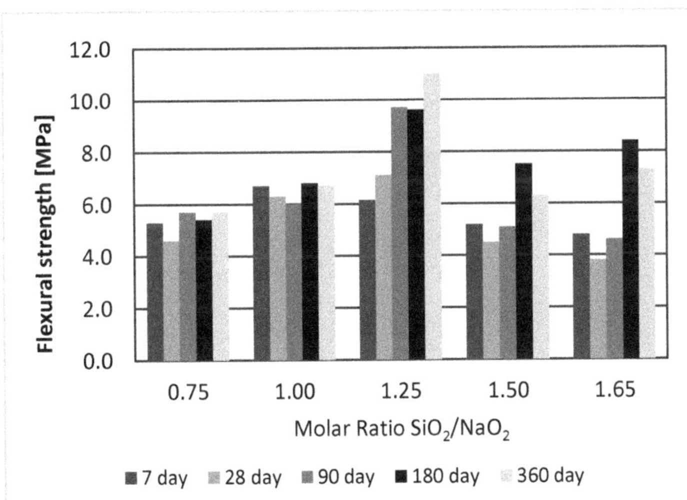

(a)

Figure 5. *Cont.*

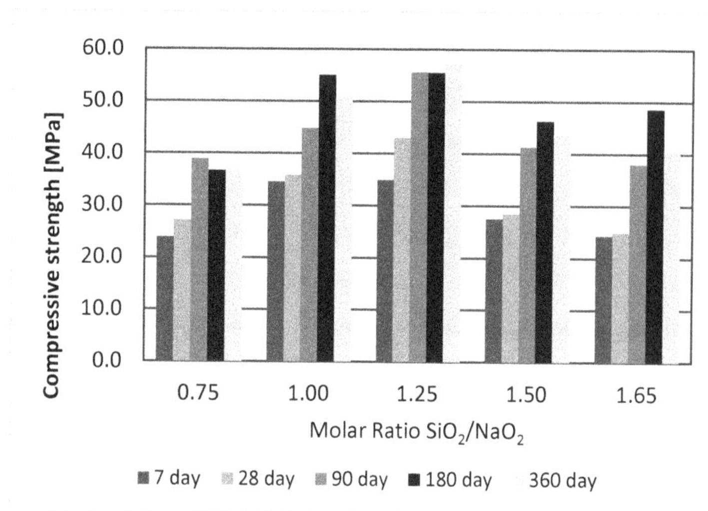

(**b**)

Figure 5. Flexural strengths (**a**) and compressive strengths (**b**) of samples with different SiO_2/Na_2O molar ratio.

As the SiO_2/Na_2O molar ratio was increased to 1.25, the compressive strength increased, but beyond that value, compressive strength decreased. In this case, only one sample obtained the highest strength after 360 days, and the highest value reached 57.1 MPa. The sample's (1.25) strength did not increase with time, since after 90 days, the strength was 55.6 MPa, meaning the compressive strength remained roughly the same.

The water absorption test was examined over 24 h. These results are presented in Figure 6. Water absorption reflected compressive strength results, with samples with higher compressive strengths achieving better and lower water absorption. The lowest percentage was 6.84% (12.5 sample).

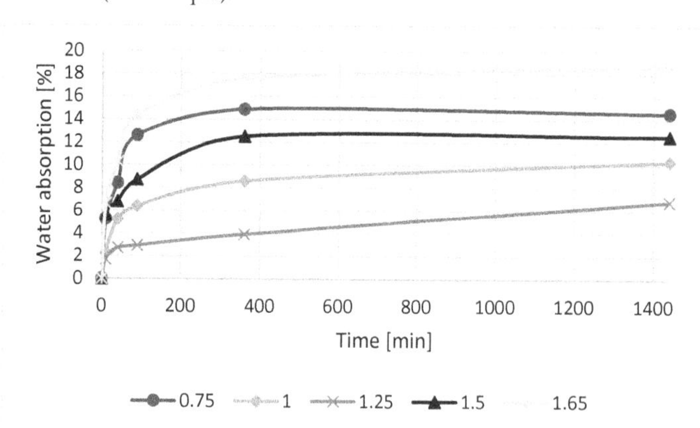

Figure 6. Water absorption results with different SiO_2/Na_2O molar ratio.

3.3. *Effect of HDFA/Water Ratio*

In this series of experiments, the effect of the HDFS/water ratio in the mixture on the resulting properties of geopolymers was studied. The amount of water (w) in the mixture ranged from 30 to 35%. The results of the flexural and compressive strengths are in Figure 7 and water absorption in Figure 8.

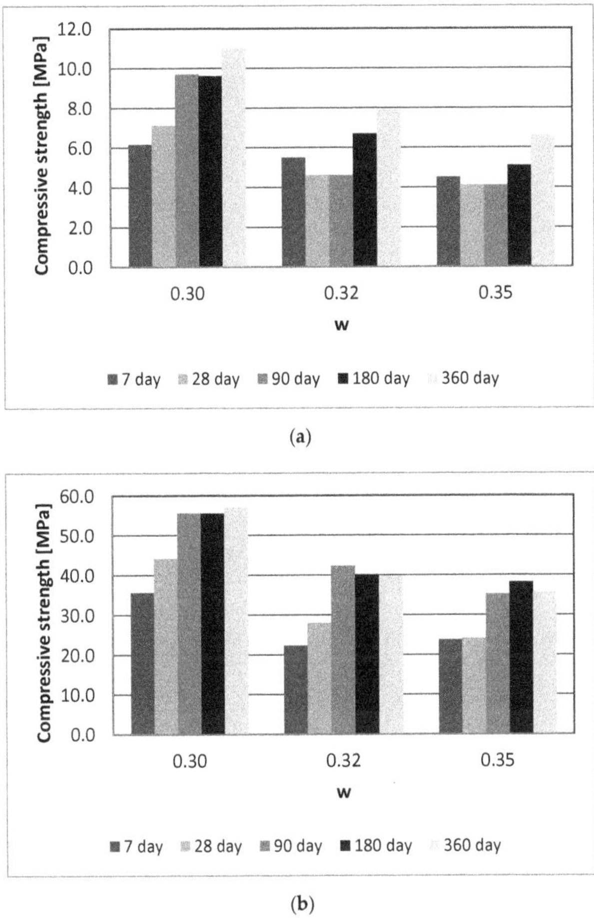

Figure 7. Flexural strengths (**a**) and compressive strengths (**b**) strength results with different HDFS/water ratios in the mixture.

An increase in the amount of water in the geopolymer mixture has a negative effect on the development of the strength of the geopolymer samples. With a constant amount of Na_2O in the mixture, the molarity of the sodium hydroxide used decreases, and the mechanical strength of the geopolymer samples decreases. Another factor contributing to the weakening of geopolymer samples is porosity, which is caused by a greater number of water-filled pores in a product with higher initial water content. The logical consequence of increasing the porosity of the synthesized geopolymers is their reduced strength. The geopolymer sample with the least amount of water in the alkaline activator solution (w = 30%) had the greatest mechanical strength. Its flexural strength was 11.0 MPa and compressive strength was 57.1 MPa after 360 days of storage in laboratory conditions.

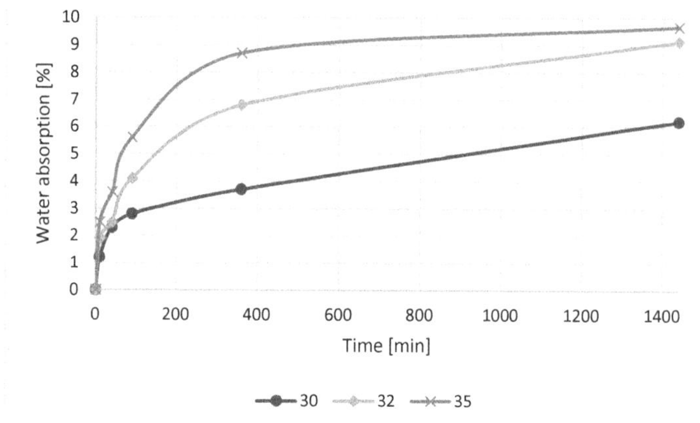

Figure 8. Water absorption results with different HDFS/water ratios in the mixture.

The development of the water absorption of the formed geopolymer samples is also affected by an increase in the amount of water in the geopolymer mixture. The sample with the lowest water absorption, 6.21 percent, had w = 30%. When the amount of water in the geopolymer mixture was increased, the water absorption of the geopolymer samples increased as well. The sample with w = 35% had the highest absorbency of up to 9.68%.

Although geopolymers are an extensively researched topic, there is a constant need to improve the mechanical performance and properties of geopolymers due to the relentless search for environmentally friendly solutions. In this context, it is generally known that fly ash-based geopolymers have very good properties such as good fire and chemical resistance, high compressive strength development, low thermal contraction, and conductivity [11,24,45,46].

In this study, the composition of the alkaline activator significantly affected the resulting properties of the geopolymers. Increasing the amount of Na_2O in the mixture, at a constant SiO_2/Na_2O molar ratio and amount of water, resulted in an increase in strength and a decrease in the water absorption of the geopolymers. Similar results were revealed by [47], who confirmed that higher compressive strength can be obtained with higher Na_2O content, but also with higher curing temperatures and longer pre-curing periods at relatively low temperature. These findings indicate that Na_2O can be considered as the major alkali oxide required for strength development [48]. The amount of sodium hydroxide in the activating solution mainly affects the rate of release of silicon and aluminum from the fly ash into the solution [41], because a partial role of the base is to increase the solubility of fly ash. In the base environment, the acidic hydroxyl groups on the surface of the quartz are neutralized, which allows for the breaking of Si-O-Si bonds [42]. Increasing the Na_2O amount in previous work shows the positive effect on mechanical strengths [48,49]. The mechanical strengths of the prepared geopolymers improved as the amount of sodium water glass in the alkaline activator solution increased while the amount of Na_2O and water in the mixture remained constant. This is consistent with the findings of [50], who revealed that the amount of sodium–water glass had effects on the properties of their samples. At the molar ratio $SiO_2/Na_2O = 1.25$, the maximum compressive strength and lowest water absorption are achieved. The strengths of the geopolymers decreased as the volume of sodium water glass was increased, while the water absorption of the geopolymers increased. Higher SiO_2/Na_2O molar ratios have been shown to reduce the strengths due to sodium silicate diluting the alkalinity, resulting in less precursor dissolution during geopolymerization [34]. The amount of water in the mixture had a huge effect on the mechanical strength of the geopolymer materials that were produced. As the volume of water in the mixture was increased, the mechanical strength of the mixture decreased and the water absorption of the geopolymers increased. A 2% increase in the volume of water in the mixture resulted in a 30% reduction in mechanical strength and a 30% increase in water absorption. The degradation of geopolymer mechanical properties may be, in part, due to the fact that as the amount of water in the mixture increases while the amount of NaOH remains unchanged, the concentration of NaOH decreases, decreasing the ash particle dissolution rate. Another explanation may be porosity, which is caused by a greater number of water-filled pores in a substance with higher water content. Increasing the porosity of geopolymers has the logical result of lowering their strengths [51].

Overall, fly ash-based geopolymers prove to be a promising material due to the obtained compressive strength, environmental, and economic aspects. It is possible to speak of both environment protection and cost savings [52]. Currently, some geopolymers are already replacing naturally occurring materials due to their unique properties, but the production of carbon dioxide from OPC remains an open and persistent problem, as does the way fly ash is processed [4,53]. Over time, much more stringent emission reduction measures can be expected to slow down global warming, as it is impossible to stop it altogether [52]. Geopolymers help to reduce the carbon footprint and address sustainable growth issues. The role of policy makers is therefore to promote the use of geopolymers in practice and to encourage civil engineers to favor this still little-known alternative in Slovakia. The results of this study provide them with a valuable platform. It should be borne in mind that the sustainable development of new materials can ensure the industry's competitiveness, its stable economic growth, and a healthy environment [39]. Before effectively implementing environmentally friendly approaches, it is first necessary to know the possibilities, and this study offers one of them. The presented findings show the possibility of efficient use of fly ash for the production of useful materials, i.e., geopolymers, which contribute to greener solutions. Based on the results, geopolymers can be better used in practice in Slovakia. This approach is fully in line with the principles of sustainable development.

4. Conclusions

The current state of the environment requires countries to be more proactive in using green approaches, but this cannot be done without experimental evidence. The main objective of this study was to examine the utilization of fly ash deposited in a heap as a potential raw material for the geopolymerization process. An experimental analysis of the mechanical properties of geopolymer prepared using heap-deposited fly ash and various alkaline activators were carried out and described in this study. The following observations can be drawn based on the experimental findings.

The geopolymeric constructions and infrastructure sectors have undergone a revolution as a result of the recent development of unique and ultramodern geopolymer technology and applicable advanced research. As a result, a new age of environmentally friendly construction materials has begun thanks to the practical substitution of geopolymer cement (GPC) for regular Portland cement (OPC). In addition, compared to current OPC manufacturing, this novel, user- and eco-friendly inorganic family of geopolymeric materials has a minimal carbon footprint and can be made using a simple technique at a low temperature and with less energy.

1. The mechanical strengths of the geopolymers improved as the volume of Na_2O in the mixture increased. After 360 days, the compressive strengths have the highest values, but after 90 days, the strength did not improve much.
2. The geopolymer strengths improved as the SiO_2/Na_2O molar ratio increased to 1.25, but a further increase was accompanied by a decline in strength. After 180 days, several samples showed a decrease in strength.
3. The mechanical strength of the geopolymers was decreased as the volume of water in the mixture was increased.
4. A water absorption experiment revealed that an increase in the amount of water in the geopolymer mixture has a negative effect on the development of the strength of

the geopolymer samples. The mechanical strengths of geopolymer samples have an impact on final water absorption.

5. The geopolymer has the greatest mechanical properties when it was prepared with an alkaline activator solution with a SiO_2/Na_2O molar ratio of 1.25, 8% Na_2O in the mixture, and 30% water in the mixture. This alkaline activator solution's formulation was ideal for the HDFA geopolymers from our investigation. This mixture was used to produce geopolymer, which was cured for 6 h with 80 °C. It had a flexural strength of 11 MPa and a compressive strength of 57.1 MPa after 360 days, and its water absorption was 6.68%.

This study supports the use of fly ash to produce geopolymers in line with reducing environmental degradation and promoting sustainable development. Ultimately, it can be claimed that geopolymers as a material have potential, as they are completely resilient, sustainable, and affordable.

In Slovakia, nearly all the heating plant waste, primarily fly ash, is not utilized and is discarded on in the form of heaps. It can be concluded that little attention is paid to this concern in Slovakia, which is reflected in the lack of evidence-based practices. According to Slovakian legislation, heaps are sources of dangerous wastes, which is the main environmental problem. This issue has long been neglected and underestimated in practice in Slovakia, and therefore this study can be a valuable platform for proactive and evidence-based solutions. In this paper, the utilization of fly ash from Slovakia, deposited on a heap near the city Kosice which creates an environmental problem, was used as a potential aluminosilicate material for the geopolymerization process. As we proved from the results obtained from alkali activation, the heap near Kosice can be utilized, and the environmental problem can be removed by starting to use new methods of discarding fly ash.

Author Contributions: Conceptualization, M.M.; methodology, M.M and M.S. (Martin Sisol); investigation, M.M. and M.S. (Michaela Sud'ová); data curation, M.S. (Martin Sisol) and K.D.; writing—original draft preparation, M.M.; writing—review and editing, M.M. and V.I.; supervision, M.S. (Martin Sisol). All authors have read and agreed to the published version of the manuscript.

Funding: This research was funded by the European Institute of Innovation and Technology (EIT), a body of the European Union, under the Horizon 2020, the EU Framework Programme for Research and Innovation (Project BioLeach: Innovative Bio-treatment of RM, grant number: 18259). This research was funded by the European Institute of Innovation and Technology (EIT), a body of the European Union, under the Horizon Europe, the EU Framework Programme for Research and Innovation (Project DYNOSORT Dynamic ore sorting of polymetallic stockpiles, grant number: 21050).

Conflicts of Interest: The authors declare no conflict of interest.

References

1. European Commission. 2050 Long-Term Strategy. Available online: https://ec.europa.eu/clima/eu-action/climate-strategies-targets/2050-long-term-strategy_en (accessed on 25 May 2022).
2. European Commission. A European Green Deal. Available online: https://ec.europa.eu/info/strategy/priorities-2019-2024/european-green-deal_en (accessed on 25 May 2022).
3. Sachs, J.; Schmidt-Traub, G.; Kroll, C.; Lafortune, G.; Fuller, G.; Woelm, F. *Sustainable Development Report 2020: The Sustainable Development Goals and COVID-19 Includes the SDG Index and Dashboards*; University Press Cambridge: Cambridge, UK, 2021. [CrossRef]
4. Nodehi, M.; Taghvaee, V.M. Alkali-activated materials and geopolymer: A review of common precursors and activators addressing circular economy. *Circ. Econ. Sust.* **2022**, *2*, 165–196. [CrossRef]
5. John, S.K.; Nadir, Y.; Girija, K. Effect of source materials, additives on the mechanical properties and durability of fly ash and fly ash-slag geopolymer mortar: A review. *Constr. Build. Mater.* **2021**, *280*, 122443. [CrossRef]
6. Lee, N.K.; Jang, J.G.; Lee, H.K. Shrinkage characteristics of alkali-activated fly ash/slag paste and mortar at early ages. *Cem. Concr. Compos.* **2014**, *53*, 239–248. [CrossRef]
7. Gunasekara, C.; Law, D.W.; Setunge, S.; Sanjayan, J.G. Zeta potential, gel formation and compressive strength of low calcium fly ash geopolymers. *Constr. Build. Mater.* **2015**, *95*, 592–599.
8. Duxson, P.; Fernández-Jiménez, A.; Provis, J.L.; Lukey, G.C.; Palomo, A.; van Deventer, J.S. Geopolymer technology: The current state of the art. *J. Mater. Sci.* **2007**, *42*, 2917–2933. [CrossRef]

9. Khale, D.; Chaudhary, R. Mechanism of geopolymerization and factors influencing its development: A review. *J. Mater. Sci.* **2007**, *42*, 729–746. [CrossRef]
10. Duxson, P.; Provis, J.L.; Lukey, G.C.; Van Deventer, J.S. The role of inorganic polymer technology in the development of 'green concrete'. *Cem. Concr. Res.* **2007**, *37*, 1590–1597. [CrossRef]
11. Palomo, A.; Grutzeck, M.W.; Blanco, M.T. Alkali-activated fly ashes: A cement for the future. *Cem. Concr. Res.* **1999**, *29*, 1323–1329. [CrossRef]
12. Davidovits, J. Geopolymers and geopolymeric materials. *J. Therm. Anal.* **1989**, *35*, 429–441. [CrossRef]
13. Wong, C.L.; Mo, K.H.; Alengaram, U.J.; Yap, S.P. Mechanical strength and permeation properties of high calcium fly ash-based geopolymer containing recycled brick powder. *J. Build. Eng.* **2020**, *32*, 101655. [CrossRef]
14. Baran, P.; Nazarko, M.; Włosińska, E.; Kanciruk, A.; Zarębska, K. Synthesis of geopolymers derived from fly ash with an addition of perlite. *J. Clean. Prod.* **2021**, *293*, 126112. [CrossRef]
15. Kumar, S.; Kumar, R. Mechanical activation of fly ash: Effect on reaction, structure and properties of resulting geopolymer. *Ceram. Int.* **2011**, *37*, 533–541. [CrossRef]
16. Barbosa, V.F.; MacKenzie, K.J. Thermal behaviour of inorganic geopolymers and composites derived from sodium polysialate. *Mater. Res. Bull.* **2003**, *38*, 319–331. [CrossRef]
17. Kaur, M.; Singh, J.; Kaur, M. Synthesis of fly ash based geopolymer mortar considering different concentrations and combinations of alkaline activator solution. *Ceram. Int.* **2018**, *44*, 1534–1537. [CrossRef]
18. Patankar, S.V.; Ghugal, Y.M.; Jamkar, S.S. Mix design of fly ash based geopolymer concrete. *Adv. Struct. Eng.* **2015**, *3*, 1619–1634. [CrossRef]
19. Dvořák, K.; Hájková, I. The Effect of High-Speed Grinding Technology on the Properties of Fly Ash. *Mater. Tehnol.* **2016**, *50*, 683–687. [CrossRef]
20. Gollakota, A.R.; Volli, V.; Shu, C.M. Progressive utilisation prospects of coal fly ash: A review. *Sci. Total Environ.* **2019**, *672*, 951–989. [CrossRef]
21. Sisol, M.; Kudelas, D.; Marcin, M.; Holub, T.; Varga, P. Statistical evaluation of mechanical properties of slag based alkali-activated material. *Sustainability* **2019**, *11*, 5935. [CrossRef]
22. Marcin, M.; Sisol, M.; Brezani, I. Effect of slag addition on mechanical properties of fly ash based geopolymers. *Procedia Eng.* **2016**, *151*, 191–197. [CrossRef]
23. Davidovits, J. Geopolymer cement. A review. *Geopolymer Inst.* **2013**, *21*, 1–11.
24. Ryu, G.S.; Lee, Y.B.; Koh, K.T.; Chung, Y.S. The mechanical properties of fly ash-based geopolymer concrete with alkaline activators. *Constr. Build. Mater.* **2013**, *47*, 409–418. [CrossRef]
25. Görhan, G.; Kürklü, G. The influence of the NaOH solution on the properties of the fly ash-based geopolymer mortar cured at different temperatures. *Compos. Eng.* **2014**, *58*, 371–377. [CrossRef]
26. Aliabdo, A.A.; Abd Elmoaty, M.; Salem, H.A. Effect of water addition, plasticizer and alkaline solution constitution on fly ash based geopolymer concrete performance. *Constr. Build. Mater.* **2016**, *121*, 694–703. [CrossRef]
27. Pavithra, P.E.; Reddy, M.S.; Dinakar, P.; Rao, B.H.; Satpathy, B.K.; Mohanty, A.N. A mix design procedure for geopolymer concrete with fly ash. *J. Clean. Prod.* **2016**, *133*, 117–125. [CrossRef]
28. Xie, T.; Ozbakkaloglu, T. Behavior of low-calcium fly and bottom ash-based geopolymer concrete cured at ambient temperature. *Ceram. Int.* **2015**, *41*, 5945–5958. [CrossRef]
29. Reddy, M.S.; Dinakar, P.; Rao, B.H. A review of the influence of source material's oxide composition on the compressive strength of geopolymer concrete. *Microporous Mesoporous Mater.* **2016**, *234*, 12–23. [CrossRef]
30. Shi, C.; Qu, B.; Provis, J.L. Recent progress in low-carbon binders. *Cem. Concr. Res.* **2019**, *122*, 227–250. [CrossRef]
31. Wang, H.; Li, H.; Yan, F. Synthesis and mechanical properties of metakaolinite-based geopolymer. *Colloids Surf. Physicochem. Eng. Asp.* **2005**, *268*, 1–6. [CrossRef]
32. Ozer, I.; Soyer-Uzun, S. Relations between the structural characteristics and compressive strength in metakaolin based geopolymers with different molar Si/Al ratios. *Ceram. Int.* **2015**, *41*, 10192–10198. [CrossRef]
33. Yuan, J.; He, P.; Jia, D.; Yang, C.; Yan, S.; Yang, Z.; Zhou, Y. Effect of curing temperature and SiO_2/K_2O molar ratio on the performance of metakaolin-based geopolymers. *Ceram. Int.* **2016**, *42*, 16184–16190. [CrossRef]
34. da Silva Rocha, T.; Dias, D.P.; França, F.C.C.; de Salles Guerra, R.R.; de Oliveira, L.R.D.C. Metakaolin-based geopolymer mortars with different alkaline activators (Na^+ and K^+). *Constr. Build. Mater.* **2018**, *178*, 453–461. [CrossRef]
35. Lee, W.H.; Lin, K.L.; Chang, T.H.; Ding, Y.C.; Cheng, T.W. Sustainable development and performance evaluation of marble-waste-based geopolymer concrete. *Polymers* **2020**, *12*, 1924. [CrossRef] [PubMed]
36. Le, V.S.; Louda, P.; Tran, H.N.; Nguyen, P.D.; Bakalova, T.; Ewa Buczkowska, K.; Dufkova, I. Study on Temperature-Dependent Properties and Fire Resistance of Metakaolin-Based Geopolymer Foams. *Polymers* **2020**, *12*, 2994. [CrossRef] [PubMed]
37. La Scalia, G.; Saeli, M.; Adelfio, L.; Micale, R. From lab to industry: Scaling up green geopolymeric mortars manufacturing towards circular economy. *J. Clean. Prod.* **2021**, *316*, 128164. [CrossRef]
38. De Azevedo, A.R.G.; Teixeira Marvila, M.; Barbosa de Oliveira, L.; Macario Ferreira, W.; Colorado, H.; Rainho Teixeira, S.; Mauricio Fontes Vieira, C. Circular economy and durability in geopolymers ceramics pieces obtained from glass polishing waste. *Int. J. Appl. Ceram. Technol.* **2021**, *18*, 1891–1900. [CrossRef]

39. Behun, M.; Kascak, P.; Hrabcak, M.; Behunova, A.; Knapcikova, L.; Sofranko, M. Investigation of sustainable geopolymer composite using automatic identification technology. *Sustainability* **2020**, *12*, 6377. [CrossRef]
40. Kanuchova, M.; Drabova, M.; Sisol, M.; Mosej, J.; Kozakova, L.; Skvarla, J. Influence of mechanical activation of fly ash on the properties of geopolymers investigated by XPS method. Environ. *Prog. Sustain. Energy* **2016**, *35*, 1338–1343. [CrossRef]
41. Assi, L.N.; Deaver, E.E.; Ziehl, P. Effect of source and particle size distribution on the mechanical and microstructural properties of fly Ash-Based geopolymer concrete. *Constr. Build. Mater.* **2018**, *167*, 372–380. [CrossRef]
42. Temuujin, J.; Williams, R.P.; Van Riessen, A. Effect of mechanical activation of fly ash on the properties of geopolymer cured at ambient temperature. *J. Mater. Process. Technol.* **2009**, *209*, 5276–5280. [CrossRef]
43. Paruthi, S.; Husain, A.; Alam, P.; Khan, A.H.; Hasan, M.A.; Magbool, H.M. A review on material mix proportion and strength influence parameters of geopolymer concrete: Application of ANN model for GPC strength prediction. *Constr. Build. Mater.* **2022**, *356*, 129253. [CrossRef]
44. Adak, D.; Mandal, S. Strength and durability performance of fly ash-based process-modified geopolymer concrete. *J. Mater. Civ. Eng.* **2019**, *31*, 04019174. [CrossRef]
45. Komljenović, M.; Baščarević, Z.; Bradić, V. Mechanical and microstructural properties of alkali-activated fly ash geopolymers. *J. Hazard. Mater.* **2010**, *181*, 35–42. [CrossRef] [PubMed]
46. Rashidian-Dezfouli, H.; Rangaraju, P.R. Comparison of strength and durability characteristics of a geopolymer produced from fly ash, ground glass fiber and glass powder. *Mater. De Constr.* **2017**, *67*, e136. [CrossRef]
47. Cho, Y.-K.; Yoo, S.-W.; Jung, S.-H.; Lee, K.-M.; Kwon, S.-J. Effect of Na_2O content, SiO_2/Na_2O molar ratio, and curing conditions on the compressive strength of FA-based geopolymer. *Constr. Build. Mater.* **2017**, *145*, 253–260. [CrossRef]
48. Leong, H.Y.; Ong, D.E.L.; Sanjayan, J.G.; Nazari, A. The effect of different Na_2O and K_2O ratios of alkali activator on compressive strength of fly ash based-geopolymer. *Constr. Build. Mater.* **2016**, *106*, 500–511. [CrossRef]
49. Fernández-Jiménez, A.; Palomo, A. Mid-infrared spectroscopic studies of alkali-activated fly ash structure. *Microporous Mesoporous Mater.* **2005**, *86*, 207–214. [CrossRef]
50. Feng, J.; Zhang, R.; Gong, L.; Li, Y.; Cao, W.; Cheng, X. Development of porous fly ash-based geopolymer with low thermal conductivity. *Mater. Des.* **2015**, *65*, 529–533. [CrossRef]
51. Prochon, P.; Zhao, Z.; Courard, L.; Piotrowski, T.; Michel, F.; Garbacz, A. Influence of Activators on Mechanical Properties of Modified Fly Ash Based Geopolymer Mortars. *Materials* **2020**, *13*, 1033. [CrossRef]
52. Culkova, K.; Khouri, S.; Straka, M.; Rosova, A. Ecological and economic savings of fly ash using as geopolymer. *Rocz. Ochr. Srodowiska* **2018**, *20*, 73–88.
53. Ahmaruzzaman, M. A review on the utilization of fly ash. *Prog. Energy Combust. Sci.* **2010**, *36*, 327–363. [CrossRef]

buildings

Article

Comparing Mechanical Characterization of Carbon, Kevlar, and Hybrid-Fiber-Reinforced Concrete under Quasistatic and Dynamic Loadings

Yeou-Fong Li [1,*], Kun-Han Yang [2], Pei-Yao Hsu [1], Jin-Yuan Syu [1], Shea-Jue Wang [2], Wen-Shyong Kuo [3] and Ying-Kuan Tsai [4]

1 Department of Civil Engineering, National Taipei University of Technology, Taipei 10608, Taiwan; qtim24681012@yahoo.com.tw (P.-Y.H.); t9679010@ntut.org.tw (J.-Y.S.)
2 Department of Materials, National Taipei University of Technology, Taipei 10608, Taiwan; lucky90305@gmail.com (K.-H.Y.); sjwang@ntut.edu.tw (S.-J.W.)
3 Department of Aerospace and Systems Engineering, Feng Chia University, Taichung 40724, Taiwan; wskuo@fcu.edu.tw
4 Department of Environmental Information and Engineering, Chung Cheng Institute of Technology, National Defense University, Taoyuan 33551, Taiwan; jeremytsai0406@gmail.com
* Correspondence: yfli@mail.ntut.edu.tw

Abstract: Concrete is a brittle material due to its poor tensile strength; consequently, concrete tends to crack or peel under an applied external load. Previous studies have investigated the effect of incorporating fiber into concrete, which can improve its tensile strength. In this study, the static and dynamic mechanical characteristics of three types of fiber-reinforced concrete (FRC) were examined: carbon-fiber-reinforced concrete (CFRC); Kevlar-fiber-reinforced concrete (KFRC); and a combination of both, known as carbon/Kevlar-hybrid-fiber-reinforced concrete (HFRC). This study created concrete specimens by pneumatically dispersing carbon and Kevlar fibers and mixing them with cement to comprise 1% of the weight. The mixture was then combined with aggregates and water to form the concrete specimens. When compared with the benchmark concrete specimens, it was found that the compressive strength of the CFRC, KFRC, and HFRC specimens increased by about 19% to 50%, the bending strength increase by about 8% to 32%, and the splitting strength increased by about 4% to 36%. Specifically, the HFRC made with the 24 mm carbon and Kevlar fibers displayed the most significant mechanical strength in a static state. Furthermore, the HFRC showed superior resistance to impact compared to the benchmark concrete specimens across various impact energies, with the 24 mm carbon and Kevlar fiber HFRC showing the highest resistance. The inclusion of fibers in the split Hopkinson pressure bar (SHPB) test demonstrated a notable increase in the maximum strength, particularly in the case of the 12 mm carbon fiber combined with the 24 mm Kevlar fiber in the HFRC specimen.

Keywords: carbon fiber; aramid fiber; hybrid-fiber-reinforced concrete; mechanical strength; impact loading; split Hopkinson pressure bar

Citation: Li, Y.-F.; Yang, K.-H.; Hsu, P.-Y.; Syu, J.-Y.; Wang, S.-J.; Kuo, W.-S.; Tsai, Y.-K. Comparing Mechanical Characterization of Carbon, Kevlar, and Hybrid-Fiber-Reinforced Concrete under Quasistatic and Dynamic Loadings. *Buildings* **2023**, *13*, 2044. https://doi.org/10.3390/buildings13082044

Academic Editor: Tomáš Dvorský

Received: 24 July 2023
Revised: 7 August 2023
Accepted: 9 August 2023
Published: 10 August 2023

1. Introduction

Concrete is prone to damage due to its inherent brittleness and low tensile strength, particularly when subjected to overloading or repeated dynamic forces. In an effort to overcome these shortcomings, fiber-reinforced concrete (FRC) has been engineered. This composite material is developed by integrating fibrous elements into the concrete. FRC offers several advantages, including an improved impact strength, crack limitation, enhanced durability, resistance to freeze–thaw cycles, mitigation of plastic shrinkage, increased toughness, and enhanced fatigue strength. Consequently, FRC finds applications in scenarios where high tensile strength, reduced cracking, or conventional reinforcement placement is impractical. Examples of applications for this material include use in concrete pavements, airplane runways, spillways of dams, the expansion joints of highway bridges, and tunnel constructions.

The mechanical attributes and ease of use of FRC can be adjusted by a number of variables, such as the kind of fiber used, the ratio of fiber to cement, and the length of the fibers. To enhance the mechanical performance of concrete, fibers such as chopped carbon, steel, aramid, glass, polyvinyl alcohol (PVA), and polyethylene are frequently used. Both single-typed fiber-reinforced concrete and hybrid-fiber-reinforced concrete (HFRC) are frequently used to improve concrete's mechanical properties [1–3]. Table 1 provides an overview of the material properties of popular fiber types [4–8].

Table 1. Material properties of various fibers.

Property \ Fiber	Carbon	Aramid	Glass	Steel	Polypropylene
Density (g/cm^3)	1.65~2.18	1.44	2.48~2.76	7.8	1.18
Tensile strength (MPa)	1700~6000	2500~3100	1400~2500	280~2800	300~770
Elastic modulus (GPa)	220~880	60~120	70~80	200~250	3~9
Elongation (%)	0.3~2.0	2.1~4.5	2.5~3.5	<15	15~25

Prior research has indicated that the fiber-to-cement ratio significantly influences the mechanical characteristics of fiber-reinforced concrete [9–15]. For example, steel-fiber-reinforced concrete demonstrates maximum tensile strength when cured for 28 days at a volume percentage of 1.5% [16]. Conversely, an increase in the glass fiber weight percentage beyond 0.1% results in a decline in the compressive strength of the fiber-reinforced concrete [17]. Research conducted on basalt-fiber-reinforced concrete (BFRC) has evaluated the effects of varying the basalt-fiber-to-cement ratios on its mechanical strength. BFRC shows an optimal strength increment at a weight percentage of 0.2%; any further increase in the fiber content causes a decline in the strength increment [18–23]. Studies on cement and concrete's mechanical attributes and microstructures have been performed by utilizing different carbon fibers, fiber lengths, and fiber ratios [24–30].

The length of fibers notably affects FRC's mechanical properties. Bijo and Unnikrishnan found that hybrid FRC incorporating 12 mm polypropylene fiber and 50 mm coconut fiber at a 0.2% volume displayed maximum impact resistance [31]. Yoo and colleagues tested ultrahigh-performance fiber-reinforced concrete (UHPFRC) with steel fibers of different lengths (16 mm, 17.5 mm, and 19.5 mm) and found that the 19.5 mm steel fiber yielded the greatest flexural strength [32]. Mastali et al. [33] utilized recycled carbon fiber of various lengths from a carbon-fiber-reinforced polymer (CFRP), concluding that CFRC with 30 mm of recycled carbon fiber showed superior compressive and flexural strength. Additionally, Li et al. [34] found that 6 mm basalt fiber displayed an enhanced compressive strength, while 24 mm basalt fiber showed a greater splitting tensile and flexural strength.

Carbon fiber, known for its high density, strength, light weight, chemical and high-temperature resistance, and low thermal expansion, is widely used in aerospace engineering, racing, military, and sports products. When added to concrete, it hinders the crack propagation, thereby improving FRC's mechanical properties [35–37]. Conversely, Kevlar fiber, a strong and lightweight organic fiber with a tensile strength five times that of steel and a density about one fifth of steel [38], improves FRC's mechanical properties and impact resistance when incorporated [39,40].

Hybrid-fiber-reinforced concrete (HFRC), a blend of at least two fiber types, boosts FRC's mechanical properties through fiber synergy [41,42]. Steel–polypropylene HFRC, for example, shows excellent synergistic behavior, particularly at lower hybrid-fiber-to-cement-volume fractions [43]. The introduction of polypropylene and basalt fibers to concrete enhances its acid–alkali and crack propagation resistance [44]. Studies on hybrid-fiber-reinforced concrete (HFRC) with Kevlar and carbon fibers revealed that a 1% fiber-to-cement-weight percentage maximizes the mechanical strength [45].

While FRC improves the static mechanical performance and dynamic loading resistance compared to regular concrete, a high static strength does not always correspond to a high impact resistance as the fiber type plays a crucial role [46,47]. The Hopkinson pressure bar, proposed by Hopkinson in 1914 [48] and later refined by Kolsky in 1949 [49], is often used to measure high strain rates [50–52]. Previous studies [53–55] have played a vital role in advancing the study of the mechanical behavior of rocks and concrete subjected to dynamic loading and high strain rates. However, the standard split Hopkinson pressure bar (SHPB) test configuration does not suit studies on high strain rates' effect on soft materials due to repeated loading pulse issues. Nemat-Nasser et al. [56] devised a momentum trapping method for a single stress pulse, involving an incident tube and a reaction mass for a tensile pulse after the compression pulse. Later research has successfully applied similar concepts [57–59]. The dynamic stress–strain relationship of the cementitious material from the SHPB test can contribute to further studies on explosive-induced air-blast loadings [60].

This study focuses on carbon and aramid fibers, which exhibit a higher tensile strength and elastic modulus than other fibers. The goal is to explore the static and dynamic mechanical properties of carbon-fiber-reinforced concrete (CFRC), Kevlar-fiber-reinforced concrete (KFRC) with varying length ratios, and carbon/Kevlar-hybrid-fiber-reinforced concrete (HFRC).

2. Materials

2.1. Carbon Fiber (CF)

Carbon fiber (CF) (TC36P, Tairyfil, Formosa Plastics Corporation, Taipei, Taiwan) was employed to produce the CFRC and HFRC specimens. The CF exhibited tensile strength, tensile modulus, and elongation properties of 4900 MPa, 250 GPa, and 2.0%, respectively. Furthermore, potential interference with the mechanical attributes of FRC was anticipated from the sizing present on the CF surface [61,62].

To mitigate this interference, the CF was placed in a crucible and subjected to a temperature of 550 °C for a duration of 3 h to remove the sizing [63]. Upon completion of this process, the CF was dispersed by utilizing a pneumatic dispersion method, as illustrated in Figure 1.

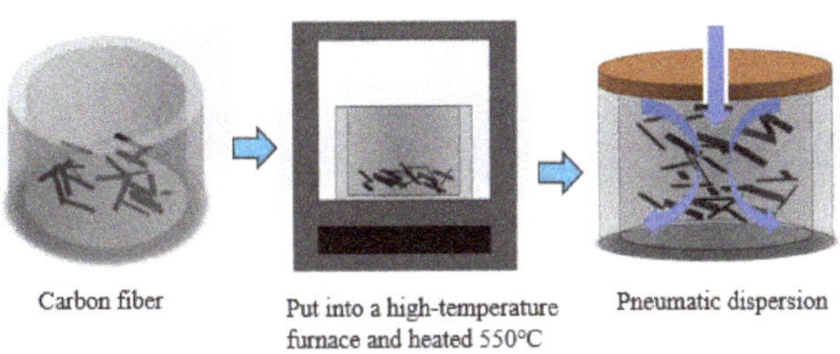

Figure 1. Carbon fiber pretreatment eliminates the interface coupling agent.

In the present study, a scanning electron microscope (SEM, JEOL, JSM-6510LV, Tokyo, Japan) was utilized to examine the fiber surface imagery. Figure 2a depicts the cross-sectional view of the original carbon fiber, having a diameter of 8.0 μm. Meanwhile, Figure 2b,c provide SEM images of the original chopped CF and the sizing-removed chopped CF, respectively, both captured at a 1000-fold magnification.

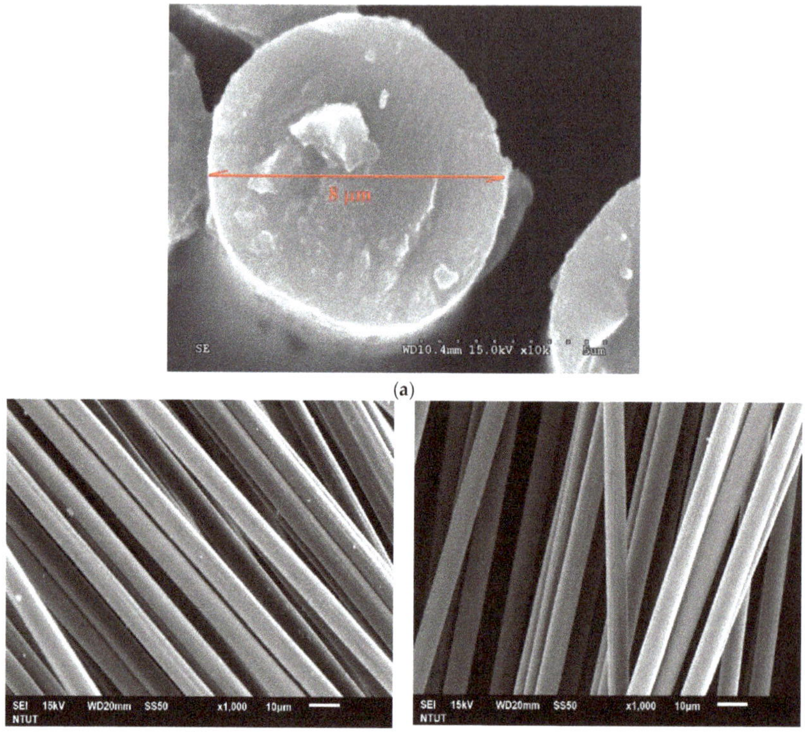

(**a**)

(**b**) (**c**)

Figure 2. SEM image: (**a**) cross-section of the original carbon fiber; (**b**) side profile of the original carbon fiber; (**c**) side profile of the sizing-removed chopped carbon fiber.

The single filament tensile strength of the original carbon fiber was conducted per ASTM D3822-07 [64]. The single filament tensile test results of the original carbon fiber showed that the average tensile force was 13.22 ± 1.39 gf and the average fracture displacement was 0.356 ± 0.024 mm. Figure 3 illustrates the failure force and displacement for the original carbon fiber filaments.

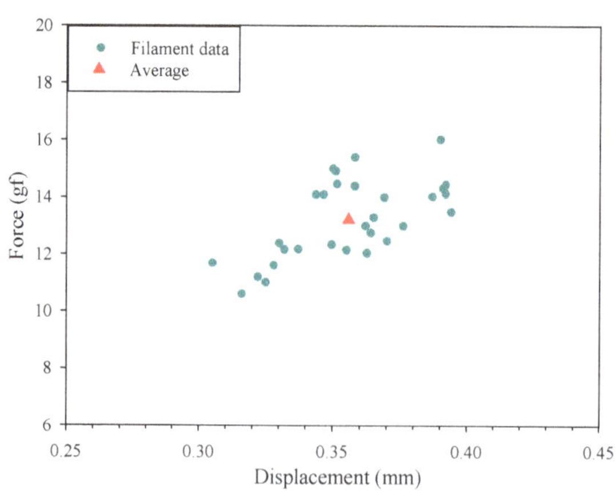

Figure 3. Filament tensile test result of the original carbon fiber.

2.2. *Kevlar Fiber (KF)*

DuPont's Kevlar® 29 fiber was used in this study to improve the strength of the concrete. The chemical formula of Kevlar is –[-CO-C_6H_4-CONH-C_6H_4-NH-]-. KF is thermally stable and chemically resistant, especially in acetic acid, hydrochloric acid, and other acidic chemicals. The tensile strength, tensile modulus, and elongation of KF were measured at 2920 Mpa, 70 Gpa, and 3.6%, respectively; two lengths of KF, namely 12 mm and 24 mm, were used.

Figure 4a,b present images of the chopped KF and its corresponding SEM representation at a 1000-fold magnification, respectively. Figure 4c illustrates the SEM-acquired cross-sectional view of the Kevlar fiber. The KF was dispersed in a manner similar to the CF, utilizing the pneumatic dispersion method.

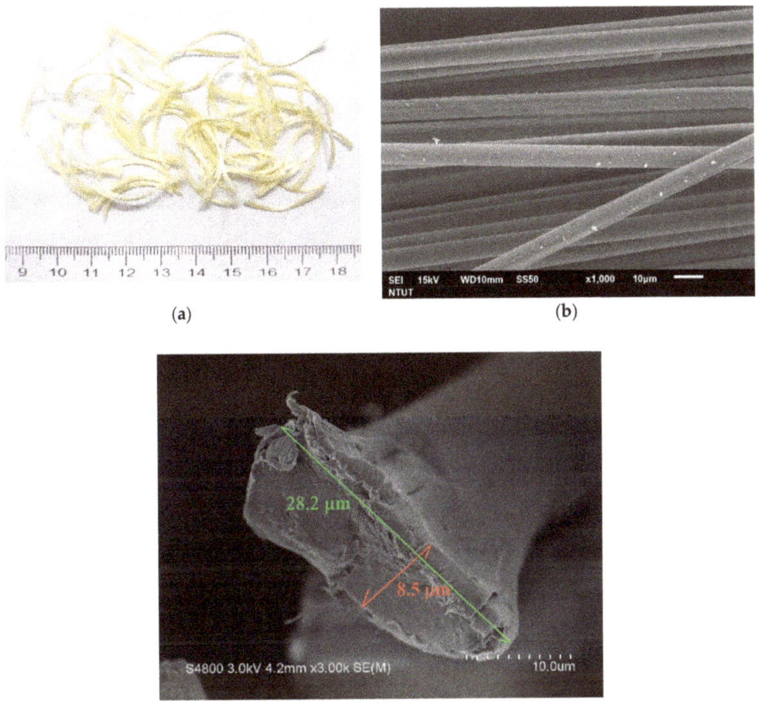

Figure 4. The Kevlar fiber: (**a**) chopped KF; (**b**) side profile by SEM; (**c**) cross-section by SEM.

The single filament tensile strength of the Kevlar fiber was also conducted per ASTM D3822-07 [64]. The single filament tensile test results of the Kevlar fiber showed that the average tensile force was 51.27 ± 8.54 gf and the average fracture displacement was 0.660 ± 0.087 mm. Figure 5 illustrates the failure force and displacement for the Kevlar fiber filaments.

2.3. *Cement and Aggregate*

The main binder used in this study was Type I Portland cement, supplied by Taiwan Cement Ltd (Taipei, Taiwan). The fineness modulus (F.M.) of both the fine and coarse aggregates was determined following the ASTM C33/C33M-18 standard [65]. The calculated F.M. values for the fine and coarse aggregates were 4.71 and 7.17, respectively. Furthermore, the overall F.M. value for all the aggregates combined was measured to be 5.96.

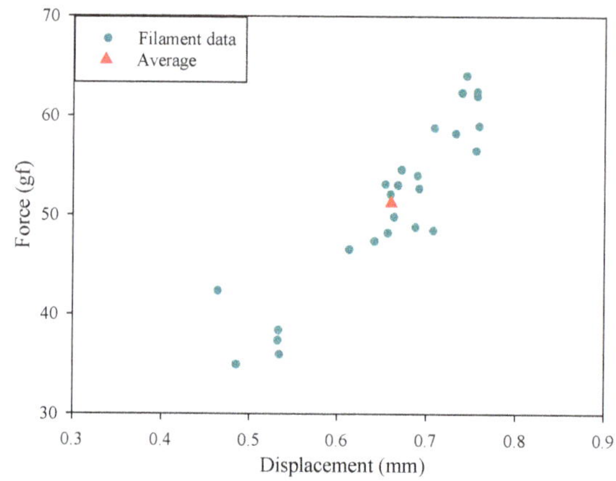

Figure 5. Filament tensile test result of the Kevlar fiber.

3. Experimental Methods

There were nine different specimens in this experiment, including the benchmark, CFRC, KFRC, and HFRC. The HFRC specimens were prepared in different proportions including two fiber types with two fiber lengths that were 1% of the weight percentage of the total fiber-to-cement ratio. The mechanical properties, including those explored with the slump test, compressive test, three-point bending test, splitting tensile test, and impact test, were tested in this study based on ACI and ASTM standards.

3.1. Specimen Nomenclature

The specimen-naming convention is structured as follows: The first letter designates the type of mechanical test, such as a compression test (C), flexural test (F), splitting test (S), and impact test (I). The second letter denotes the type of fiber used, where "C" represents the carbon fiber and "K" represents the Kevlar fiber. The subsequent number indicates the length of the fiber; for instance, "12" signifies a length of 12 mm. For example, C-C12 refers to the CFRC specimen with a 12 mm carbon fiber used in the compressive test, while S-C12K12 represents the HFRC specimen with both 12 mm carbon and 12 mm Kevlar fibers used in the splitting test.

3.2. Slump Test

The workability of the control, CFRC, KFRC, and HFRC samples was evaluated by using a slump test as per ASTM C143/C143M-20 [66] specifications, designed to gauge the flowability and consistency of the concrete.

3.3. Compressive Test

The compressive strength of the control, CFRC, KFRC, and HFRC was determined in adherence to the ASTM C39/C39M-01 [67] guidelines, utilizing a universal testing machine (HT-9501 Series, Hong-Ta, Taipei, Taiwan). Cylindrical specimens (ϕ 10 cm × 20 cm) were subjected to a steady load rate ranging between 900 and 1800 N/s.

3.4. Three-Point Bending Test

The universal testing machine was again used to determine the flexural strength of the control, KFRC, CFRC, and HFRC as per ASTM C293-02 [68]. The three-point bending test was conducted on a specimen with dimensions of 28 cm × 7 cm × 7 cm (length × width × height) at a loading rate of 20 kPa/s, as shown in Figure 6. The rupture modulus or flexural strength of the specimen was calculated by using the formula below:

$$R = \frac{3PL}{2bd^2} \tag{1}$$

where R stands for the rupture modulus (flexural strength) in Mpa, P for the maximum load as per the testing machine in N, L for the span length in mm, b for the average specimen width at rupture in mm, and d for the average specimen thickness at rupture in mm.

Figure 6. Three-point bending test setup.

3.5. Splitting Tensile Test

The universal testing machine was employed to conduct the splitting tensile strength test on the control, KFRC, CFRC, and HFRC samples in accordance with ASTM C496/C496M-17 [69]. The test was performed on a cylindrical specimen with dimensions ϕ 10 cm × 20 cm at a loading rate of 4300 N/min. The splitting tensile strength of the specimen was computed by using the following equation:

$$T = \frac{2P}{\pi LD} \quad (2)$$

where T denotes the splitting tensile strength (Mpa), P stands for the maximum load applied as indicated by the testing machine (N), L stands for the length (mm), and d stands for the diameter (mm).

3.6. Impact Test

The impact test specimen, measuring ϕ 15.2 cm × 6.35 cm, was placed on a fixture with a hardened steel ball, as depicted in Figure 7a. The impact test was conducted following the guidelines of ACI 544.2R-89T [70] by using an impact testing machine (SP-006, Shengpeng, Yunlin, Taiwan). The specimen was positioned in a fixture and subsequently set on a sandbox's top, as displayed in Figure 7b. The potential energy (E) can be computed as E = m × g × h, where m stands for the hammer's mass (kg), g stands for the gravitational acceleration (m/s^2), and h stands for the hammer's setting height (m). The impact energies tested for both the control and HFRC specimens varied between 100 J and 300 J.

3.7. Split Hopkinson Pressure Bar Test

The device used in this study is located in the Materials Laboratory of the Environmental Information and Engineering Department of Chung Cheng Institute of Technology, National Defense University. The equipment was improved during the experiment, the original equipment was used in the experiment on the CFRC, and the HFRC was improved. Figure 8 illustrates the original SHPB devices.

During the SHPB examination, a striker bar was accelerated from the gas pressure chamber under a prescribed pressure, which impacted the incident bar and generated stress waves. These stress waves propagated from the incident bar to the specimen's interface, leading to the creation of reflective and transmissive waves. Strain gauges were attached to the incident bar and transmission bar to record the strain–time signals of the incident wave $\varepsilon_i(t)$, reflected wave $\varepsilon_r(t)$, and transmitted wave $\varepsilon_t(t)$. The specimen's stress $\sigma(t)$, strain $\varepsilon(t)$, and strain rate $\dot{\varepsilon}(t)$ obtained from the three-wave method are represented as follows [71]:

$$\varepsilon(t) = \frac{C_e}{L_S}\int_0^t [\varepsilon_i(t) - \varepsilon_r(t) - \varepsilon_t(t)]t \tag{3}$$

$$\dot{\varepsilon}(t) = \frac{C_e}{L_S}[\varepsilon_i(t) - \varepsilon_r(t) - \varepsilon_t(t)] \tag{4}$$

$$\sigma(t) = \frac{A_e E}{2A_S}[\varepsilon_i(t) + \varepsilon_r(t) + \varepsilon_t(t)] \tag{5}$$

where C_e is the wave speed; E is the elastic modulus of the incident bar; A_e is the cross-sectional area of the incident bar; and L_s and A_S are the length and cross-sectional area of the specimen, respectively.

(**a**) (**b**)

Figure 7. Impact test. (**a**) Test fixture and FRC specimen; (**b**) instrumented drop weight system.

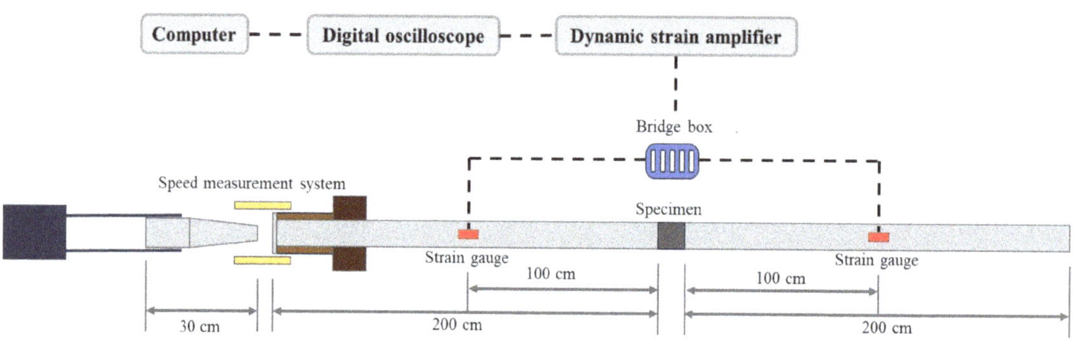

Figure 8. SHPB testing devices.

4. Results and Discussion

This section outlines the findings from the experimental work conducted for this study. CF and KF, with sizing removed, were employed in the creation of the CFRC, KFRC, and HFRC specimens, which had varying fiber lengths and mixing proportions, and each maintained a 1% fiber-to-cement weight ratio. The specimens' workability was evaluated via the slump test, while the ASTM and ACI standards were applied to ascertain the compressive strength, flexural strength, splitting tensile strength, and impact resistance.

4.1. Slump Test

Table 2 displays the slump values of the benchmark, CFRC, KFRC, and HFRC specimens at different weight percentages. Compared to the benchmark concrete, the CFRC, KFRC, and HFRC exhibited a significantly reduced workability, with slump values ranging from 20 mm to 30 mm, while the benchmark had a slump value of 78 mm.

Table 2. Slump of benchmark, CFRC, KFRC, and HFRC.

Specimens	Benchmark	C12	C24	K12	K24
Slump (mm)	78	24	23	25	27
Specimens	-	**C12K12**	**C12K24**	**C24K12**	**C24K24**
Slump (mm)	-	24	22	25	21

4.2. Compression Test

Table 3 presents the compressive strengths of the C-C12, C-C24, C-K12, and C-K24 specimens. The average compressive strengths of these specimens were 30.01 MPa, 31.81 MPa, 27.73 MPa, and 31.49 MPa, respectively. Compared to the benchmark specimen, the average compressive strengths increased by 19.3%, 26.4%, 10.2%, and 25.2%, respectively.

Table 3. Compressive strengths of benchmark, CFRC, and KFRC specimens.

Specimen	C-B	CFRC and KFRC			
		C-C12	C-C24	C-K12	C-K24
Compressive strength (MPa)	24.87 25.27 25.34	30.59 29.71 29.73	31.92 32.03 31.47	26.56 27.19 29.45	29.97 32.36 32.15
Average compressive strength (MPa)	25.16	30.01	31.81	27.73	31.49
Increase (%)	-	19.3	26.4	10.2	25.2

The compressive strengths of the C-C12K12, C-C12K24, C-C24K12, and C-C24K24 specimens are presented in Table 4. The respective mean compressive strengths were measured to be 32.63 MPa, 33.63MPa, 35.32 MPa, and 37.86 MPa. In comparison to the benchmark specimen, these signify respective average compressive strength enhancements of 29.7%, 33.7%, 40.4%, and 50.5%.

Table 4. Compressive strengths of benchmark and HFRC specimens.

Specimen	C-B	HFRC			
		C-C12K12	C-C12K24	C-C24K12	C-C24K24
Compressive strength (MPa)	24.87 25.27 25.34	32.98 32.12 32.78	33.87 33.09 33.94	36.40 35.39 34.16	37.20 37.82 38.56
Average compressive strength (MPa)	25.16	32.63	33.63	35.32	37.86
Increase (%)	-	29.7	33.7	40.4	50.5

Figure 9 illustrates the average compressive strengths of the benchmark, CFRC, KFRC, and HFRC specimens. The CFRC exhibited a higher compressive strength than the KFRC, and the length of the fiber had a positive correlation with the compression strength. The C-C24K24 specimen had the highest compressive strength among all the specimens.

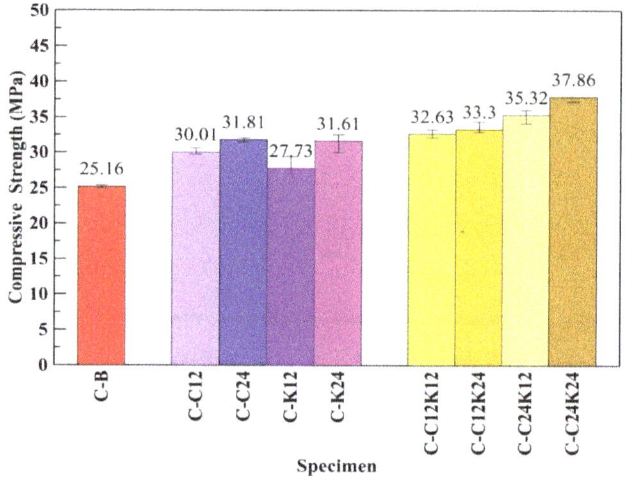

Figure 9. Compressive strength of benchmark, CFRC, KFRC, and HFRC specimens.

4.3. Three-Point Bending Test

The flexural strengths of the F-C12, F-C24, F-K12, and F-K24 specimens are detailed in Table 5. They manifested average flexural strengths of 5.76 MPa, 5.91 MPa, 6.19 MPa, and 6.34 MPa, respectively. When compared to the benchmark specimen, these averages demonstrate enhancements in the flexural strength by 8.3%, 11.0%, 16.2%, and 19.0%, respectively.

Table 5. Flexural strengths of benchmark, CFRC, and KFRC specimens.

Specimen	F-B	CFRC and KFRC			
		F-C12	F-C24	F-K12	F-K24
Flexural strength (MPa)	5.34	5.78	5.89	6.12	6.34
	5.24	5.76	5.88	6.22	6.33
	5.39	5.75	5.95	6.22	6.34
Average flexural strength (MPa)	5.32	5.76	5.91	6.19	6.34
Increase (%)	-	8.3	11.0	16.2	19.0

Table 6 presents the flexural strengths of the F-C12K12, F-C12K24, F-C24K12, and F-C24K24 specimens. The average flexural strengths of these specimens were 6.41 MPa, 6.49 MPa, 6.89 MPa, and 7.02 MPa, respectively. Compared to the benchmark specimen, the average flexural strengths increased by 20.5%, 22.0%, 29.5%, and 32.0%, respectively.

Table 6. Flexural strengths of benchmark and HFRC specimens.

Specimen	F-B	HFRC			
		F-C12K12	F-C12K24	F-C24K12	F-C24K24
Flexural strength (MPa)	5.34	6.41	6.50	6.81	7.17
	5.24	6.46	6.49	6.92	6.96
	5.39	6.35	6.49	6.92	6.92
Average flexural strength (MPa)	5.32	6.41	6.49	6.89	7.02
Increase (%)	-	20.5	22.0	29.5	32.0

Figure 10 portrays the mean flexural strengths of the benchmark, CFRC, KFRC, and HFRC specimens. It can be observed that the KFRC samples exhibited an enhanced flexural strength compared to the CFRC samples. Additionally, the fiber length displayed a beneficial association with the flexural strength. Remarkably, the F-C24K24 specimen stood out with the highest flexural strength among all the specimens.

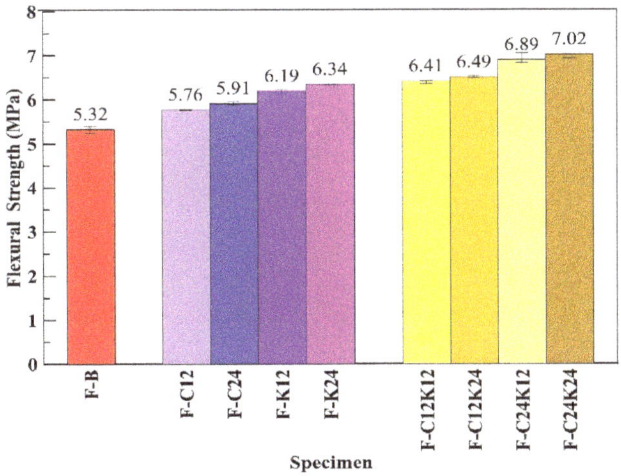

Figure 10. Flexural strengths of benchmark, CFRC, KFRC, and HFRC specimens.

4.4. Splitting Tensile Test

Table 7 shows the splitting strengths of the S-C12, S-C24, S-K12, and S-K24 specimens. The average splitting strengths of these specimens were 3.27 MPa, 3.38 MPa, 3.04 MPa, and 3.09 MPa, respectively. Compared to the benchmark specimen, the average splitting strengths increased by 12.1%, 15.8%, 4.0%, and 5.9%, respectively.

Table 7. Splitting strengths of concrete benchmark, CFRC, and KFRC specimens.

Specimen	S-B	CFRC and KFRC			
		S-C12	S-C24	S-K12	S-K24
Splitting strength (MPa)	3.04	3.21	3.37	3.10	3.12
	2.82	3.27	3.44	3.01	3.06
	2.90	3.34	3.33	3.00	3.10
Average splitting strength (MPa)	2.92	3.27	3.38	3.04	3.09
Increase (%)	-	12.1	15.8	4.0	5.9

Table 8 presents the splitting strengths of the S-C12K12, S-C12K24, S-C24K12, and S-C24K24 specimens. The average splitting strengths of these specimens were 3.49 MPa, 3.55 MPa, 3.80 MPa, and 3.99 MPa, respectively. Compared to the benchmark specimen, the average splitting strengths increased by 19.5%, 21.7%, 30.3%, and 36.5%, respectively.

Table 8. Splitting strengths of benchmark and HFRC specimens.

Specimen	S-B	HFRC			
		S-C12K12	S-C12K24	S-C24K12	S-C24K24
Splitting strength (MPa)	3.04	3.52	3.54	3.77	3.99
	2.82	3.47	3.57	3.81	4.00
	2.90	3.48	3.55	3.83	3.97
Average splitting strength (MPa)	2.92	3.49	3.55	3.80	3.99
Increase (%)	-	19.5	21.7	30.3	36.5

Figure 11 presents the average splitting strengths of the benchmark, CFRC, KFRC, and HFRC specimens. As seen in Figure 11, it can be deduced that the CFRC exhibited a superior splitting strength compared to the KFRC, and a trend can be observed where an increase in the fiber length corresponds to a higher splitting strength. Notably, the S-C24K24 specimen had the highest peak splitting strength amongst all the samples.

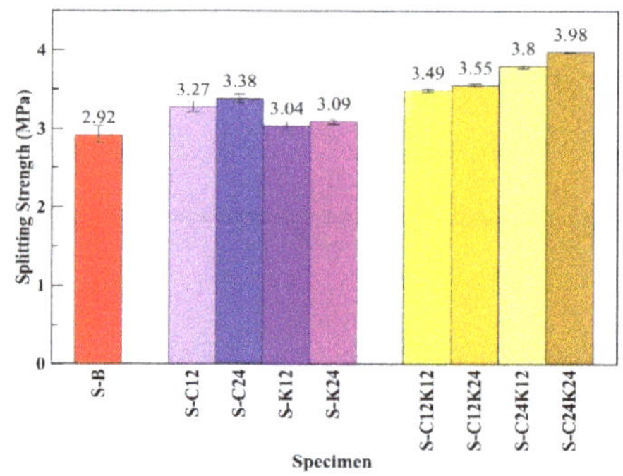

Figure 11. Splitting strength of benchmark, CFRC, KFRC, and HFRC specimens.

According to the above quasistatic test results, the HFRC demonstrated highly improved mechanical performances compared with the CFRC and KFRC. Based on Table 1, the elongation of the carbon fiber ranged from 0.3 to 0.2% and of aramid ranged from 2.1 to 4.5%. When a force is applied to the FRC specimen, the concrete and the fibers would experience the same amount of strain, and since carbon fiber and Kevlar fiber have distinct elongation, it is suspected that the carbon fiber would break first while the Kevlar fiber would still maintain its overall structural integrity due to the synergistic effect of the two fibers [72].

4.5. Impact Test

Based on the previous test results, the HFRC exhibited greater strength compared to the single-type FRC. Therefore, the impact test was conducted on the HFRC and benchmark specimens by using impact energies of 100 J, 150 J, 200 J, 250 J, and 300 J. Table 9 presents the impact energies and corresponding impact numbers for the HFRC and benchmark specimens. For example, at 100 J of impact energy, the average impact numbers for the benchmark, I-12K12, I-C12K24, I-C24K12, and I-C24K24 HFRC specimens were 18.0, 37.5, 46.8, 56.5, and 79.0, respectively.

Figure 12 illustrates the average impact resistance of the benchmark and HFRC specimens. Specimen I-C24K24 exhibited the highest impact number under different impact energies. It was observed that the HFRC with longer fibers demonstrated greater impact resistance.

4.6. Split Hopkinson Pressure Bar Test

During the SHPB test on the HFRC, the stress–strain relationship was obtained with various ratios under the impact of different gas pressures (from 0.6 to 1.5 kgf/cm^2). Since the breaking point of the machine cannot be set in the SHPB test unlike in the compression test, in this test, the breaking point is always set as 80% of the maximum stress. The stress–strain relationship from the SHPB test are shown in Figure 13. As seen from this figure, the higher the gas pressure, the larger the maximum strength. Additionally, the addition of fiber will also increase the maximum strength of the concrete, especially in the C12K24 HFRC specimen.

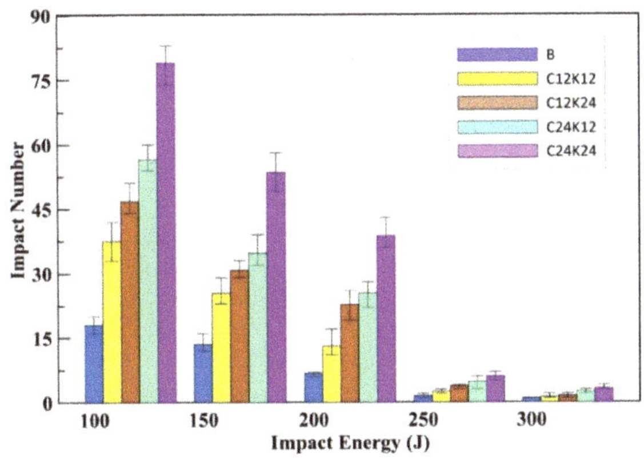

Figure 12. Impact energy–number relationships of benchmark and HFRC specimens.

Table 9. Impact energy and the corresponding impact number of benchmark and HFRC specimens.

Specimen	Impact Energy (J)	Impact Number				
		1	2	3	4	Average
I-B	300	1	1	1	1	1.0
	250	2	2	1	1	1.5
	200	7	7	6	7	6.8
	150	13	12	16	13	13.5
	100	17	20	16	19	18.0
I-C12K12	300	1	2	1	1	1.3
	250	3	2	2	3	2.5
	200	11	17	13	11	13.0
	150	23	27	29	23	25.5
	100	37	42	38	33	37.5
I-C12K24	300	1	2	2	1	1.5
	250	4	3	4	4	3.8
	200	24	26	22	19	22.8
	150	29	33	30	31	30.8
	100	45	44	47	51	46.8
I-C24K12	300	2	3	3	2	2.5
	250	3	5	6	5	4.5
	200	25	22	27	28	25.5
	150	35	33	39	32	34.8
	100	60	54	55	57	56.5
I-C24K24	300	3	4	3	3	3.3
	250	7	5	6	6	6.0
	200	37	39	43	36	38.8
	150	49	54	58	53	53.5
	100	79	83	74	80	79.0

The strain energy, which encapsulates the energy stored within a specimen, is calculated by determining the integral of the area beneath the stress–strain curve and subsequently multiplying this value by the specimen's volume. Figure 14 displays a bar chart of the strain energy, indicating a higher concentration of strain energy in the HFRC specimens composed of 12 mm carbon fiber and 24 mm Kevlar fiber relative to the other tested samples.

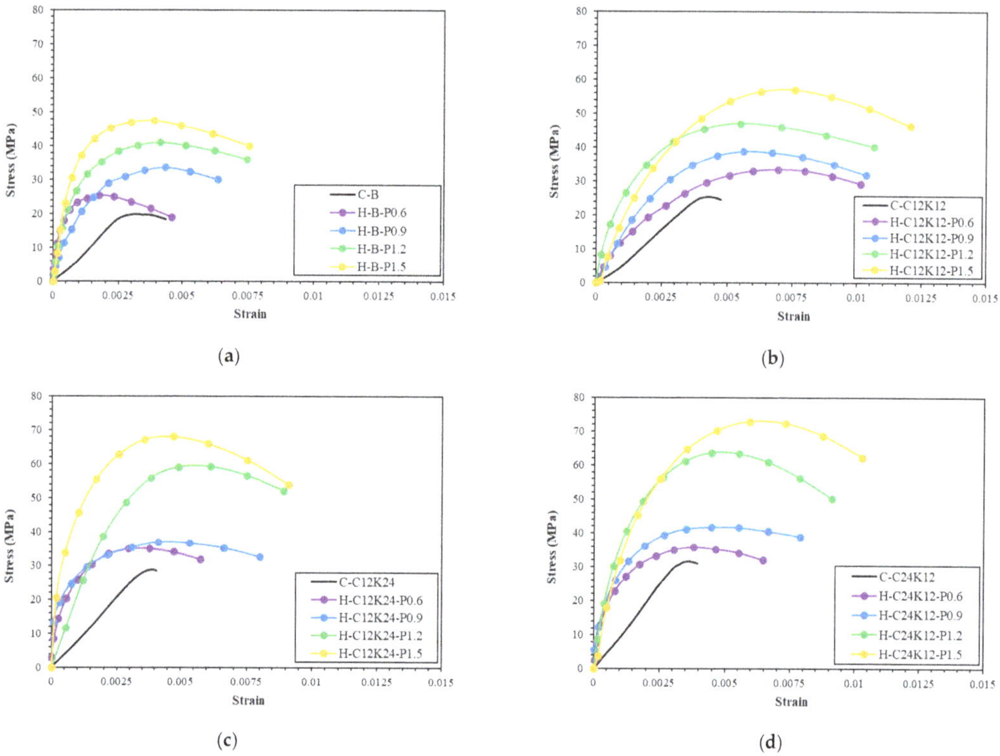

Figure 13. Figure 13. Stress–strain relationship from SHPB test: (**a**) benchmark; (**b**) specimen C12K12; (**c**) specimen C12K24; (**d**) specimen C24K12.

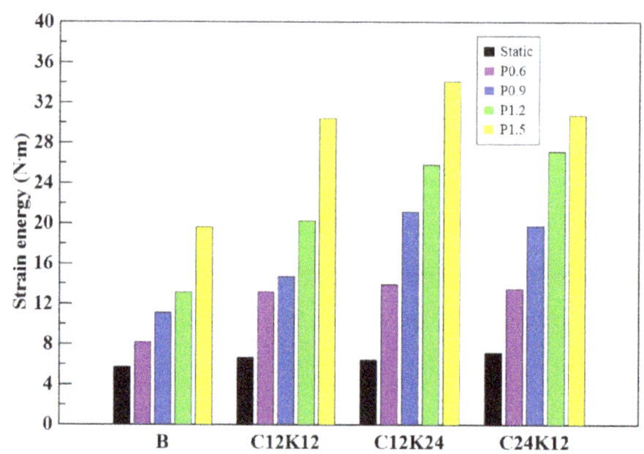

Figure 14. SHPB of benchmark and HFRC specimens.

5. Conclusions

This research embarked on a rigorous examination of the mechanical characteristics of HFRC synthesized through the fusion of carbon and Kevlar fibers and each endowed with distinct material attributes. This study brings to light compelling distinctions when juxtaposing HFRC with singular-fiber-reinforced concretes, specifically CFRC and KFRC, culminating in a series of key insights:

1. An assessment of the slumps of the CFRC, KFRC, and HFRC mixtures indicated minimal variation. Despite this, these measures were consistently less than those of conventional concrete, underscoring the impact of fiber incorporation on the workability of the mix.
2. A significant contrast was observed in the compressive, flexural, and splitting strengths between the CFRC and KFRC specimens, with the CFRC demonstrating superior compressive and splitting strengths. Conversely, the KFRC specimens exhibited an enhanced flexural strength, a crucial attribute affecting the concrete's response under stress.
3. The introduction of disparate fiber types into the HFRC specimens yielded an overall improvement in the compressive strength by a maximum increment of 50.5%, flexural strength by 32%, and splitting tensile strength by 36.5%, all surpassing those of the CFRC and KFRC specimens.
4. The HFRC variant composed of longer fibers (I-C24K24) exhibited remarkable impact resistance, surpassing that of its shorter counterpart. Compared with the benchmark, the impact number under 100 J of load increased from 18 times to 79 times. This heightened resistance can be traced back to the greater bonding force achieved with longer fibers, which facilitated the effective energy absorption from shear and tensile forces, subsequently mitigating concrete fracture.
5. The empirical data acquired through a split Hopkinson pressure bar test demonstrated an increase in the maximum strength correlating with a higher gas pressure. The addition of fibers led to the further enhancement of this maximum strength. In particular, the HFRC specimens equipped with 12 mm carbon fiber and 24 mm Kevlar fiber manifested higher strain energies than the other specimens, promising more robust and enduring constructions.

Author Contributions: Conceptualization, Y.-F.L. and S.-J.W.; data curation, K.-H.Y. and P.-Y.H.; formal analysis, K.-H.Y. and P.-Y.H.; funding acquisition, Y.-F.L. and W.-S.K.; investigation, K.-H.Y. and J.-Y.S.; methodology, Y.-F.L. and Y.-K.T.; project administration, Y.-F.L. and S.-J.W.; supervision, Y.-F.L. and Y.-K.T.; writing—original draft, K.-H.Y. and J.-Y.S.; writing—review and editing, Y.-F.L. and W.-S.K. All authors have read and agreed to the published version of the manuscript.

Funding: This research was funded by the Ministry of Science and Technology of the Taiwan government under contract No. MOST 110-2221-E-027-031 and the "Research Center of Energy Conservation for New Generation of Residential, Commercial, and Industrial Sectors" from the Ministry of Education in Taiwan under contract No. L7121101-19.

Institutional Review Board Statement: Not applicable.

Informed Consent Statement: Not applicable.

Data Availability Statement: Not applicable.

Conflicts of Interest: The authors declare no conflict of interest.

References

1. Tran, T.K.; Tran, N.T.; Nguyen, D.L.; Kim, D.J.; Park, J.K.; Ngo, T.T. Dynamic fracture toughness of ultra-high-performance fiber-reinforced concrete under impact tensile loading. *Struct. Concr.* **2021**, *22*, 1845–1860. [CrossRef]
2. Conforti, A.; Minelli, F.; Plizzari, G.A.; Tiberti, G. Comparing test methods for the mechanical characterization of fiber reinforced concrete. *Struct. Concr.* **2018**, *19*, 656–669. [CrossRef]
3. Zheng, D.; Song, W.; Fu, J.; Xue, G.; Li, J.; Cao, S. Research on mechanical characteristics, fractal dimension and internal structure of fiber reinforced concrete under uniaxial compression. *Constr. Build. Mater.* **2020**, *258*, 120351. [CrossRef]
4. Chen, B.; Wu, K.; Yao, W. Conductivity of carbon fiber reinforced cement-based composites. *Cem. Concr. Compos.* **2004**, *26*, 291–297. [CrossRef]
5. Jung, S.H.; Kishimoto, H.; Nakazato, N.; Nakata, D.; Park, J.S.; Kohyama, A. Effect of the fabrication process on the microstructural evolution of carbon fibers and flexural property on C/SiC composites by the NITE method. *Ceram. Int.* **2022**, *48*, 32712–32722. [CrossRef]
6. Ravichandran, D.; Prem, P.R.; Kaliyavaradhan, S.K.; Ambily, P.S. Influence of fibers on fresh and hardened properties of Ultra High-Performance Concrete (UHPC)—A review. *J. Build. Eng.* **2022**, *57*, 104922. [CrossRef]

7. Li, C.; Xian, G. Experimental and modeling study of the evolution of mechanical properties of PAN-based carbon fibers at elevated temperatures. *Materials* **2019**, *12*, 724. [CrossRef]
8. Pertuz-Comas, A.D.; Díaz, J.G.; Meneses-Duran, O.J.; Niño-Álvarez, N.Y.; León-Becerra, J. Flexural fatigue in a polymer matrix composite material reinforced with continuous Kevlar fibers fabricated by additive manufacturing. *Polymers* **2022**, *14*, 3586. [CrossRef]
9. El Naggar, H.; Abu Abdo, A.M. Properties and behavior of rubberized concrete enhanced with PVA fibers. *Building* **2023**, *13*, 1681. [CrossRef]
10. Li, Y.F.; Kadagathur Ramanathan, G.; Syu, J.Y.; Huang, C.H.; Tsai, Y.K. Mechanical behavior of different fiber lengths mix-proportions carbon fiber reinforced concrete subjected to static, impact, and blast loading. *Int. J. Prot. Struct.* **2022**, 20414196221138596. [CrossRef]
11. Babaei, A.; Mortezaei, A.; Salehian, H. Experimental study on seismic performance of steel fiber reinforced concrete wall piers. *Struct. Concr.* **2021**, *22*, 1363–1377. [CrossRef]
12. Li, Y.F.; Wang, H.F.; Syu, J.Y.; Kadagathur Ramanathan, G.; Tsai, Y.K. Investigating the mechanical performance on static and shock wave loading of aramid fiber-reinforced concrete. *Fibers* **2022**, *10*, 82. [CrossRef]
13. Torrijos, M.C.; Giaccio, G.; Zerbino, R. Glass macrofiber self-compacting concrete: Fiber distribution and mechanical properties in thin walls and slabs. *Struct. Concr.* **2019**, *20*, 798–807. [CrossRef]
14. Li, Y.F.; Huang, Y.R.; Syu, J.Y.; Tsai, Y.K.; Huang, C.H. A study on mechanical behavior of Kevlar fiber reinforced concrete under static and high-strain rate loading. *Int. J. Prot. Struct.* **2022**, *14*, 20414196221118596. [CrossRef]
15. Yang, L.; Xie, H.; Fang, S.; Huang, C.; Yang, A.; Chao, Y.J. Experimental study on mechanical properties and damage mechanism of basalt fiber reinforced concrete under uniaxial compression. *Structures* **2021**, *31*, 330–340. [CrossRef]
16. Köksal, F.; Altun, F.; Yiğit, İ.; Şahin, Y. Combined effect of silica fume and steel fiber on the mechanical properties of high strength concretes. *Constr. Build. Mater.* **2008**, *22*, 1874–1880. [CrossRef]
17. Tibebu, A.; Mekonnen, E.; Kumar, L.; Chimdi, J.; Hailu, H.; Fikadu, N. Compression and workability behavior of chopped glass fiber reinforced concrete. *Mater. Today Proc.* **2022**, *62*, 5087–5094. [CrossRef]
18. Dong, J.Q.; Du, Y.T.; Wen, B.L.; Sun, L.X.; Wang, J.Q. Research on mechanical properties and toughening mechanism of basalt fiber reinforced concrete. *Ind. Concr.* **2011**, 638–641.
19. Sun, X.; Gao, Z.; Cao, P.; Zhou, C. Mechanical properties tests and multiscale numerical simulations for basalt fiber reinforced concrete. *Constr. Build. Mater.* **2019**, *202*, 58–72. [CrossRef]
20. Wang, D.; Ju, Y.; Shen, H.; Xu, L. Mechanical properties of high performance concrete reinforced with basalt fiber and polypropylene fiber. *Constr. Build. Mater.* **2019**, *197*, 464–473. [CrossRef]
21. Tumadhir, M. Thermal and mechanical properties of basalt fibre reinforced concrete. *Int. J. Civ. Environ. Eng.* **2013**, *7*, 334–337.
22. Jalasutram, S.; Sahoo, D.R.; Matsagar, V. Experimental investigation of the mechanical properties of basalt fiber-reinforced concrete. *Struct. Concr.* **2017**, *18*, 292–302. [CrossRef]
23. Wang, Y.; Hu, S.; Sun, X. Experimental investigation on the elastic modulus and fracture properties of basalt fiber–reinforced fly ash geopolymer concrete. *Constr. Build. Mater.* **2022**, *338*, 127570. [CrossRef]
24. Park, S.-J.; Seo, M.-K.; Shim, H.-B.; Rhee, K.Y. Effect of different cross-section types on mechanical properties of carbon fibers-reinforced cement composites. *Mater. Sci. Eng. A* **2014**, *366*, 348–355. [CrossRef]
25. Tabatabaei, Z.S.; Volz, J.S.; Keener, D.I.; Gliha, B.P. Comparative impact behavior of four long carbon fiber reinforced concretes. *Mater. Des.* **2014**, *55*, 212–223. [CrossRef]
26. Han, B.; Zhang, L.; Zhang, C.; Wang, Y.; Yu, X.; Ou, J. Reinforcement effect and mechanism of carbon fibers to mechanical and electrically conductive properties of cement-based materials. *Constr. Build. Mater.* **2016**, *125*, 479–486. [CrossRef]
27. Rangelov, M.; Nassiri, S.; Haselbach, L.; Englund, K. Using carbon fiber composites for reinforcing pervious concrete. *Constr. Build. Mater.* **2016**, *126*, 875–885. [CrossRef]
28. Li, Y.F.; Li, J.Y.; Syu, J.Y.; Yang, T.H.; Chang, S.M.; Shen, M.Y. Mechanical Behaviors of Microwave-Assisted Pyrolysis Recycled Carbon Fiber-Reinforced Concrete with Early-Strength Cement. *Materials* **2023**, *16*, 1507. [CrossRef]
29. Li, Y.F.; Li, J.Y.; Kadagathur Ramanathan, G.; Chang, S.-M.; Shen, M.-Y.; Tsai, Y.-K.; Huang, C.-H. An Experimental study on mechanical behaviors of carbon fiber and microwave-assisted pyrolysis recycled carbon fiber-reinforced concrete. *Sustainability* **2021**, *13*, 6829. [CrossRef]
30. Li, Y.F.; Lee, K.F.; Kadagathur Ramanathan, G.; Cheng, T.-W.; Huang, C.-H.; Tsai, Y.-K. Static and dynamic performances of chopped carbon-fiber-reinforced mortar and concrete incorporated with disparate lengths. *Materials* **2021**, *14*, 972. [CrossRef]
31. Bijo, M.D.; Unnikrishnan, S. Mechanical strength and impact resistance of hybrid fiber reinforced concrete with coconut and polypropylene fibers. *Mater. Today Proc.* **2022**, *65*, 1873–1880.
32. Yoo, D.Y.; Zi, G.; Kang, S.T.; Yoon, Y.S. Biaxial flexural behavior of ultra-high-performance fiber-reinforced concrete with different fiber lengths and placement methods. *Cem. Concr. Compos.* **2015**, *63*, 51–66. [CrossRef]
33. Mastali, M.; Dalvand, A.; Sattarifard, A. The impact resistance and mechanical properties of the reinforced self-compacting concrete incorporating recycled CFRP fiber with different lengths and dosages. *Compos. Part B Eng.* **2017**, *112*, 74–92. [CrossRef]
34. Li, Y.F.; Hung, J.Y.; Syu, J.Y.; Chang, S.M.; Kuo, W.S. Influence of sizing of basalt fiber on the mechanical behavior of basalt fiber reinforced concrete. *J. Mater. Res. Technol.* **2022**, *21*, 295–307. [CrossRef]

35. Hossain, M.Z.; Awal, A.A. Flexural response of hybrid carbon fiber thin cement composites. *Constr. Build. Mater.* **2011**, *25*, 670–677. [CrossRef]
36. De Souza Abreu, F.; Ribeiro, C.C.; da Silva Pinto, J.D.; Nsumbu, T.M.; Buono, V.T.L. Influence of adding discontinuous and dispersed carbon fiber waste on concrete performance. *J. Clean. Prod.* **2020**, *273*, 122920. [CrossRef]
37. Sharma, M.; Gao, S.; Mäder, E.; Sharma, H.; Wei, L.Y.; Bijwe, J. Carbon fiber surfaces and composite interphases. *Compos. Sci. Technol.* **2014**, *102*, 35–50. [CrossRef]
38. Rolland, A.; Chataigner, S.; Benzarti, K.; Quiertant, M.; Argoul, P.; Paul, J.M. Experimental investigation and modeling of the bond between aramid fiber-reinforced polymer bars and concrete. *Mater. Infrastruct.* **2016**, *1*, 115–128.
39. Karthik, K.; Rajamani, D.; Manimaran, A.; Udayaprakash, J. Evaluation of tensile properties on Glass/Carbon/Kevlar fiber reinforced hybrid composites. *Mater. Today Proc.* **2021**, *39*, 1655–1660. [CrossRef]
40. Krishna, A.; Kaliyaperumal, S.R.M.; Kathirvel, P. Compressive strength and impact resistance of hybrid fiber reinforced concrete exposed to elevated temperatures. *Struct. Concr.* **2022**, *23*, 1611–1624. [CrossRef]
41. *ASTM-C1371*; Standard Test Method for Determination of Emittance of Materials Near Room Temperature Using Portable Emissometers. American Society for Testing and Materials: West Conshohocken, PA, USA, 2004.
42. Huang, H.; Yuan, Y.; Zhang, W.; Zhu, L. Experimental study on the mechanical properties and the microstructure of hybrid-fiber-reinforced concrete under an early stage. *Struct. Concr.* **2020**, *21*, 1106–1122. [CrossRef]
43. Singh, N.K.; Rai, B. Assessment of synergetic effect on microscopic and mechanical properties of steel-polypropylene hybrid fiber reinforced concrete. *Struct. Concr.* **2021**, *22*, 516–534. [CrossRef]
44. Deng, Z.; Liu, X.; Yang, X.; Liang, N.; Yan, R.; Chen, P.; Xu, Y. A study of tensile and compressive properties of hybrid basalt-polypropylene fiber-reinforced concrete under uniaxial loads. *Struct. Concr.* **2021**, *22*, 396–409. [CrossRef]
45. Li, Y.F.; Wang, H.F.; Syu, J.Y.; Kadagathur Ramanathan, G.; Tsai, Y.K.; Lok, M.H. Mechanical properties of aramid/carbon hybrid fiber-reinforced concrete. *Materials* **2021**, *14*, 5881. [CrossRef]
46. Isabai, B.; Nurzhan, S.; Yerlan, A. Strength properties of various types of fiber-reinforced concrete for production of driven piles. *Buildings* **2023**, *13*, 1733. [CrossRef]
47. Anas, S.M.; Shariq, M.; Alam, M.; Yosri, A.M.; Mohamed, A.; AbdelMongy, M. Influence of supports on the low-velocity impact response of square rc slab of standard concrete and ultra-high performance concrete: FEM-Based Computational Analysis. *Buildings* **2023**, *13*, 1220. [CrossRef]
48. Hopkinson, B. A method of measuring the pressure produced in the detonation of high, explosives or by the impact of bullets. *Philos. Trans. R. Soc. Lond. Ser. A* **1914**, *213*, 437–456.
49. Kolsky, H. An investigation of the mechanical properties of materials at very high rates of loading. *Proc. Phys. Soc. Sect. B* **1949**, *62*, 676. [CrossRef]
50. Malvar, L.J.; Crawford, J.E. Dynamic Increase Factors for Concrete. In Proceedings of the 28th DDESB Seminar, Orlando, FL, USA, 10–12 August 1998.
51. Xiong, B.; Demartino, C.; Xu, J.; Simi, A.; Marano, G.C.; Xiao, Y. High-strain rate compressive behavior of concrete made with substituted coarse aggregates: Recycled crushed concrete and clay bricks. *Constr. Build. Mater.* **2021**, *301*, 123875. [CrossRef]
52. Yang, H.; Song, H.; Zhang, S. Experimental investigation of the behavior of aramid fiber reinforced polymer confined concrete subjected to high strain-rate compression. *Constr. Build. Mater.* **2015**, *95*, 143–151. [CrossRef]
53. Ross, C.A.; Tedesco, J.W. Split-Hopkinson pressure-bar tests on concrete and mortar in tension and compression. *Mater. J.* **1989**, *86*, 475–481.
54. Tedesco, J.W.; Powell, J.C.; Ross, C.A.; Hughes, M.L. A strain-rate-dependent concrete material model for ADINA. *Comput. Struct.* **1997**, *64*, 1053–1067. [CrossRef]
55. Zhu, W.C.; Bai, Y.; Li, X.B.; Niu, L.L. Numerical simulation on rock failure under combined static and dynamic loading during SHPB tests. *Int. J. Impact Eng.* **2012**, *49*, 142–157. [CrossRef]
56. Nemat-Nasser, S.; Isaacs, J.B.; Starrett, J.E. Hopkinson techniques for dynamic recovery experiments. *Proc. R. Soc. London. Ser. A* **1991**, *435*, 371–391.
57. Prot, M.; Cloete, T.J. A tandem momentum trap for dynamic specimen recovery during split Hopkinson pressure bar testing of cancellous bone. *J. Dyn. Behav. Mater.* **2016**, *2*, 50–58. [CrossRef]
58. Song, B.; Chen, W. Loading and unloading split Hopkinson pressure bar pulse-shaping techniques for dynamic hysteretic loops. *Exp. Mech.* **2004**, *44*, 622–627. [CrossRef]
59. Anas, S.M.; Alam, M.; Umair, M. Experimental and numerical investigations on performance of reinforced concrete slabs under explosive-induced air-blast loading: A state-of-the-art review. *Structures* **2021**, *31*, 428–461. [CrossRef]
60. Wang, S.; Xu, Y.; Xia, K.; Tong, T. Dynamic fragmentation of microwave irradiated rock. *J. Rock Mech. Geotech. Eng.* **2021**, *13*, 300–310. [CrossRef]
61. Yu, K.; Shi, Q.; Dunn, M.L.; Wang, T.; Qi, H.J. Carbon fiber reinforced thermoset composite with near 100% recyclability. *Adv. Funct. Mater.* **2016**, *26*, 6098–6106. [CrossRef]
62. Karthik, K.; Rajamani, D.; Raja, T.; Subramani, K. Experimental investigation on the mechanical properties of Carbon/Kevlar fibre reinforced epoxy LY556 composites. *Mater. Today Proc.* **2022**, *52*, 668–674. [CrossRef]
63. Li, Y.-F.; Yang, T.-H.; Kuo, C.-Y.; Tsai, Y.-K. A Study on Improving the mechanical performance of carbon-fiber-reinforced cement. *Materials* **2019**, *12*, 2715. [CrossRef] [PubMed]

64. *ASTM D3822-07*; Tensile Properties of Single Textile Fibers. ASTM: West Conshohocken, PA, USA, 2010.
65. *ASTM C33/C33M-18*; Standard Specification for Concrete Aggregates. ASTM: West Conshohocken, PA, USA, 2018.
66. *ASTM C143/C143M−20*; Standard Test Method for Slump of Hydraulic-Cement Concrete. ASTM: West Conshohocken, PA, USA, 2020.
67. *ASTM C39/C39M-01*; Standard Test Method for Compressive Strength of Cylindrical Concrete Specimens. ASTM: West Conshohocken, PA, USA, 2017.
68. *ASTM C293/C293M-16*; Standard Test Method for Flexural Strength of Concrete. ASTM: West Conshohocken, PA, USA, 2016.
69. *ASTM C496/C496M–17*; Standard Test Method for Splitting Tensile Strength of Cylindrical Concrete Specimens. ASTM: West Conshohocken, PA, USA, 2017.
70. *ACI 544.2R-89*; Measurement of Properties of Fiber Reinforced Concrete. America Concrete Institute: Farmington Hills, MI, USA, 1999.
71. Lindholm, U.S. Some experiments with the split Hopkinson pressure bar. *J. Mech. Phys. Solids* **1964**, *12*, 317–335. [CrossRef]
72. Pakravan, H.R.; Latifi, M.; Jamshidi, M. Hybrid short fiber reinforcement system in concrete: A review. *Constr. Build. Mater.* **2017**, *142*, 280–294. [CrossRef]

buildings

MDPI

Article

Effects of Fly Ash Composition to Mitigate Conversion of Calcium Aluminate Cement Composites

Thwe Thwe Win [1], Chinnapat Panwisawas [2], Pitcha Jongvivatsakul [3], Withit Pansuk [3] and Lapyote Prasittisopin [1,*]

[1] Architectural Technology Research Unit, Department of Architecture, Faculty of Architecture, Chulalongkorn University, Bangkok 10330, Thailand
[2] School of Engineering and Materials Science, Queen Mary University of London, London E1 4NS, UK
[3] Centre of Excellence in Innovative Construction Materials, Department of Civil Engineering, Faculty of Engineering, Chulalongkorn University, Bangkok 10330, Thailand
* Correspondence: lapyote.p@chula.ac.th

Abstract: Calcium aluminate cement (CAC) is one of the alternative cements that is widely used for special applications. However, during the hydration process degradation of CAC microstructure, the so-called hydrate conversion process, hexagonal calcium aluminate hydrate (CAH_{10}) transforms into a cubic (C_3AH_6) phase, resulting in increased porosity and reduced strengths. It is known that alternative means for stabilizing the CAC conversion are conducted by introducing fly ash (FA) in CAC, where its microstructure is attributed to aluminosilicates. However, no study has yet been conducted on different FA compositions influencing CAC performance. This study aims to evaluate the effects of different compositions of FA on CACs' fresh and hardened characteristics. Results revealed that the microstructure was denser when CAC was mixed with FA. Regarding reactivity, CAC with calcium-rich FA systems is 13% faster than the silica-rich one. The higher the density and the lower the porosity of calcium-rich FA mixtures were found compared with silica-rich FA in both micro- and macro structures. As seen in the microscopic structure, this is due to the calcium-rich phase formation.

Keywords: fly ash; calcium aluminate cement; conversion; hydration; performance; composition

check for updates

Citation: Win, T.T.; Panwisawas, C.; Jongvivatsakul, P.; Pansuk, W.; Prasittisopin, L. Effects of Fly Ash Composition to Mitigate Conversion of Calcium Aluminate Cement Composites. *Buildings* **2023**, *13*, 2453. https://doi.org/10.3390/buildings13102453

Academic Editor: Tomáš Dvorský

Received: 4 September 2023
Revised: 19 September 2023
Accepted: 22 September 2023
Published: 27 September 2023

1. Introduction

Calcium aluminate cement (CAC) or high-alumina cement is widely used and usually adopted in engineering applications with special purposes, such as refractory, biomaterial, ability to resist acid exposure, fast-setting, durable, and additive manufacturing concretes [1]. Several research programs were being determined, especially for the purposes of property improvement, durability, and more eco-friendly production processes. In addition, during the manufacturing process of CAC, depending on the availability of the source, it is reported that there is a significant reduction in carbon dioxide emissions generated in the atmosphere from lower sintering temperatures compared to portland cement, so CAC can be considered an eco-cement material.

One of the primary applications of CAC is in the construction sector. It is widely used in refractory applications due to its ability to withstand extreme temperatures, making it invaluable in the manufacturing of furnaces, kilns, and foundry linings. Moreover, CAC is employed in high-performance concrete formulations, where its rapid setting and exceptional strength gain are critical, particularly in precast concrete elements and underwater construction [2].

However, the widespread adoption of CAC-based materials is not without challenges. Regulatory constraints governing the use of CAC can vary significantly from one country to another. These constraints may pertain to environmental considerations and quality control measures. Understanding these variations and how they impact the use of CAC-based materials is essential for engineers and researchers.

One of the primary concerns associated with CAC is the hydrate conversion process during hydration, which can impact the material's performance. The CA phase is the fundamental hydraulic phase in all CAC systems. Regarding the affecting factors, the most significant factor is the curing temperature, since the hydration product directly relies on the curing temperature [3]. CAH_{10} and C_2AH_8 phases are regarded as metastable at ambient temperature [4] and ultimately change to the more stable C_3AH_6 and AH_3 phases due to thermodynamic stabilization. The conversion from the metastable to stable phases leads to porosity gains, more permeability, and hence strength reduction [5]. The conversion reactions are presented in Equations (1) and (2). The hydration reaction of CA produced is dependent on the temperature, which is given in Table 1. The strength progression of CAC concrete at room temperature is shown in Figure 1. The strength first rises rapidly to a high early strength (corresponding to CAH_{10} and C_2AH_8), then gradually drops to a minimum (during the transformation to C_3AH_6 and AH_3) before rising once more from prolonged hydration.

$$2CAH_{10} \longrightarrow C_2AH_8 + AH_3 + 9H \quad (1)$$

$$3C_2AH_8 \longrightarrow 2C_3AH_6 + AH_3 + 9H \quad (2)$$

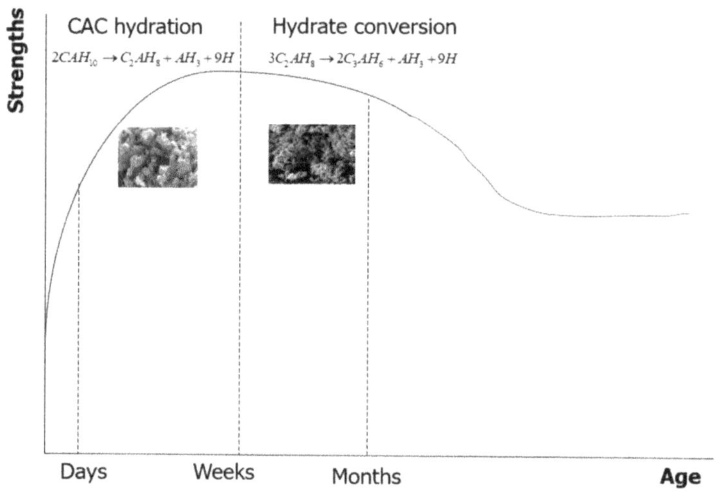

Figure 1. Schematic strength development of CAC concrete at ambient temperature.

Boris et al. [6] investigated the conversion process, especially in the presence of nanomaterials. These nanomaterials vary in chemical composition and grain size, and their effects on CAC depend on their type and quantity. Nano-silica was found to accelerate the CAC hydration process, shortening the induction period and intensifying the secondary heat release effect. Despite similar types of hydration products in all CAC pastes, the quantitative composition could vary [6]. Due to this conversion effect, the implementation of CAC in structural engineering practice is restricted. Although CAC is no longer a novel material, the conversion raises questions about how well it performs under harsh loading and the stability of its microstructure. The processes that transform the metastable phases of CAH_{10} and C_2AH_8 into stable, long-lasting, and dense C_3AH_6 hydrates are responsible for this thermodynamic instability [7,8]. Many researchers have attempted to identify the chemical reactions and the effects of the conversion on the overall performance of CAC concretes [8,9]. Some researchers [3–5] have investigated whether replacing some of the

CAC with pozzolanic materials such as ground granulated blast furnace slag (GGBFS), metakaolin, silica fume, and rice husk ash is a proper technique to mitigate the hydrate conversion process and strength reduction.

Therefore, utilizing coal fly ash (FA) as pozzolan in CAC is carried out to overcome the conversion process. The incorporation of FA could improve the hydrogarnet due to an aluminate hydrate reacting alternatively to silica in the C-A-S-H phase or stratlingite (C_2ASH_8) [2,10] as revealed by the subsequent equations:

$$2CAH_{10} + S \longrightarrow C_2ASH_8 + AH_3 + 9H \quad (3)$$

$$C_2AH_8 + S \longrightarrow C_2ASH_8 \quad (4)$$

Majumdar and Singh [11] found that replacing CAC with pozzolanic materials such as micro-silica, GGBFS, and metakaolin reduced the hydrate conversion while also increasing its strengths. Its hydrate conversion of hexagonal hydrated structures could be prevented by the pozzolanic reaction in the CAC system, which has been observed in the reaction of CA. Lastly, this reaction could prevent hexagonal (C_2AH_8) formation and C_3AH_6 transformation [11,12].

FA is an industrial by-product with very fine particles that is manufactured from the combustion of coal in power plants. FA has a spherical shape that has a high silica and calcium composition [13]. Although the CAC hydration process may receive some attention for inclusion with pozzolanic materials by many researchers, the mechanical and microstructural performances of such systems have not been systematically investigated in engineering applications. In addition, the effect of pozzolanic materials on the hydration reaction of CAC and its mechanical performance have not been completely investigated. Similarly, a study that examines how FA impacts the mechanical and microstructural characteristics of CAC systems is lacking.

Based on the alumina and calcium contents, various CAC systems are remarkably different, with low silica contents. The mechanical properties of this CAC concrete can be changed by variations in the alumina content [14]. Although it has been researched how the water-to-cement ratio (w/c) affects the fracture performance of CAC concrete, the effect of adopting different silica contents in the CAC system has not been reviewed yet [15]. According to the review of the literature, the novelty of this work includes:

- The effect of replacing FA with CAC has not been thoroughly considered.
- The effect of two types of FA (class F and class C) on the performance of this type of cement has not been investigated.

The experiment explores various aspects of CAC composites incorporating FA, entailing initial setting time, evolution of FA reactivity, workability, densities at 7, 28, and 56 days, compressive strengths at 7, 28, and 56 days, and porosities at 7, 28, and 56 days. To examine the pore structures and chemical composition of CAC, including FA, nitrogen adsorption and thermogravimetric analysis (TGA) have been performed. Furthermore, microstructural analysis using Scanning Electron Microscopy (SEM) with Energy Dispersive Spectrometer (EDS) has been employed to gain a deeper understanding of the transformation of CAC containing FA.

Table 1. Summary of the temperature history of conversion reaction of CAC from literature.

Reference	Temperature Range (°C)		
	CAH_{10}	C_2AH_8	C_3AH_6
Scrivener et al. [16]	<15	15–70	>70
Adams et al. [17]	<15	15–27	>27
Khaliq and Khan [18]	<15	15–27	>27
Zapata et al. [19]	<15	15–35	>35

Table 1. *Cont.*

Reference	Temperature Range (°C)		
	CAH_{10}	C_2AH_8	C_3AH_6
Son et al. [20]	<20	20–40	40–60
Antonovič et al. [7]	5	20	40
Ukrainczyk and Matusinović [21]	20	30	>55
Vafaei and Allahverdi [14]	15–25	25–40	40–60
Zapata et al. [9]	≥20	~30	>55
Mean	<16	28	>45
Median	<15	15–27	>40

2. Materials and Methods

2.1. Materials

The CAC used in the preparation of the mixture was Secar 71, manufactured by Kerneos in Paris, France. Standard sand, in accordance with ASTM C33 [22], was utilized as fine aggregate for making CAC mortar. The standard sand, which has a specific gravity of 2.61, a fineness modulus of 2.96, and a water absorption of 0.66%, was obtained from Saraburi Province, Thailand [23]. Deionized water was used throughout the experiment. Two types of as-received FA were procured from different sources. SEM pictures of aluminosilicate sources display that both class F (FA1) and class C FA (FA2) have spherical shapes, and CAC has angular and asymmetrical shapes (in Figure 2). Figure 3 shows the sieve analysis of fine aggregate obtained from the sieve test and the upper and lower limits according to ASTM C33 [22]. Their grading meets the limits specified by ASTM standards. The aggregate was used in this investigation under saturated surface dry (SSD) conditions. It should be mentioned that the used FA was categorized into two types: low calcium content FA (class F, max: 18%) and high calcium content FA (class C > 18%) according to the ASTM C618 specification on pozzolanic materials [24]. FA employed here was supplied for class F and class C FA from sources in Map Ta Phut Industrial Estate, Rayong Province, Thailand, and Mae Moh, Lampang Province, Thailand, respectively. The chemical composition of CAC and FA analyses was performed using X-ray Fluorescence Spectroscopy (XRF) (BRUKER S8 TIGER instrument, Bruker Corporation, Bremen, Germany). In addition, the Blaine-specific surface areas of CAC and FA were determined in conformity with ASTM C204 [25]. The specific gravity test was carried out to determine the weight per unit volume of the CAC and FA particles according to ASTM C188 [26]. The chemical composition and physical properties of CAC and FA are presented in Table 2.

Figure 2. SEM micrographs of CAC, FA1, and FA2.

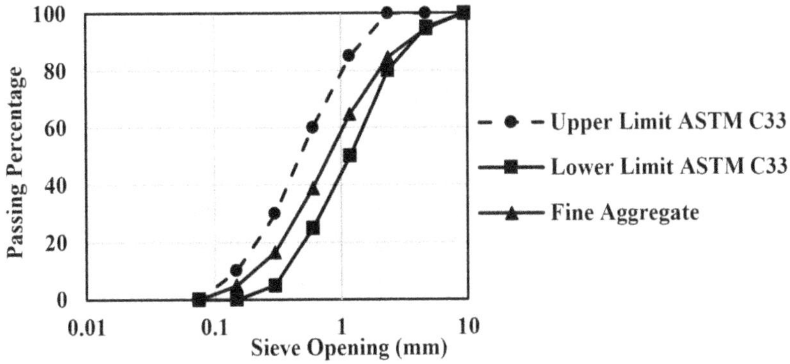

Figure 3. Sieve analysis of fine aggregate.

Table 2. Chemical composition and physical properties of CAC and different FA.

Composition (%)	CAC	FA1	FA2
SiO_2	0.19	73.80	27.40
Al_2O_3	68.9	17.70	15.80
Fe_2O_3	0.13	1.94	12.10
CaO	29.6	0.80	21.60
MgO	0.21	0.31	2.26
SO_3	4.14	0.19	6.92
Na_2O	0.33	0.39	1.70
$SiO_2 + Al_2O_3 + Fe_2O_3$	-	93.44	55.30
LOI	0.5	1.8	0.2
Grade/Class	High alumina	Class F	Class C
Color	White	Gray	Tan
Specific gravity	3.21	2.11	2.40
Blaine surface area (cm^2/g)	3889	4654	4131
Average particle size (μm)	15.6	7.32	7.64

2.2. Preparation of Samples

All mortar mixes produced with two types of FA, namely, FA1M (silica-rich FA) and FA2M (calcium-rich FA), have a 20% replacement level by weight with a binder-sand ratio of 1:2.75 by weight (per ASTM C109) [27] and a water-to-binder ratio (w/b) value of 0.55. Based on previous research [13], a FA substitution level was used as the optimal replacement level for cement. Therefore, this substitution level was selected in the research, as it is a well-established and commonly used substitution level in the industry.

The mortar mix design is given in Table 3. The required amount of water was immediately poured after mixing the materials homogenously in a mechanical mixer. The mortar specimens containing FA were produced according to ASTM C305 [28]. The mixing continued until a uniform mixture was obtained. After the end of mixing, the workability of the fresh mortars was analyzed using the flow table test according to ASTM C1437 [28]. The fresh mixtures were then placed lightly in oiled plastic cubic molds with a dimension of 50 mm for testing the density, compressive strength, and porosity [29]. To maintain the moisture loss, cast specimens were covered with a polyethylene film. After 24 h, the cured specimens were demolded and kept in a water-curing tank at ambient temperature until the designated test age [30]. The SEM/EDS (JSM-IT300LV instrument, JEOL Ltd., Tokyo,

Japan) was used to investigate the morphology of the microstructure, pore structure, and chemical composition of the samples. In addition, the EDS analysis was semi-quantitatively conducted to determine the elemental components.

Table 3. Proportions of mortar mixtures (in kg/m^3).

Mix	CAC	FA	Sand	Water	w/b
CACM	525.00	0.00	1443.75	288.75	0.55
FA1M	420.00	105.00	1443.75	288.75	0.55
FA2M	420.00	105.00	1443.75	288.75	0.55

2.3. Experimental Methods

The initial setting times of CAC pastes (CACP) with different compositions of FA were investigated according to the ASTM C191 standard [31]. Various curing temperatures (30, 40, 60, and 80 °C) were applied to evaluate the reactivity of FA. These curing temperatures were selected based on a combination of factors, including the specific properties of CAC and FA phases, the intended application of the samples, and previous research performed in the field. These curing temperatures were selected to simulate a range of practical conditions. At lower temperatures (30 and 40 °C), the intent was to replicate conditions where the material might experience milder curing environments, such as ambient temperature curing in certain applications. As the temperature increases to 60 and 80 °C, the focus shifts towards investigating the material's behavior under more accelerated curing conditions, which can be relevant for time-sensitive applications. The higher temperatures were selected to assess how the material's properties evolve as exposed to elevated temperature conditions. The observed temperature gradient inconsistencies may arise from complex interactions between the cement and FA phases, heat distribution, and reaction kinetics. In addition, to assess the reactivity of cement paste systems containing class F FA (FA1P) and class C FA (FA2P), the experiment was applied thermodynamically using the Arrhenius Equation as follows:

$$k = k_0\ \exp(-E_a/RT) \tag{5}$$

where k is the rate coefficient (s^{-1}), k_0 is the pre-exponential constant (s^{-1}); E_a is the activation energy (J·mol^{-1}), R is the gas constant (=8.314 J·K^{-1}·mol^{-1}), and T is the temperature (K).

Additionally, dry densities of CAC mortars containing different FA at the curing ages of 7, 28, and 56 days have been examined, per ASTM C642 [32]. The cast mortars were made in 50 × 50 × 50 mm^3 cube molds. The measured densities were conducted based on three tests.

The 7-, 28-, and 56-day compressive strengths of CAC mortars containing FA (CACM) were carried out according to ASTM C109/C109M [27]. After casting, mortar specimens were kept in molds at an ambient temperature of 23 ± 2 °C and saturated relative humidity (≥95% RH) for 24 h and then demolded. Demolded specimens were kept in a curing tank until the designated test age. The variety of 7, 28, and 56-day durations is based on industry standards and common practices for assessing the long-term performance of construction materials, including CAC-based materials. These test durations evaluate both early-age properties and the material's evolution over time. The 7-day compressive strength insights into early-age properties are particularly relevant for early strength development. The 28-day compressive strength assesses the material's performance at a more intermediate stage, while the 56-day compressive strength provides insights into the longer-term behavior and durability of the CAC-FA mixtures.

This study considered the porosity and density of CAC mortars affected by the conversion reaction. As mentioned above, the porosity of a cement system is determined by its vital microstructural performances relating to its durability. To determine the porosity of CAC containing FA, their hardened porosity is determined by evaluating the total weight of water in the SSD samples, which have a dimension of 50 × 50 × 50 mm^3. The cubed specimens were cast and then cured for 7, 28, and 56 days. After that, the specimens were placed in an oven at 105 °C for 24 h, the dried specimens were kept in a water tank to completely penetrate water into their voids. It is noted that at a temperature of 105 °C, all the physically bound and capillary water in the cement paste had completely evaporated. A weight scale with an accuracy of 0.01 g was used to measure the mass of saturated and dried samples. The three specimens were carried out for each test, and the average data and error bars were presented. A similar method was utilized in previous studies by Shen et al. [29]. Equation (6) was determined for calculating the porosity values.

$$p = \frac{M_w - M_d}{v \times \rho} \tag{6}$$

where p is the porosity in %, M_w is the weight of the saturated sample in g, M_d is the weight of the dried sample in g, v is the volume of the sample in cm^3, and ρ is the density of water at 20°C in g/cm^3.

The Brunauer-Emmet-Teller (BET) isotherm is a promising method for assessing the surface area and adsorption properties of a variety of materials [33]. The results obtained from BET analysis can reveal the pore characteristics of CAC-FA composites, which are closely linked to their strengths. In this approach, the sample was tested in vacuum conditions, causing the cement matrix to absorb nitrogen gas using a nitrogen adsorption measurement instrument (Micromeritics 3Flex, Micromeritics Instrument Corporation, Boulder, CO, USA). The quantity of adsorbed nitrogen gas was then quantified to express the surface properties and adsorption capabilities. Using the BET theory, the pore structures of CAC composites ranging from 1 to 500 nm were evaluated. The results presented are the specific surface area and pore volume of the CAC composites.

In addition, powdered specimens cured for 56 days were obtained by extracting material from the central portion of the samples. TGA was carried out using the Perkin Elmer Pyris 1 instrument (PerkinElmer, Shelton, CT, USA) to quantify the extent of CAC conversion. In the TGA procedure, the temperature was incrementally increased from 25 to 1000 °C at a controlled rate of 10 °C per minute while maintaining a continuous nitrogen flow of 70 mL/min. The content of C_3AH_6 was calculated and reported.

For SEM-EDS analysis, the CAC composite's microstructural arrangement was examined. A fracture of hardened CAC was carefully procured, dehydrated, purified, and coated with a layer of gold before being fixed to copper mounts using carbon tape. The SUPRA 66VP ultra-high-resolution equipment(Carl Zeiss AG, Aalen, Germany) was utilized for analyzing the surface morphology and chemical composition of the samples at the microscale.

3. Results

3.1. Chemical Composition of CAC and FA

The chemical compositions of CAC, FA1, and FA2 are illustrated in Table 2. As observed, the principal constituents of CAC are Al_2O_3 > CaO > SiO_2. FA1 and FA2 were largely made up of 73.8% and 27.4% SiO_2, respectively, with other components of Al_2O_3 and CaO. In addition, the CaO content of FA2 (21.6%) was 27 times higher than that of FA1 (0.8%). The silica composition in FA was considered to have pozzolanic properties.

3.2. Initial Setting Time

Figure 4 shows the effect of replacing CAC with a different FA on their initial setting times. Results indicated that the initial setting time of CACP was longer than FA1P and FA2P set at ambient temperature. FA significantly changed the setting time of the CAC. Bensted [2] reported that CAC was moderately gradual stiffening; however, rapid hardening occurred thereafter. Therefore, CAC can provide slow stiffening characteristics at a suitable time for concrete placement without using any retarders. The beginning of steady development of hydrates can also permit this characteristic with a load of time, which enables high early strength development. The results also indicated that control and

CAC binders with silica-rich FA had longer setting times than those with Ca-rich FA for all temperature levels tested at 30, 40, 50, and 60 °C. However, the setting time of CAC without FA noticeably declined at elevated temperatures. Additionally, the process of solidification has been significantly improved by the partial replacement of CAC with FA. The tested results apparently showed that CAC with FA could significantly reduce the setting time, representing accelerated hydration.

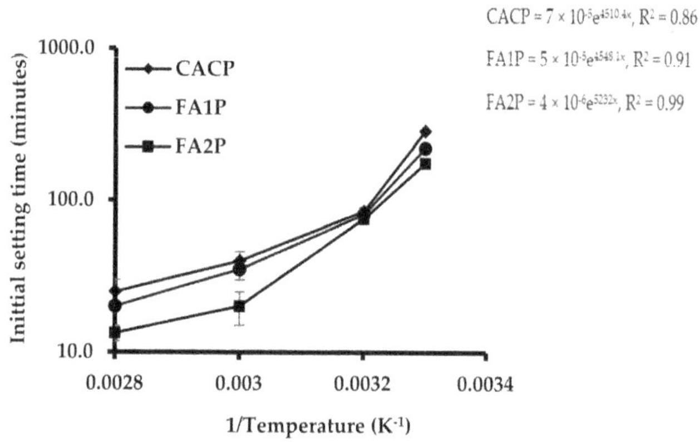

Figure 4. Plot of the effect of FA on the initial setting times of CAC systems (minutes in log scale) as a function of 1/temperature (K^{-1}) and fitted Arrhenius equations (right).

3.3. Assessment of the Reactivity of FA

The Arrhenius equation, which can be used to calculate FA reactivity, is given in Equation (5) and is measured by the initial setting times of the pastes [34]. The activation energy can be denoted by the slope of a straight line. The energy during the stiffening stage is represented as the activation energy, which is calculated here from the initial setting time of CAC paste containing various FA. To change the phase from liquid to solid, higher slopes are associated with a higher activation energy requirement [34]. Temperature is represented as an inverse function of the initial setting time, as illustrated in Figure 4. The findings revealed that the CACP had a steeper slope than the CAC paste with various FAs. This indicated that the CAC paste with FA demanded less energy for paste stiffening than the CAC system. The CAC had more reactive components that rapidly and well reacted with water, leading to a higher rate of heat release and requiring more energy for hydration. In contrast, FA slowly reacted with $Ca(OH)_2$, gradually releasing heat and requiring less energy for hydration.

Therefore, the partial replacement of CAC by FA was used in the conversion reactions of CAC due to the lower consumption of energy. Furthermore, the FA1P had a lower reactivity value than the FA2P, around 13%. Meaningfully, using high-calcium-content FA or class C FA can offer faster reactivity than low-calcium-content FA or class F FA. This contributes to calcium reacting faster than silica at an early age. Although silica offers siliceous reactions at later ages, this should be discussed in Section 3.6.

3.4. Flowability

The flowability test was conducted on fresh mortars per ASTM C1437. Figure 5 demonstrates the workability value of CAC mortar with and without 20% FA. Results revealed that the flow values of all mixes ranged from 76 to 102%. The incorporation of FA in the CAC system led to an increase in flow value. This is generally due to the unique spherical shape of FA particles, which can enhance the workability of concrete by filling in voids between cement particles and lubricating the mixture. Their spherical shape creates a ball-bearing effect, allowing them to move more freely within the mixture, reducing the friction between particles and improving the flowability of the concrete. This led to a reduction in the amount of water needed to achieve the desired workability.

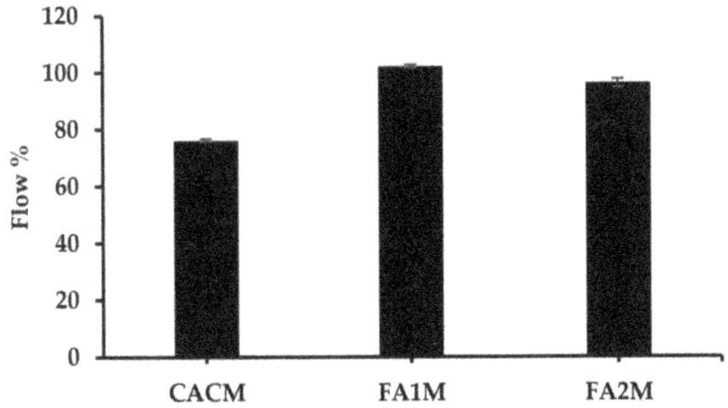

Figure 5. Flow value of fresh CACM with different FA.

The FA1M had the largest flow value of 102 mm, while the CACM exhibited the lowest flow value of 76 mm. The flow of FA2M was approximately 5% lower than FA1M. These FA-based mixes had higher cohesiveness than the control mix. This cohesiveness diminishes segregation all through the mixing, transporting, placing, and compacting processes [35].

3.5. Dry Density

Measuring the densities of mortars is similarly vital as porosity due to their effectiveness in filling gaps in pastes with solid cement components [36]. The plot of the dry density of CACM containing FA for different ages is illustrated in Figure 6. Results exhibited that the dry densities of the CACM samples ranged from 2.34 to 2.42 g/cm^3. The density of CACM cured for 7 days was about 3% higher than that of CACM cured for 28 and 56 days. Moreover, the 7-day densities of these samples were 5% and 9% greater than those containing FA (FA1M and FA2M, respectively). Because FA had a lower specific gravity than CAC, the dry density of the specimens with FA declined. Moreover, the sample density decreased slightly for FA1M. It is implied that the specific gravity of FA1 was smaller than that of CAC and FA2. As a result, the 28- and 56-day densities of CACM without FA had the lowest density values, while the 28- and 56-day densities of FA2M had the highest values. This was because their densities were different, with conversion phases leading to the development of porosity (which will be discussed in Section 3.7). The densities of intermediate phases CA and C_2AH_8 are 2.98 and 1.97, respectively, whereas the converted phases C_3AH_6 and AH_3 are 2.52 and 2.44 [2]. For that reason, the conversion of the CAC hydrates to C_3AH_6 and AH_3 permitted porosity enhancement. Moreover, the formation of density is consistent with the mechanical properties of CACM, including FA.

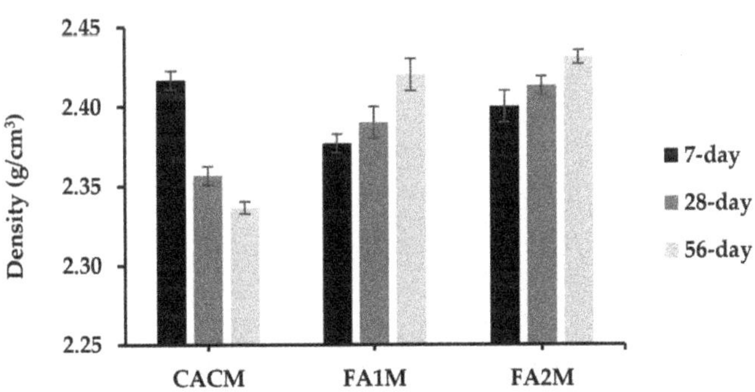

Figure 6. Dry density of CACM with different types of FA cured for 7, 28, and 56 days.

3.6. Compressive Strength

Figure 7 shows 7-, 28-, and 56-day mortar compressive strengths having FA. The 7-, 28-, and 56-day compressive strengths of CACM were 28.3, 24.6, and 22.9 MPa, respectively. According to the observed results, the 7-day compressive strengths of the CACM were 15% and 24% higher than those of the CACM cured for 28 and 56 days, respectively. Moreover, the 7-day compressive strength of the CACM was 30% and 11.5% higher than the FA1M and FA2M. The strength improvement of CAC concrete at ambient temperature has the phase formations illustrated in Figure 1 [16], with an initial improvement sharply increasing to an elevated early strength (corresponding to CAH_{10} and C_2AH_8).

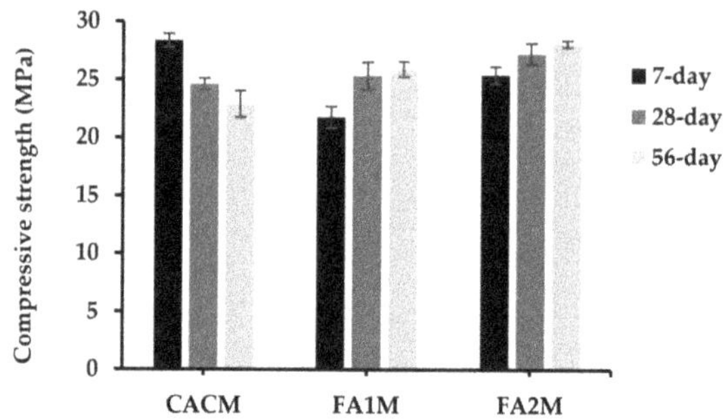

Figure 7. Compressive strengths of CACM containing different FA cured for 7, 28, and 56 days.

Conversely, the presence of FA in the CAC system offers higher 28- and 56-day compressive strengths when compared to the control. Previous studies [37,38] investigated a similar figure and considered that one of the key factors determining the compressive strength and durability of the FA-cement system was the pozzolanic reactivity. Additionally, results found that pozzolanic materials with larger silica contents reacted with $Ca(OH)_2$ more readily and thus produced C–(A)–S–H at higher contents. The chemical and physical effects of FA determined their compressive strength development. The pozzolanic reactions between the amorphous silica in FA and $Ca(OH)_2$, which were produced by the hydration reaction of cement to create C–S–H gel, were mostly responsible for the chemical effect. The physical effect, known as the "filler effect," was that FA particles improved the packing of solid materials by filling the spaces between the cement grains, similarly to how the cement particles filled the spaces between fine aggregate particles.

In addition, FA2M led to one of the highest compressive strength developments (25.91 and 28.76 MPa) in CAC-FA systems in both 28- and 56-day compressive strengths. Results found that there were 5% and 8.5% greater compressive strengths of the FA2M compared with the FA1M at 28- and 56-day curing periods, respectively. Although the compressive strength of FA2M was greater than that of FA1M at all testing ages, the trend was likely to be the same in both FA1M and FA2M. As the CAC was replaced by FA2, the compressive strength increased to lesser extents, which was likely because of the high silica content in FA2 that was assessable with $Ca(OH)_2$ formed from the hydrate reaction. It was reported that once the cement systems contained huge amounts of silica during hydration, because they could not release enough $Ca(OH)_2$ to hydrate with all the available silica, some silica remained present without causing any chemical reaction [39]. Later, a substantial portion of silica was absent from the chemical reactions [39].

In addition, after 7 days of curing, the conversion of all CACM occurred, which resulted in a suddenly reduced compressive strength. It is inferred due to the conversion of CAC. Nonetheless, using FA in CAC diminished the conversion process from metastable phases (CAH_{10} and C_2AH_8) to stable phases (C_3AH_6 and AH_3). It can be explained that the C_2ASH_8 development in the case of loss of compressive strength could be lessened if there was FA content. By generating a favorable situation for the creation of C_2ASH_8 instead of C_3AH_6 by inducing calcium ions in the pore solution to be consumed, the insertion of FA in this work was seen to prevent the phase evolution to a calcium-rich phase, C_3AH_6 [20]. It is noticeable that at room temperature, the primary hydrates of CAC having siliceous materials have been described to form as C_3AH_{10}, C_2AH_8, C-(A)-S-H, and AH_3 gel phases. Therefore, this resulted in a decrease in mechanical strength since the phase shift in CAC accompanied an increase in porosity in a rigid state [40,41]. This phenomenon is detailed in Section 3.7.

3.7. Porosity

Porosity is a crucial factor influencing the strength, durability, and mechanical effectiveness of concrete. The porosity of the hydrated CACM increased, as shown in Figure 8. As a result, the 7-day porosity of CACM was 13% and 19% lower than the CACM at 28- and 56-day curing ages, where the values ranged between 12.1% and 15.0%. In addition, the porosity of the CACM cured for 7 days was about 19% and 5% lower than the FA1M and FA2M. On the other hand, the 28- and 56-day porosities of CACM were significantly higher than FA1M and FA2M. Therefore, replacing CAC with FA resulted in reduced porosity by approximately 8% to 48% at longer curing ages. The figures of FA1M and FA2M exhibited lower porosity values in the CAC-FA systems over time, whereas the conversion took place in CACM at longer ages. It was apparently seen that FA2M stood as one of the lowest porosity values (11.23% and 7.8%) at both 28 and 56 days, respectively. Since the pozzolanic reaction of FA reacts for a longer time, called the second reaction, the FA cannot provide a satisfactory reaction for the first 7 days. Based on the results, the converted CACM increases porosity considerably when compared to FA1M and FA2M. This indicates that conversion significantly influences the morphology of CAC.

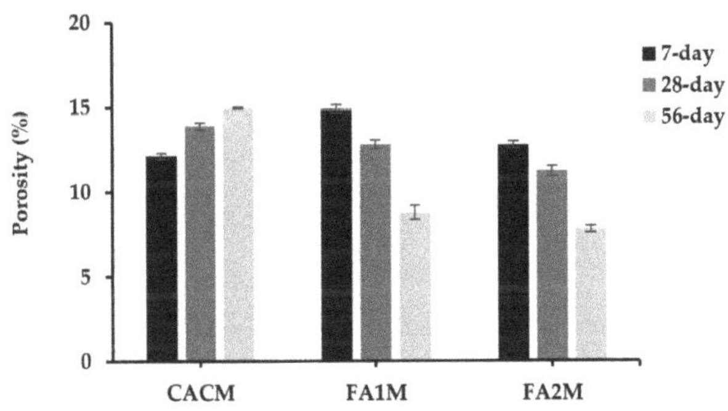

Figure 8. Porosity of CACM with different FA cured for 7, 28, and 56 days.

Bensted [2] suggested that the increase in porosity could be attributed to the formation effects resulting from conversion processes. This was observed due to the higher density found in the converted phase (C_3AH_6) compared to the metastable phase. While the formation of stable C_3AH_6 and AH_3 increased porosity in the CAC conversion, which allowed for strength reduction. The formation of metastable phases resisted and controlled the shrinkage of the hydrated CAC [42]. There will be a higher ingress of CAC with more porosity, which will worsen its durability. According to Equations (1) and (2), the aforementioned conversion reactions cause CAC to become more porous, permeable, and weaker as a result of the release of water from the conversion structure and the production of a packed C_3AH_6 (which results in a reduction in the molar volume of the solids) [43,44].

3.8. BET Isotherm Analysis

The nitrogen sorption isotherms for all mixtures are shown in Figure 9. During the early phases of sorption (at p/p° values between 0 and 0.1), there is rapid initial molecular coverage on the surfaces of the pores as the gas pressure rises. Subsequently, the rate of sorption volume increase diminishes considerably, resulting in a distinct bend known as a 'knee' [45]. The knee characteristics are essential for distinguishing between monolayer and multilayer sorption. Less overlap exists between monolayer coverage and the start of multilayer adsorption if the curvature is more pronounced (indicating a more distinct turning point). The isotherms show that during the initial sorption phase, all of the knees have very strong, clearly defined turning points (shown by the arrow in Figure 9). This means that the sorption is changing from monolayer coverage to multilayer. As the relative gas pressure rises to 0.7, the adsorption volume increases gradually and progressively. The IU-PAC type II pattern [45] is approximately followed by the adsorption branches of the isotherms, indicating that the tested materials possess macropores. When the relative gas pressure drops from 1 (0.98 in actual testing) to 0.45, the desorption and adsorption branches move away from each other. This causes hysteresis loops to form in this range of gas pressure (p/p° between 0.45 and 1). According to IUPAC standards [45], these hysteresis loops are of the H3 type, which can be attributed to the presence of macro-porous structures. Notably, the hysteresis loop observed in the CACP sample was slightly larger than that observed in the other samples, presumably due to the CAC system's conversion process. The FA1P and FA2P mixtures exhibited comparable isotherm geometries and hysteresis loops, indicating that their pore structures can be comparable.

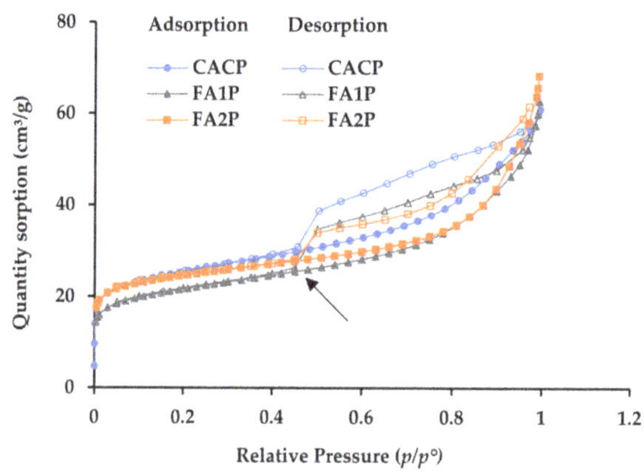

Figure 9. Nitrogen sorption isotherms of all mixes.

Furthermore, the hysteresis loops observed in the FA1P and FA2P gradually diminished in size and became less prominent when compared to the hysteresis loop of the CACP, which primarily occurred at higher pressures ($p/p^\circ > 0.6$). This observation suggests a decrease in the porosity of the FA1P and FA2P, indicating smaller pore volumes compared to the CACP, where its pronounced hysteresis loop is evident. This phenomenon can be attributed to the limited number of pores introduced by the incorporation of FA. Consequently, the conversion process of CAC containing FA resulted in lower strength loss when compared to CACP, as detailed in Table 4. The analysis revealed that the surface area of CACP was notably higher than that of FA1P and FA2P. As a result, the inclusion of FA led to a reduction in surface area. Regarding total pore volume, the findings showed that incorporating FA2P into the CAC system resulted in an 8.4% reduction as compared with CACP. The incorporation of FA effectively promotes the development of CAC composites with finer pore structures.

Table 4. Pore structure of CAC composites containing different FA.

	CACP	FA1P	FA2P
Surface area (m^2/g)	82.824	78.666	70.633
Total pore volume (cm^3/g)	0.083	0.077	0.076

3.9. TGA

Figure 10 presents the TGA analyses of hydrated CAC systems containing FA. As shown in Equations (7)–(10), when a heated sample of hydrated CAC breaks down, chemically bound water is released from its different aluminate hydrates. This investigation reveals that AH_3 releases the bound water at 250 °C, and the C_3AH_6 phase releases the bound water at 300 °C. These findings can also be found elsewhere [20,46–48].

$$CAH_{10} \longrightarrow CaAl_2O_4 + 10H_2O\ (100\text{–}160\ ^\circ C) \tag{7}$$

$$C_2AH_8 \longrightarrow Ca_2Al_2O_5 + 8H_2O\ (140\text{–}200\ ^\circ C) \tag{8}$$

$$AH_3 \longrightarrow Al_2O_3 + 3H_2O\ (210\text{–}300\ ^\circ C) \tag{9}$$

$$C_3AH_6 \longrightarrow Ca_3Al_2O_6 + 6H_2O\ (240\text{–}370\ ^\circ C) \tag{10}$$

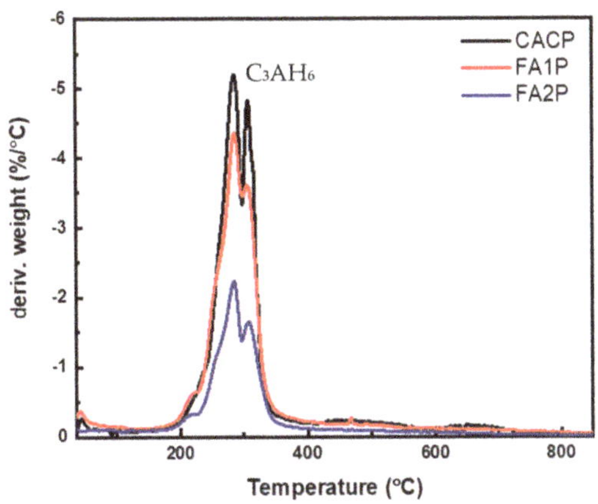

Figure 10. The DTG curve of CAC composites including FA.

The results showed that TGA analysis only found AH_3 and C_3AH_6 hydrates. There were no signs of additional CAH_{10} or C_2AH_8 phases in the samples. The reason behind this absence is that a significant portion of the C_2AH_8 transformed into AH_3 and C_3AH_6. These results are consistent with prior research findings regarding the transformation of metastable phases into C_3AH_6. Equation (11) was used to determine the quantity of C_3AH_6.

$$C_3AH_6 = WL_{C_3AH_6} \times \frac{m_{C_3AH_6}}{6 \times m_{H_2O}} \tag{11}$$

where C_3AH_6 is the amount of measured C_3AH_6 present in the mixture (%), and WL_{C3AH6} is the weight loss percentage due to water evaporation within the range of 240 to 370 °C. Moreover, m_{C3AH6} is the molecular weight of C_3AH_6 (378 g/mol), and m_{H2O} is the molecular weight of water (18 g/mol). The outcomes are given in Table 5, illustrating the calculated percentages of transformed C_3AH_6 within the CAC system. Obviously, CACP yields a notable expansion in the corresponding area, indicating heightened conversion and a greater presence of C_3AH_6. Conversely, when employing FA, notably FA2P, the formation of C_3AH_6 is reduced. This can be attributed to the CASH phase formation of FA1P and FA2P.

Table 5. Percentage of C_3AH_6 of CAC containing FA from TGA analysis.

	CACP	FA1P	FA2P
C_3AH_6 (%)	57.7	45.7	44.1

3.10. Microstructure Assessment

SEM micrographs of CAC/FA specimens are illustrated in Figure 11. Some of the pores caused by the conversion hydration in CACP were very obvious, and due to the lack of FA, the highly porous structure of the CACP was obvious. Moreover, there was a larger size and number of pores in CACP compared with FA1P and FA2P. The larger size and increased number of pores can result in the mechanical properties of CAC deteriorating. In contrast, CAC containing FA1 and FA2 showed a more uniform and compact microstructure in the cement matrix. In addition, while comparing the micrographs of both FA1P and FA2P samples, it became obvious that FA2P had a denser and lower inter-particle space than FA1P. It is evident that by replacing FA2 in CAC, the surface of these samples' morphology showed a radical transition from a non-compact, porous structure to a compact one. This could also happen because FA1 has a larger specific surface area than FA2 and CAC, resulting in a higher rate of hydration due to its high pozzolanic reactivity. According to some researchers [49,50], the replacement of pozzolans has also led to an improvement in the mechanical properties of concrete. Chousidis et al. [51] observed that the reinforced concrete structure containing FA improved compressive strength and decreased porosity.

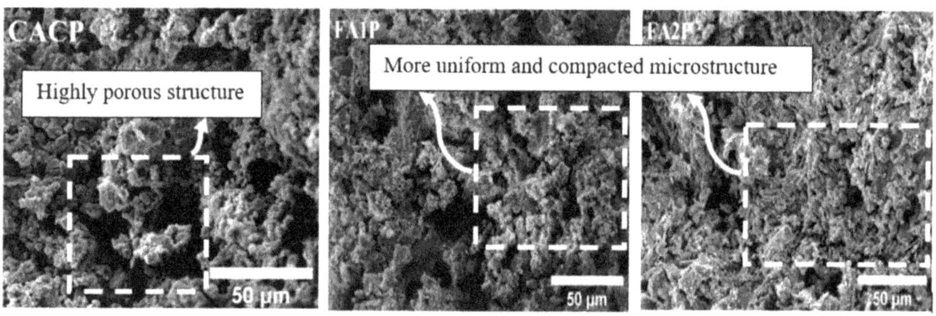

Figure 11. SEM micrographs of CACP, FA1P, and FA2P.

4. Discussion

As aforementioned, the performance of FA in CAC systems is strongly influenced by its physical and chemical properties. The ternary diagram of CAC systems containing different FAs is depicted in Figure 12. Regarding the early-age properties, different activation energies of CAC systems with FA contribute to calcium reacting faster than silica at a very young age. Different FAs can offer different [52] activities. These reactions thereafter change hydration reactions. When CAC mortar is converted, it undergoes a process of mineralogical development, and the porosity has increased. Therefore, the conversion method could avoid the reaction that could be formed in the subsequent way [2,3,10,11,44]. The mineral admixtures containing silica would react first with the CA, avoiding the formation of C_2AH_8 and, subsequently, the conversion into C_3AH_6. Consequently, instead of this C_3AH_6 phase, the C_2ASH_8 phase was proposed to be formed. This implies that samples containing FA exhibit reduced levels of stable phases and encounter comparatively lesser degrees of strength loss (see Table 5).

Based on SEM-EDS images as shown in Figure 13 and Table 6 of CAC specimens prepared from fractured surfaces after 56-day water curing, The main elements involve Al, Si, and Ca, and other elements C and O. By using FA, a pozzolan with a high specific surface area and strong reactivity, in relation to the CACP sample, the concentration of the remaining $Ca(OH)_2$ and silicon dioxide (SiO_2) among the compounds was reduced, according to data from EDS elemental compositions for each system. This showed that these components had a higher likelihood of forming C-(A)-S-H gel when FA was present in the specimens, which consequently improved the strength and decreased porosity. It is believed that, compared to the control sample, CAC specimens containing FA exhibit enhanced mechanical characteristics. In addition, it is likely that the specimen lacking FA will perform poorly in terms of mechanical properties given the quantity and size of pores among the samples in this work.

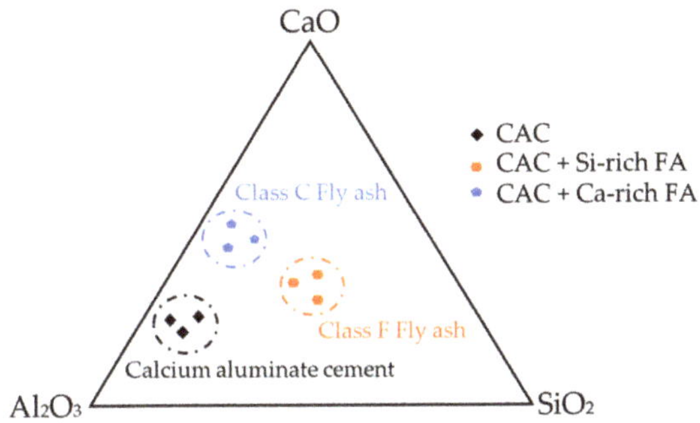

Figure 12. Ternary diagram of CAC, CAC + Ca-rich FA, and CAC + Si-rich FA.

The partial utilization of FA in CAC systems, especially for class-C FA, improved the quality of resultant products from very early to later ages. It can be mainly adopted for increasing sustainability in our manufacturing industries such as construction, refractory, and rapid repair applications, as well as reducing material costs since FA is an industrial by-product that is less expensive, and using it can increase reactivity so that the CAC can be used in smaller fractions.

In summary, the incorporation of FA into CAC offers numerous advantages, including improved environmental sustainability and enhanced material performance. By introducing FA into CAC, we contribute to the cement industry's efforts to combat climate change. Our study has shown that FA addition leads to a range of beneficial effects on CAC properties. Firstly, it results in a reduced setting time, enhancing construction efficiency and reducing construction time. Additionally, FA improves workability and positively influences microstructural characteristics. This includes the increased formation of C-(A)-S-H phases and the reduction of inter-particle space, indicating the pozzolanic action of FA. These changes lead to improved microstructure, density, and compressive strength, which are highly desirable attributes in the construction industry. Furthermore, the presence of FA in CAC alters the kinetics of hydration reactions, lowering the activation energy required for the conversion process. This has significant implications for the material's setting time, strength development, and overall performance. While our study highlights the positive impacts of FA on CAC, further research is needed to fully understand the mechanisms underlying FAs' influence on the conversion process of CAC. This will enable us to optimize the use of FA for specific applications in the construction industry, ultimately advancing both environmental sustainability and construction practices.

Table 6. EDS elemental composition (mass%) shown in SEM micrographs.

Sample	Element	C	O	Al	Si	Ca
CACM	Wt. %	-	52.46	23.25	-	24.28
	At. %	-	69.08	18.16	-	12.76
FA1M	Wt. %	4.25	49.53	22.43	3.35	20.44
	At. %	7.21	63.05	16.93	2.43	10.39
FA2M	Wt. %	-	53.36	28.16	-	18.49
	At. %	-	68.91	21.56	-	9.53

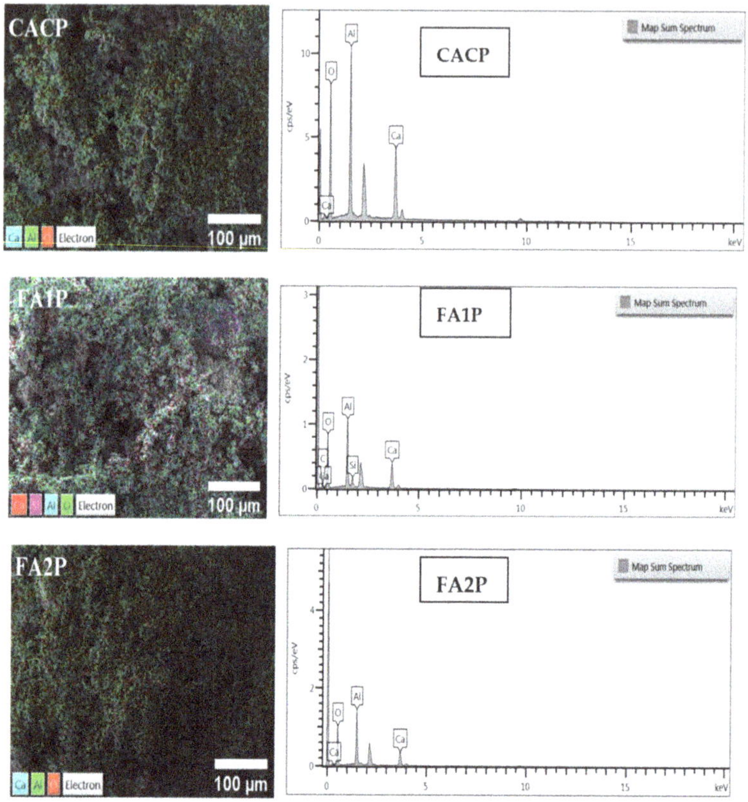

Figure 13. SEM micrographs (**left**) and EDS elemental spectra (**right**) of CACP, FA1P, and FA2P after 56-day water curing.

5. Conclusions

This work examined the utilization of different compositions of FA affecting the conversion of CAC. When the CAC materials considered in this research were mixed with FA, the conclusions were made as follows:

- CAC with fly ash can be satisfied by minimizing its conversion process.
- Adding FA resulted in a reduction in setting time and thus reduced the required activation energy contributed by the decrease in the slope of the FA-CAC system. Therefore, FA could be utilized to improve the hydration of CAC.
- CAC containing FA influenced the fresh properties through increasing workability. This behavior was due to the high surface area and spherical shape of the particles, and thus the high reactivity of FA particles with $Ca(OH)_2$.
- Increased gel formation C-(A)-S-H, reduced inter-particle space, and lowered porous surfaces were found in the microstructural CAC containing FA. This resulted in improved microstructure, density, and compressive strength due to the pozzolanic action.
- Utilizing FA in the BET test resulted in alterations to the pore structure, yielding a matrix characterized by reduced mesopore volume.
- TGA analysis revealed that when FA was presented, the mitigation of CAC conversion and the yield of the lower stable C_3AH_6 phase were seen.
- CAC with calcium-rich FA resulted in denser and lower mesopores than the silica-rich FA, leading to higher compressive strength. This is caused by the larger CaO formation.

These findings investigate that replacing the CAC with FA not only promotes cost efficiency but also paves the way for environmentally friendly, sustainable development. Additionally, this method reduces greenhouse gas emissions that the cement industry releases into the atmosphere, which is advantageous for preventing climate change. The effects of FA on CAC with different chemical compositions and different micropore structures on both early- and later-age performance should be further investigated. Other by-products containing a high calcium content should also be assessed.

Author Contributions: Conceptualization, L.P.; methodology, T.T.W. and L.P.; investigation, T.T.W.; writing—original draft preparation, T.T.W.; writing—review and editing, C.P., P.J., W.P. and L.P.; visualization, L.P.; supervision, L.P.; project administration, L.P., P.J. and W.P.; funding acquisition, L.P. All authors have read and agreed to the published version of the manuscript.

Funding: This project is funded by the National Research Council of Thailand (NRCT) and Chulalongkorn University (Grant No. N42A660629). This research is also funded by the Thailand Science Research and Innovation Fund at Chulalongkorn University (SOC66250010).

Data Availability Statement: Some or all data, models, or code that support the findings of this study are available from the corresponding author upon reasonable request.

Acknowledgments: The authors would like to thank Siam Refractory Industry Co., Ltd. for donating the material, "Chulalongkorn Academic Advancement into its Second Century Fund (C2F) for Postdoctoral Fellowship (T.T.W.), Chulalongkorn University, and British Council—Thai-UK World-Class University Consortium 2022".

Conflicts of Interest: The authors report that there are no competing interests to declare.

References

1. Abolhasani, A.; Samali, B.; Aslani, F. Physicochemical, Mineralogical, and Mechanical Properties of Calcium Aluminate Cement Concrete Exposed to Elevated Temperatures. *Materials* **2021**, *14*, 3855. [CrossRef]
2. Bensted, J. Calcium Aluminate Cements. *Struct. Perform. Cem.* **2002**, *2*, 114–138. Available online: https://books.google.co.th/books?hl=en&lr=&id=6wPpkyrWE5oC&oi=fnd&pg=PA114&dq (accessed on 10 March 2022).
3. Zapata, J.F.; Gomez, M.; Colorado, H.A. Cracking in Calcium Aluminate Cement Pastes Induced at Different Exposure Temperatures. *J. Mater. Eng. Perform.* **2019**, *28*, 7502–7513. [CrossRef]
4. Zapata, J.F.; Gomezc, M.; Colorado, H.A. *Characterization of Two Calcium Aluminate Cement Pastes*; John Wiley & Sons: Hoboken, NJ, USA, 2017; Volume 263, pp. 491–503. [CrossRef]
5. Zapata, J.F.; Gomez, M.; Colorado, H.A. Calcium Aluminate Cements Subject to High Temperature. *Adv. Mater. Sci. Environ. Energy Technol. VI* **2017**, *262*, 97. Available online: https://books.google.co.th/books?hl=en&lr=&id=Nlg7DwAAQBAJ&oi=fnd&pg=PA97&dq=Calcium+aluminate+cements+subject+to+high+temperature (accessed on 28 June 2022).
6. Boris, R.; Wilińska, I.; Pacewska, B.; Antonovič, V. Investigations of the Influence of Nano-Admixtures on Early Hydration and Selected Properties of Calcium Aluminate Cement Paste. *Materials* **2022**, *15*, 4958. [CrossRef]
7. Antonovič, V.; Kerienė, J.; Boris, R.; Aleknevičius, M. The Effect of Temperature on the Formation of the Hydrated Calcium Aluminate Cement Structure. *Procedia Eng.* **2013**, *57*, 99–106. [CrossRef]
8. Midgley, H. Quantitative determination of phases in high alumina cement clinkers by X-ray diffraction. *Cem. Concr. Res.* **1976**, *6*, 217–223. [CrossRef]
9. Zapata, J.F.; Colorado, H.A.; Gomez, M.A. Effect of high temperature and additions of silica on the microstructure and properties of calcium aluminate cement pastes. *J. Sustain. Cem. Mater.* **2020**, *9*, 323–349. [CrossRef]
10. López, A.H.; Calvo, J.L.G.; Olmo, J.G.; Petit, S.; Alonso, M.C. Microstructural Evolution of Calcium Aluminate Cements Hydration with Silica Fume and Fly Ash Additions by Scanning Electron Microscopy, and Mid and Near-Infrared Spectroscopy. *J. Am. Ceram. Soc.* **2008**, *91*, 1258–1265. [CrossRef]
11. Majumdar, A.; Singh, B. Properties of some blended high-alumina cements. *Cem. Concr. Res.* **1992**, *22*, 1101–1114. [CrossRef]
12. Cong, X.; Kirkpatrick, R.J. Hydration of Calcium Aluminate Cements: A Solid-State 27 Al NMR Study. *J. Am. Ceram. Soc.* **1993**, *76*, 409–416. [CrossRef]
13. Win, T.T.; Wattanapornprom, R.; Prasittisopin, L.; Pansuk, W.; Pheinsusom, P. Investigation of Fineness and Calcium-Oxide Content in Fly Ash from ASEAN Region on Properties and Durability of Cement–Fly Ash System. *Eng. J.* **2022**, *26*, 77–90. [CrossRef]
14. Vafaei, M.; Allahverdi, A. Influence of calcium aluminate cement on geopolymerization of natural pozzolan. *Constr. Build. Mater.* **2016**, *114*, 290–296. [CrossRef]
15. Abolhasani, A.; Nazarpour, H.; Dehestani, M. The fracture behavior and microstructure of calcium aluminate cement concrete with various water-cement ratios. *Theor. Appl. Fract. Mech.* **2020**, *109*, 102690. [CrossRef]

16. Scrivener, K.L.; Cabiron, J.-L.; Letourneux, R. High-performance concretes from calcium aluminate cements. *Cem. Concr. Res.* **1999**, *29*, 1215–1223. [CrossRef]
17. Adams, M.P.; Ideker, J.H. Influence of aggregate type on conversion and strength in calcium aluminate cement concrete. *Cem. Concr. Res.* **2017**, *100*, 284–296. [CrossRef]
18. Khaliq, W.; Khan, H.A. High temperature material properties of calcium aluminate cement concrete. *Constr. Build. Mater.* **2015**, *94*, 475–487. [CrossRef]
19. Zapata, J.F.; Azevedo, A.; Fontes, C.; Monteiro, S.N.; Colorado, H.A. Environmental Impact and Sustainability of Calcium Aluminate Cements. *Sustainability* **2022**, *14*, 2751. [CrossRef]
20. Son, H.; Park, S.; Jang, J.; Lee, H. Effect of nano-silica on hydration and conversion of calcium aluminate cement. *Constr. Build. Mater.* **2018**, *169*, 819–825. [CrossRef]
21. Ukrainczyk, N.; Matusinović, T. Thermal properties of hydrating calcium aluminate cement pastes. *Cem. Concr. Res.* **2010**, *40*, 128–136. [CrossRef]
22. *ASTM C33-18*; Standard Specification for Concrete Aggregates. American Society of Testing Materials: West Conshohocken, PA, USA, 2018.
23. *ASTM C128-15*; Standard Test Method for Density, Relative Density (Specific Gravity), and Absorption of Fine Aggregate. American Society of Testing Materials: West Conshohocken, PA, USA, 2015.
24. *ASTM C618-19*; Standard Specification for Coal Fly Ash and Raw or Calcined Natural Pozzolan for Use in Concrete. American Society of Testing Materials: West Conshohocken, PA, USA, 2019.
25. *ASTM C204-18*; Standard Test Method for Fineness of Hydraulic Cement by Air-Permeability Apparatus. American Society of Testing Materials: West Conshohocken, PA, USA, 2018.
26. *ASTM C188-17*; Standard Test Method for Density of Hydraulic Cement. American Society of Testing Materials: West Conshohocken, PA, USA, 2017.
27. *ASTM C109/C109M-16*; Standard Test Method for Compressive Strength of Hydraulic Cement Mortars (Using 2-in. or [50-mm] Cube Specimens). American Society of Testing Materials: West Conshohocken, PA, USA, 2016.
28. *ASTM C1437-20*; Standard Test Method for Flow of Hydraulic Cement Mortar. American Society of Testing Materials: West Conshohocken, PA, USA, 2020.
29. Shen, J.; Xu, Q. Effect of moisture content and porosity on compressive strength of concrete during drying at 105 °C. *Constr. Build. Mater.* **2018**, *195*, 19–27. [CrossRef]
30. *ASTM C511-19*; Standard Specification for Mixing Room; Moist Cabinets; Moist Rooms; and Water Storage Tanks Used in the Testing of Hydraulic Cements and Concretes. American Society of Testing Materials: West Conshohocken, PA, USA, 2019.
31. *ASTM C191-21*; Standard Test Method for Time of Setting of Hydraulic Cement by Vicat Needle. American Society of Testing Materials: West Conshohocken, PA, USA, 2021.
32. *ASTM C642-21*; Standard Test Method for Density, Absorption, and Voids in Hardened Concrete. American Society of Testing Materials: West Conshohocken, PA, USA, 2021.
33. De Belie, N.; Kratky, J.; Van Vlierberghe, S. Influence of pozzolans and slag on the microstructure of partially carbonated cement paste by means of water vapour and nitrogen sorption experiments and BET calculations. *Cem. Concr. Res.* **2010**, *40*, 1723–1733. [CrossRef]
34. Prasittisopin, L.; Sereewatthanawut, I. Effects of seeding nucleation agent on geopolymerization process of fly-ash geopolymer. *Front. Struct. Civ. Eng.* **2017**, *12*, 16–25. [CrossRef]
35. Mehta, P.K.; Monteiro, P.J. *Concrete: Microstructure, Properties, and Materials*; McGraw-Hill: New York, NY, USA, 2014. Available online: https://www.accessengineeringlibrary.com/binary/mheaeworks (accessed on 1 August 2022).
36. Balonis, M.; Glasser, F. The density of cement phases. *Cem. Concr. Res.* **2009**, *39*, 733–739. [CrossRef]
37. Le, H.T.; Ludwig, H.-M. Effect of rice husk ash and other mineral admixtures on properties of self-compacting high performance concrete. *Mater. Des.* **2016**, *89*, 156–166. [CrossRef]
38. Hefni, Y.; El Zaher, Y.A.; Wahab, M.A. Influence of activation of fly ash on the mechanical properties of concrete. *Constr. Build. Mater.* **2018**, *172*, 728–734. [CrossRef]
39. Prasittisopin, L.; Trejo, D. Characterization of Chemical Treatment Method for Rice Husk Ash Cementing Materials. *ACI Symp. Publ.* **2013**, *294*, 1–14. [CrossRef]
40. Park, S.; Jang, J.; Son, H.; Lee, H. Stable conversion of metastable hydrates in calcium aluminate cement by early carbonation curing. *J. CO2 Util.* **2017**, *21*, 224–226. [CrossRef]
41. Juenger, M.; Winnefeld, F.; Provis, J.; Ideker, J. Advances in alternative cementitious binders. *Cem. Concr. Res.* **2011**, *41*, 1232–1243. [CrossRef]
42. Scrivener, K.; Capmas, A. Calcium Aluminate Cements. *Adv. Concr. Technol.* **2003**, 1–31. Available online: https://books.google.co.th/books?hl=en&lr=&id=IMdF-QR_8mkC&oi=fnd&pg=SA2-PA1&dq=Calcium+aluminate+cements (accessed on 12 April 2022).
43. Sio, J.D. Influence of Pozzolanic Material in the Conversion and Corrosion Behaviour of Calcium Aluminate Cement. Master's Thesis, University of Sydney, Camperdown, Australia, 2014. Available online: https://hdl.handle.net/2123/12694 (accessed on 1 August 2022).
44. Sereewatthanawut, I.; Prasittisopin, L. Effects of accelerating and retarding agents on nucleation and crystal growth of calcium aluminate cement. *Open Ceram.* **2022**, *11*, 100290. [CrossRef]

45. Thommes, M.; Kaneko, K.; Neimark, A.V.; Olivier, J.P.; Rodriguez-Reinoso, F.; Rouquerol, J.; Sing, K.S.W. Physisorption of gases, with special reference to the evaluation of surface area and pore size distribution (IUPAC Technical Report). *Pure Appl. Chem.* **2015**, *87*, 1051–1069. [CrossRef]
46. Prasittisopin, L.; Trejo, D. Performance Characteristics of Blended Cementitious Systems Incorporating Chemically Transformed Rice Husk Ash. *Adv. Civ. Eng. Mater.* **2017**, *6*, 17–35. [CrossRef]
47. Ahmed, A.A.; Shakouri, M.; Trejo, D.; Vaddey, N.P. Effect of curing temperature and water-to-cement ratio on corrosion of steel in calcium aluminate cement concrete. *Constr. Build. Mater.* **2022**, *350*, 128875. [CrossRef]
48. Win, T.T.; Prasittisopin, L.; Jongvivatsakul, P.; Likitlersuang, S. Investigating the synergistic effect of graphene nanoplatelets and fly ash on the mechanical properties and microstructure of calcium aluminate cement composites. *J. Build. Eng.* **2023**, *78*, 107710. [CrossRef]
49. Nakamura, K.; Inoue, Y.; Komai, T. Consideration of strength development by three-dimensional visualization of porosity distribution in coal fly ash concrete. *J. Build. Eng.* **2020**, *35*, 101948. [CrossRef]
50. Sereewatthanawut, I.; Panwisawas, C.; Ngamkhanong, C.; Prasittisopin, L. Effects of extended mixing processes on fresh, hardened and durable properties of cement systems incorporating fly ash. *Sci. Rep.* **2023**, *13*, 6091. [CrossRef]
51. Chousidis, N.; Rakanta, E.; Ioannou, I.; Batis, G. Mechanical properties and durability performance of reinforced concrete containing fly ash. *Constr. Build. Mater.* **2015**, *101*, 810–817. [CrossRef]
52. Chopra, D.; Siddique, R.; Kunal. Strength, permeability and microstructure of self-compacting concrete containing rice husk ash. *Biosyst. Eng.* **2015**, *30*, 72–80. [CrossRef]

buildings

Article

Variability in the Distinctive Features of Silica Sands in Central Europe

Martina Bašistová [1,*], Jiřina Vontorová [2], Simona Zlá [2], Monika Kawuloková [2], Petr Lichý [1] and Tomáš Dvorský [3]

1 Department of Metallurgical Technologies, Faculty of Materials Science and Technology, VSB–Technical University of Ostrava, 17. Listopadu 2172/15, 708 00 Ostrava, Czech Republic; petr.lichy@vsb.cz
2 Department of Chemistry and Physico-Chemical Processes, Faculty of Materials Science and Technology, VSB–Technical University of Ostrava, 17. Listopadu 2172/15, 708 00 Ostrava, Czech Republic; jirina.vontorova@vsb.cz (J.V.); simona.zla@vsb.cz (S.Z.); monika.kawulokova@vsb.cz (M.K.)
3 Department of Enviromental Engineering, Faculty of Mining and Geology, VSB–Technical University of Ostrava, 17. Listopadu 2172/15, 708 00 Ostrava, Czech Republic; tomas.dvorsky@vsb.cz
* Correspondence: martina.basistova@vsb.cz; Tel.: +420-596-994-206

Abstract: Quality quartz sand is globally utilized in construction due to its availability and economic factors, especially in the production of composite cements. Despite its positive properties, quartz sand also has several disadvantages. The dilation of quartz sand can be technologically significant for certain high-temperature applications. This dilation has a non-continuous character with sharp volume change caused by the phase transformation from β to α SiO_2 at temperatures around 573 °C. The extent of dilation depends on various factors such as compaction, grain size, the quantity of sand, as well as the shape and character of the grain and chemical purity, particularly the SiO_2 content. In this study, six types of quartz sand from different locations in Central Europe were examined, and the influence of chemical composition and grain shape was correlated with the final dilation of these samples. Evaluation methods included X-ray fluorescence spectroscopy (XRFS), scanning electron microscopy (SEM), energy-dispersive X-ray spectroscopy (EDX), differential thermal analysis (DTA), and linear thermal expansion analysis. It was found that angular grains, despite their chemical purity, may exhibit minimal dilation. Conversely, the least suitable combination in terms of dilation appears to be a high SiO_2 content and high roundness of grains with a smooth surface.

Keywords: fine quartz sand; thermal dilatation; DTA; chemical purity; grain shape; technological properties

Citation: Bašistová, M.; Vontorová, J.; Zlá, S.; Kawuloková, M.; Lichý, P.; Dvorský, T. Variability in the Distinctive Features of Silica Sands in Central Europe. *Buildings* **2024**, *14*, 279. https://doi.org/10.3390/buildings14010279

Academic Editor: Chuang Feng

Received: 30 November 2023
Revised: 12 January 2024
Accepted: 16 January 2024
Published: 19 January 2024

1. Introduction

Quartz fine sand, known for its easy availability, properties, and cost-effectiveness, is utilized not only in the glass, foundry, and electrical sectors but also extensively in the construction industry [1–3]. These sands serve as a crucial component in various construction materials, such as composite cements, concretes, and mortars, influencing the resulting characteristics of the material. Fine quartz sands typically exhibit grain sizes ranging from 0.02 to 0.5 mm. Finer grades of sands can effectively act as partial substitutes for fine aggregates or cement in concrete applications [4–6]. This fineness is the key factor allowing for the creation of smooth and compact concrete surfaces with sufficient strength [7,8]. For example, in study [9], the authors confirmed the effect of fine sand (granite rock stone powder was used) on the unconfined compressive strength of cement-treated clay mixed composites, where the addition of 50–70% rock stone powder sand led to an increase in surface area and hence higher strength compared to the control specimen and allowed for a 28.6% reduction in the amount of cement in the mix. Generally, used sands should also possess a suitable mineralogical composition and a minimal content of fine particles, such as clays and dust (particles smaller than 0.02 mm) [10]. Another significant characteristic of quartz sands that influences the properties of the resulting material is the shape and surface structure of the grains [11].

In some cases, fine sands can be replaced or mixed with appropriate additives or recycled materials. This can reduce the environmental footprint while customizing the material to specific requirements. The primary focus in such cases is on harnessing existing sand sources, with ongoing experiments involving desert sands [12,13]. As an alternative, from a more ecological point of view instead of landfilling, various kinds of sands sourced from the foundry industry, encompassing both pristine and recycled sands, are also tested [14–16]. For example, in [17], the authors studied mixtures of soil and foundry sand with the addition of cement and lime to assess their applicability as base materials for highways. The results of the study show that the winter resistance of the foundry sand-based samples is generally better than that of a typical reference base material and that there will be no environmental impact (effect on groundwater quality).

In the field of construction materials, continuous research is underway to enhance the properties of sands and their utilization in concrete. Innovations involve exploring alternative materials and optimizing compositions to achieve optimal results. Fine sands are a key component in structural concrete, especially where high strength or aesthetic properties are required [18–20]. They are also used in special types of concrete, such as self-sealing or low thermal conductivity concretes, or in concretes used for high-temperature applications [21–23]. They significantly influence the workability of concrete and can affect properties such as density, strength, and freeze–thaw resistance. The sands are not only crucial in various construction materials but also in 3D-printed rock analogs and other experimental model preparations. For those, some previous studies were suggested [24–26].

Fine sands are incorporated into construction materials with the aim of achieving the optimal physical and mechanical properties of concrete and cement, taking into account the specific requirements of particular applications [18,21]. The thermal expansion of sands, especially quartz sands, is an important factor in their use in various applications such as cement, concrete, or ceramics. Quartz sand, composed mainly of the mineral quartz (SiO_2), plays a crucial role in influencing the characteristics of the final product. Unlike other fine aggregates, it exhibits a non-continuous thermal expansion due to multiple changes in its crystal lattice when subjected to heat [27]. Some of these changes result in substantial fluctuations in volume. The extent of this dilation varies based on the specific type of quartz sand, its purity, and the morphology of its grains [28–30]. The presence of impurities (so-called mineralizers) can significantly influence thermal expansion or reduce the temperature of the phase transformation to cristobalite, which can occur under specific conditions starting from around 900 °C [29]. However, technologically more significant is the transformation from β-SiO_2 to α-SiO_2, occurring at lower temperatures, around 573 °C, accompanied by a sharp increase in volume of up to approximately 3.9% [11]. This transformation is reversible and is significantly affected by grain size (smaller particles resulted in higher dilatation) and chemical composition (sample with a SiO_2 content higher by 1.1% achieved 19.8% more dilatation compared with a sample with less SiO_2), as suggested by some studies [31–33].

The thermal expansion of quartz sand becomes a critical consideration at elevated temperatures because it can affect the overall thermal expansion of the resulting concrete or composite cement. Despite concrete and composite cement being generally considered materials with low thermal expansion compared to metal, they are suitable for structures such as bridge decks and similar constructions. The thermal expansion of concrete is in the order of tenths of millimeters per meter [34]. The thermal expansion of composite cement and concrete plays a key role in the design and construction of buildings and technological units, especially where they are exposed to significant temperature changes [35]. For instance, it is used to create heat-resistant insulation materials. These materials can be used to form thermal insulation layers or protective coatings for constructions exposed to high temperatures. This is the case, for example, in applications such as the cement sheath of heavy oil reservoirs subjected to hot steam injection reaching temperatures up to 380 °C, where there is ongoing exploration into quartz-enhanced cement [36–38]. For example, study [36] found that the proposed cement system with SiO_2 particles significantly increased the compressive strength by more than 50 MPa even under the conditions of the seventh round of steam curing. It can also involve the use of so-called refractory (high-temperature resistant) concretes and cements used in the construction of industrial furnaces, waste incinerators, fireplace linings, and other structures exposed to temperatures above 200–1000 °C. These materials must be resistant to thermal shocks and oxidation [39]. Further experimentation is being conducted in the construction industry with concretes that can be used in solar structures. For these concretes, the ability to accumulate heat during the day and release it at night is assumed, contributing to the energy efficiency of buildings.

When working with sands containing quartz for the production of composite cement, one of the key factors, besides grain size, should be the character (such as surface structure with pores and cavities and smoothness) and shape of the grain and its chemical composition (oxides). Individual quartz sands differ from each other in their origin, which then determines these grain properties and their subsequent technological parameters. One of these technological parameters is the resulting linear thermal expansion of quartz sands, which can be crucial when used at high temperatures. Therefore, understanding the influence of grain character and its chemical composition on the resulting dilation can be an important part of the decision-making process in selecting suitable raw materials, which, especially in terms of grain shape, has not received much attention, and differences in individual sands have been little explored so far. The aim of this experiment is to evaluate the impact of origin and grain shape, as well as its chemical composition, on the resulting thermal dilation. For greater variability, six high-quality quartz sands from mining sites in Central Europe were examined.

2. Materials and Methods

2.1. Sand Samples

In the context of this experiment, natural washed quartz sands were analyzed. These sands were obtained from 6 different mining locations in Central Europe, specifically from the Czech Republic (designated as ST 54), Poland (designated as GL 21, BG 21, and BK 23), and Slovakia (designated as SH 35 and SE 22). All six samples consisted of highly pure natural washed quartz sands, characterized by a high SiO_2 content (all 90% SiO_2 and above), representing sands of glassmaking or foundry quality also used in construction. Washing and purification of these sands ensured the minimal presence of clay particles and dust. Individual sand samples from different mining locations differ in their characteristic properties (shape and grain structure), as well as in chemical composition. Within this experiment, samples of sands with the same granulometric composition were used, with a declared median grain size around 0.21 ± 0.02 mm.

2.2. Testing Methods

Under the conditions of this experiment, the determination of the chemical composition, particularly the actual SiO_2 content and impurities, was carried out through X-ray fluorescence spectroscopy (XRFS) using the Rigaku Supermini 200 analyzer. The EasySken method (a non-standard method) was employed for measurements, set up and calibrated by the analyzer manufacturer. The initial state of individual samples for this measurement involved finely grinding the samples into powder (<0.1 mm) using a laboratory mill, homogenizing them, and subsequently forming tablets by pressing with a sample-to-binder ratio of 4:1. The binder used was Hoechst wax C. Loss on ignition (LOI) was carried out on dried samples (at 130 °C) in preannealed crucibles in a chamber annealing oven. The measurement temperature was set at 1000 °C and the holding time was 2 h. Three measurements were taken for each sample.

The grain shape of all six samples of quartz sand, including visual assessments of captured impurities and the presence of inclusions, was examined using images taken with the

Keyence VHX 6000 digital microscope at a magnification of 200× and coaxial illumination. Images from the digital microscope were complemented by scanning electron microscopy (SEM) measurements using the FEI QUANTA 450 FEG (FEI Czech Republic, Brno, Czech Republic), allowing for a visual assessment of impurity quantity, including grain shape and surface, at a magnification of 600×. The previous XRFS analysis was subsequently complemented by point EDX analysis. For each sample, a number of images were taken at different magnifications so that the most common grain shape and surface could be assessed and a reference grain for point EDX could be determined. The composition was evaluated only for pure quartz grains, not for inclusions or grains of other minerals.

Linear thermal expansion analysis was conducted using a special Al_2O_3 container for powdered materials on the Netzch DIL402/C (Netzch, Selb, Germany) dilatometer. Free-flowing samples, without any compaction, were placed into the container and gently leveled using 3 taps with tweezers. The size of the obtained samples for measurement in the container ranged from 10.0 ± 0.1 mm. The parameters of the measurement procedure were as follows: the heating rate of the samples was 15 °C/min until reaching the final temperature of 1130 °C. The measurement took place under an inert atmosphere of 6.0 Argon with a flow rate of 100 L/min.

Differential thermal analysis (DTA) was performed to precisely determine the temperature of the onset of the β-SiO_2 to α-SiO_2 phase transformation, as the measurement of thermal expansion captures only the temperature of the change itself, not its exact beginning. DTA analysis was conducted using the SETARAM Setsys 18TM (SETARAM, Caluire, France) instrument. The temperature calibration of the instrument was performed using the calibration material Al 5N. A sample weighing approximately 80 mg was placed in an Al_2O_3 crucible with a lid (100 μL). An isotherm was set at 20 °C (2 h), followed by heating from 20 to 1100 °C at a rate of 10 °C/minute and cooling from 1100 to 20 °C, also at a rate of 10 °C/min. The measurement was conducted under an inert atmosphere of 6.0 Argon.

All measurements were performed on at least 3 samples of each investigated material in its initial state. The results were then averaged. If any values differed by more than 5%, subsequent measurements were conducted to confirm the result.

Figure 1 provides a straightforward schematic depiction of the step-by-step process involving individual measurements and sample production. Detailed information regarding the specific measurements and the materials employed is outlined in the subsequent sections.

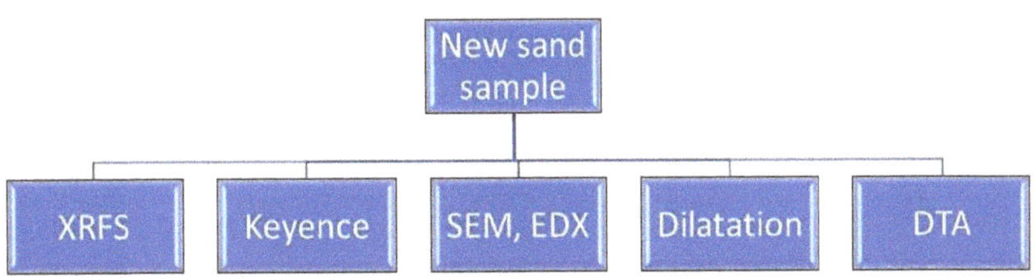

Figure 1. Illustrates the sequence of measurements in a schematic manner.

3. Results and Discussion

3.1. Chemical Composition

The chemical composition assessment, specifically the quantities of oxides, for all types of sand was conducted using XRFS analysis. It was confirmed that in all cases, these are highly chemically pure sands, designated as glassmaking quality sands, with a SiO_2 content exceeding 90%. In quartz sands, the technologically most important content is SiO_2, defining refractoriness; Al_2O_3 indicates the possible presence of clay residues, while Fe_2O_3 and potential inclusions (indicated by K_2O content) inversely reduce refractoriness. In addition to the major and technologically significant oxides, sands may contain traces of other minerals and oxides, generally in minor proportions. The comprehensive results of the chemical composition are presented for major oxides in Table 1 and for minor oxides in Table 2. The difference between the results of the individual measurements did not exceed 5%.

Table 1. Results from XRFS mineralogical analysis for 6 tested sands—major minerals and the results of loss on ignition test.

Sample	SiO_2	Al_2O_3	Fe_2O_3	TiO_2	K_2O (Mass%)	ZrO_2	CaO	P_2O_5	SO_3	LOI (%)
SH 35	91.02	3.19	0.99	-	3.75	0.01	0.44	0.05	-	0.09
Sx SH 35	0.021	0.016	0.013	-	0.017	0.001	0.018	0.004	-	0.00016
ST 54	99.15	0.50	0.10	0.13	0.04	0.02	0.02	-	0.02	0.02
Sx ST 54	0.037	0.012	0.016	0.008	0.016	0.004	0.001	-	0.006	0.00034
GL 21	98.88	0.49	0.14	0.30	0.04	0.04	0.03	0.01	0.02	0.04
Sx GL 21	0.037	0.021	0.005	0.008	0.005	0.005	0.006	0.006	0.002	0.00010
BG 21	97.80	0.78	0.23	0.83	0.08	0.15	0.04	-	0.02	0.07
Sx BG 21	0.016	0.007	0.013	0.033	0.007	0.016	0.002	-	0.002	0.00003
BK 23	94.12	2.54	0.69	0.23	1.60	0.06	0.19	0.06	0.02	0.18
Sx BK 23	0.037	0.010	0.013	0.006	0.013	0.002	0.008	0.002	0.001	0.00004
SE 22	91.46	3.52	0.79	0.15	3.15	-	0.29	0.06	0.04	0.22
Sx SE 22	0.058	0.023	0.006	0.004	0.013	-	0.010	0.002	0.003	0.00014

Table 2. Results from XRFS mineralogical analysis for 6 tested sands—minor minerals.

Sample	Na_2O	MgO	MnO	Rb_2O (Mass%)	SrO	Cl
SH 35	0.25	0.12	0.05	0.02	0.02	0.010
Sx SH 35	0.013	0.002	0.001	0.001	0.001	0.0006
ST 54						
GL 21						0.004
Sx GL 21						0.0002
BG 21						
BK 23	0.003	0.003	0.002			0.0003
Sx BK 23						
SE 22	0.18	0.08		0.02	0.02	0.007
Sx SE 22	0.003	0.002		0.002	0.002	0.0002

The highest chemical purity was achieved by sample ST 54, specifically 99.15% SiO_2, with a minimal presence of other oxides, below 0.5%. It contained no minor oxides. Sample ST 54 was closely followed in chemical purity by sample GL 21 (0.269 mass % less SiO_2 compared to ST 54) and sample BG 21 (1.35 mass % less SiO_2 compared to ST 54). In the case of GL 21 and BG 21, these were samples of sand from related deposits close to each other, where, due to the same origin of the sand, similar or nearly identical properties, including chemical composition, were expected. However, sample BG 21 exhibited almost double the content of Al_2O_3, Fe_2O_3, and K_2O, indicating a higher presence of inclusions and clay minerals, compared to sample GL 21. In comparison with other samples, the content of TiO_2 was also unusual, specifically 0.83%, approximately 3.5 times higher on average. Samples GL 21 and BG 21 did not contain minor oxides.

On the other hand, the lowest chemical purity was observed in sample SH 35 with a SiO_2 content of 91.024%, which is 8.13 mass % less compared to ST 54. This sample also showed more than 3% content of Al_2O_3 and K_2O oxides, indicating a higher level of inclusions and clays negatively affecting the refractoriness of the sand. The content of Fe_2O_3 was nine times higher than in the purest ST 54 and also compared to the other two highly pure samples. A similar representation of individual oxides was also exhibited by another sample labeled SE 22, which again was a sample from a related deposit of the same origin. This sample showed a SiO_2 content 0.44 mass % higher than sample SH 35 and a slightly lower content of impurities, including minor oxides. However, this difference can be considered negligible compared to other samples. Both samples exhibited a similar range of minor oxides. Sample BK 23 represented an intermediate between the investigated samples with high and low chemical purity in terms of SiO_2 content, which was 94.117%, 5.03 mass % less than ST 54. Major oxides were primarily represented by a higher content of Al_2O_3 (2.54 mass %) and K_2O (1.60 mass %). In the case of sample BK 23, as with sample ST 54, it was not a related sample from the same locality as any other sample examined.

3.2. Shape and Character of Grains

Using the Keyence VHX 6000 (Keyence, Osaka, Japan) digital microscope, detailed images of all sand samples were captured at a magnification of 200× (visible in Figure 2), allowing for the observation of grain shape, surface structure, and the influence of chemical composition on the appearance and color of the sand.

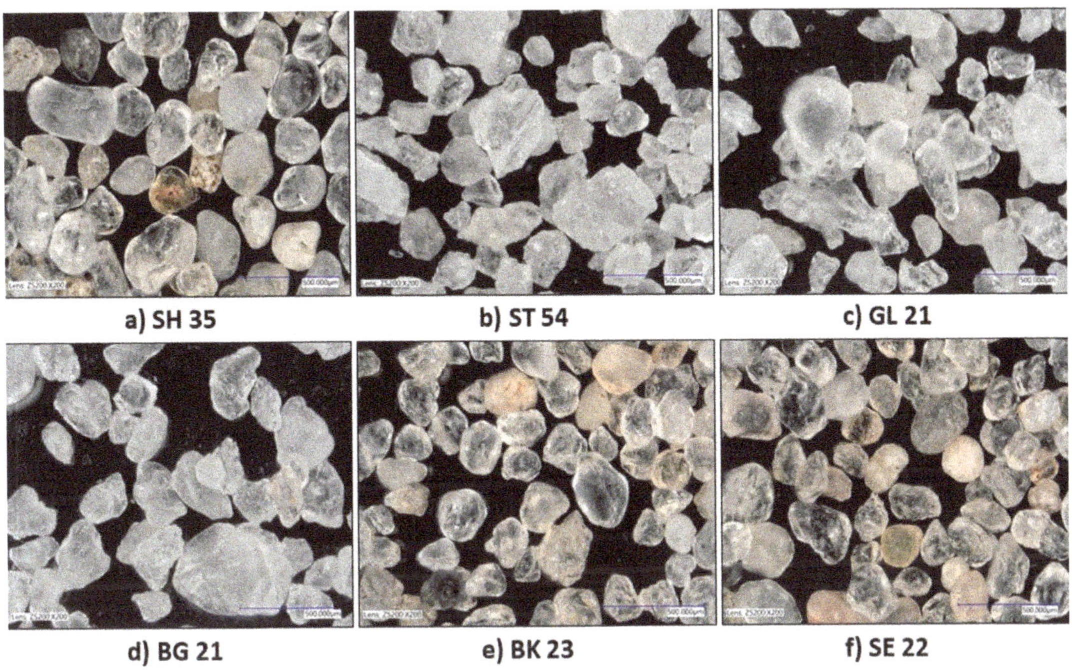

Figure 2. Photos of sand samples taken by Keyence VHX 6000 (Keyence, Osaka, Japan), 200× magnification.

Natural quartz sands occur in nature in deposits accompanied mostly by clay minerals. These sands are either supplied together with some natural clay content (illites and kaolinites) in the so-called mined state, or after extraction they are washed with water and dried, the so-called washed sands. In this way, the individual grains are separated from the clay. Most quartz sands in nature then contain varying amounts of SiO_2 in different mineralogical forms (most quartz, lesser amounts of cristobalite, and tridymite) and other minerals such as fayalite, spinels, feldspars, etc. Depending on the deposit and the origin of the deposit, their content and relative proportions vary. Highly chemically pure sands after washing (SiO_2 content above 90%) form only a part of European deposits. These sands then reach the so-called glass or foundry quality.

The examined samples shown in Figure 2 are all washed sands from deposits characterized by high SiO_2 content after washing. Nevertheless, the presence of other minerals or contamination by residual clays or quartz grains with lower SiO_2 content can also be observed according to the coloration. This is due to the formation of the deposit and it is recognized as a "nature" of the particular sand. These less pure sands generally have rounder grains, as they were transported from long distances by water or air to the present-day deposit sites (aeolian sands) at the time of formation. For this reason, they also contain more feldspars and other minerals than the more angular, so-called marine sands that were formed at the site of extraction. The aeolian sands then have generally different technological properties, such as lower refractoriness due to the presence of low-melting feldspars.

The greatest color variation, and therefore different composition, was observed in the case of samples SH 35 (Figure 2a) and SE 22 (Figure 2f), consistent with the previous results of XRFS analysis and their lower chemical purity. There were also visible grains of a different character, cracked with depressions and cavities and a dull or non-transparent surface, likely composed of inclusions. Individual grains were round to oval-shaped; in the case of transparent grains, composed mainly of SiO_2, they had a smooth surface. Another sample with round to oval-shaped grains was sample BK 23 (Figure 2e). However, this sample visibly contained more transparent, colorless, or slightly colored quartz grains with a high SiO_2 content. The proportion of impurities was noticeably smaller, corresponding to the previous chemical composition.

Sample ST 54 was without visible impurities (Figure 2b). The grains were clear and angular, with sharp edges and visible fracture surfaces. The transparency of the grains corresponds to the previously determined high SiO_2 content, where almost the majority of grains were composed mainly of this oxide. Samples GL 21 (Figure 2c) and BG 21 (Figure 2d) represented an intermediate stage between sample ST 54 and rounded-grain samples with lower purity. In both samples GL 21 and BG 21, there were occasionally visible grains with lower chemical purity, tinted slightly pink. However, these were not grains of visibly different composition and character (inclusions) as in the case of SH 35 and SE 22, but rather adhered impurities and contaminants, such as clay.

The detailed grain characteristics of each sample were evaluated based on images captured with a SEM microscope at a magnification of 600×. The focus was solely on quartz grains, thus those with a high SiO_2 content, excluding grains of other minerals, inclusions, or impurities. Evaluation primarily considered the grain surface, erosion, the quantity of cracks, and angularity. The point EDX composition of each reference pure grain, a centrally displayed grain with a marked point, was subsequently assessed. The results are evident in Figures 3 and 4. Detailed point EDX analysis results are shown in Table 3. Similar to the Keyence images, grains of samples SH 35 (Figure 3a) and SE 22 (Figure 4d) were round to oval in shape with rounded edges. The surface was smooth, without cracks, erosion, or significant porosity. There were no pronounced fracture surfaces from the crushing of sand grains. A similar character was exhibited by sample BK 23 (Figure 4e), with rather rounded oval grains. In the EDX point analysis, these three sand samples showed a high oxygen content compared to the other evaluated samples and the lowest Si content (Table 3). They also demonstrated a higher amount of Al and traces of Mg, K, and Fe, indicating the previously confirmed higher presence of other oxides besides SiO_2. The presence of K and Fe, shown in the EDX results, indicates the presence of probably residues of potassium feldspars (orthoclase) and fayalite, which are common in quartz sands. In both cases, these are minerals with lower refractoriness compared to quartz. Considering that the quartz grains were examined (high Si peak), these were likely residues of clay (Al peak) or dust remnants from the crushing of softer inclusions (Mg and K peaks).

In the case of samples GL 21 (Figure 3c) and BG 21 (Figure 4d), the larger grains were oval to slightly angular with rounded edges. The surface was mostly smooth, occasionally with shallow cracks or minor erosion. Smaller grains already showed signs of fractures, sharp unpolished edges, deeper cracks, or surface fissures, indicating grain crushing. Erosion was particularly evident in the porous structure of some grains. The Si and O content was high for both samples according to point EDX (Table 3). Additionally, trace amounts of Al were noticeable, likely from impurities adhering to the surface of the quartz sand grain. Sample ST 54 (Figure 3b) was characterized by its most angular grain shape among all six examined samples. The grain surface was very rugged, rimosed with pronounced cracks and fissures. The grains had sharp edges and distinct fracture surfaces. Some areas exhibited a porous character. EDX confirmed a high content of Si and O, consistent with the previously confirmed chemical composition with a high SiO_2 content. The Si content (Table 3) there is the highest from all tested sands, namely 46.45%. Furthermore, the grain showed a slightly higher content of Al compared to samples GL 21 and BG 21. This could be attributed to captured particles on an otherwise highly cracked grain surface.

Based on the grain shape, samples SH 35, BK 23, and SE 22 are classified as aeolian sands, which were transported to the mining site by wind or water, i.e., allochthonous in origin. This transport caused the abrasion of grain edges and their rounded shape. From a technological perspective, aeolian sands exhibit a smaller surface area compared to angular sands. The rounded shape and smooth surface directly affect the fluid absorption by the grains (e.g., water), which is lower, or the amount of material (e.g., binder) needed to coat the grain surface. Rounded sands with a smooth surface can also contribute to higher strengths in mixtures, considering the results of studies evaluating the impact of grain shape on the strengths of cured organic or inorganic molding mixtures for foundry purposes, where these sands are also used [40,41]. On the contrary, angular grains with sharp edges (ST 54) or rounded edges (GL 21 and BG 21) result from sand extraction directly at the source. These sands were not transported but formed by the erosion of original rocks or sediment beds. Angular sands are characterized by their large surface area compared to rounded aeolian sands, due to their pronounced ruggedness (considering cracks, fissures, etc.). This directly leads to the higher absorption of these grains or lower strengths of mixtures. Conversely, during compaction, angular sands fit together much better.

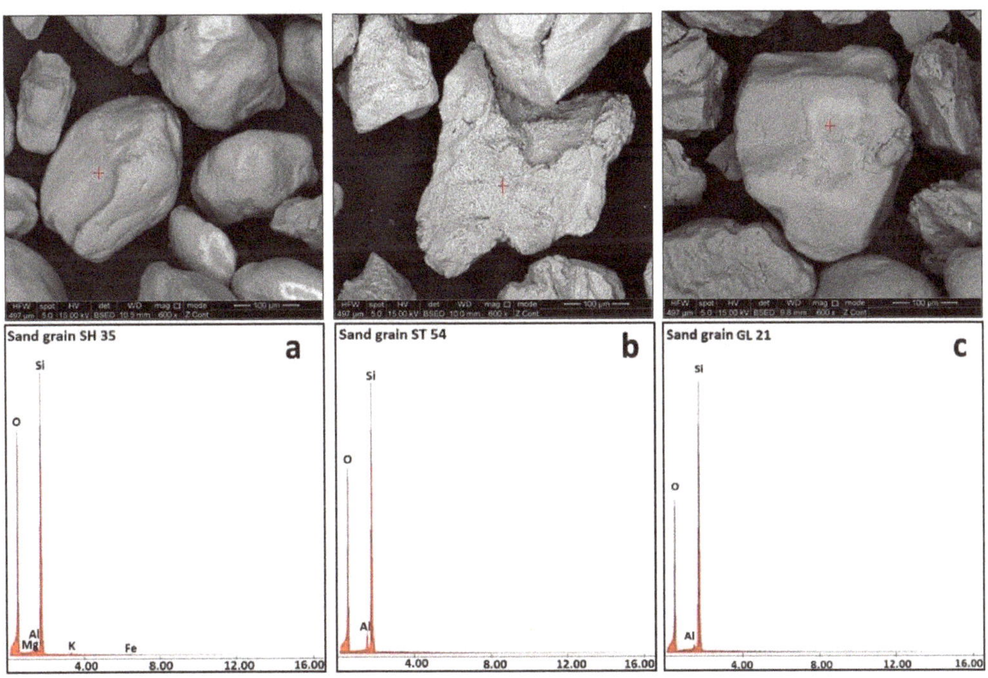

Figure 3. SEM + EDX results for sand samples: (**a**) SH 35, (**b**) ST 54, and (**c**) GL 21.

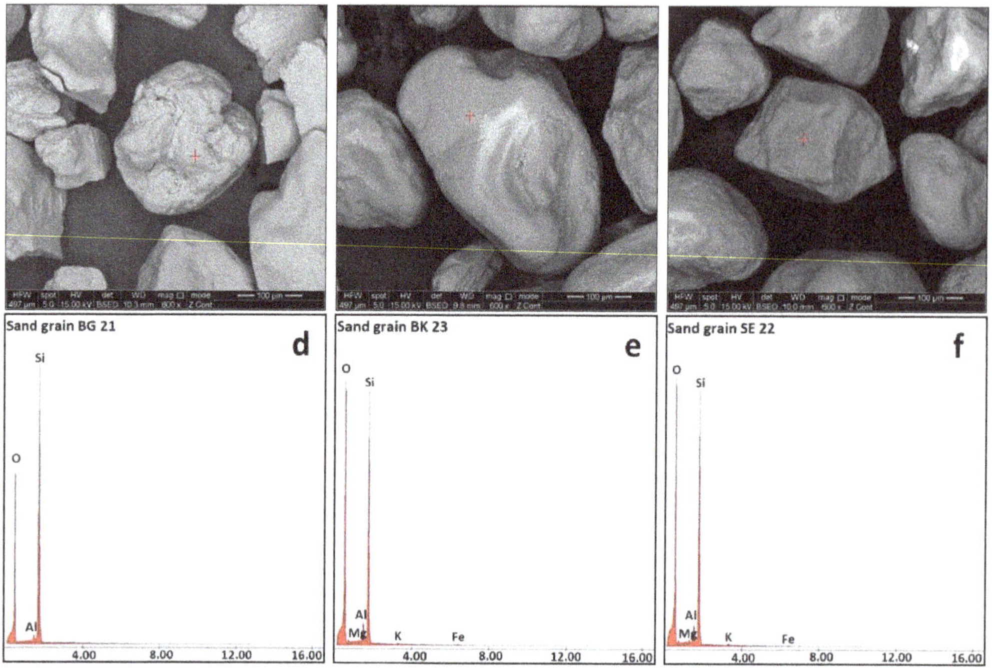

Figure 4. SEM + EDX results for sand samples: (**d**) BG 21, (**e**) BK 23, and (**f**) SE 22.

Table 3. Results of point EDX analysis of 6 tested sand samples.

Sample	O	Mg	Al	Si	K	Fe
			(Mass%)			
SH 35	52.34	0.22	1.94	43.97	1.32	0.73
Sx SH 35	5.55	0.01	1.02	3.30	0.93	0.30
ST 54	51.96	-	1.59	46.45	-	-
Sx ST 54	0.95	-	1.02	1.50	-	-
GL 21	54.25	-	1.06	44.69	-	-
Sx GL 21	2.91	-	0.22	2.97	-	-
BG 21	54.72	-	1.06	44.23	-	-
Sx BG 21	2.12	-	0.07	2.05	-	-
BK 23	55.88	0.24	3.11	38.96	0.34	1.46
Sx BK 23	2.75	0.02	0.07	2.25	0.07	0.38
SE 22	57.81	0.36	3.77	36.15	0.68	1.22
Sx SE 22	2.22	0.22	2.15	1.76	0.61	0.92

3.3. Linear Thermal Expansion and Phase Transformation

Unlike other natural sands (such as olivine) or synthetic sands (ground chamotte or other aluminosilicates) and recycled materials, quartz sands exhibit a non-linear expansion curve due to numerous changes in the crystal lattice of the main mineral, namely quartz. During thermal exposure, a phase transformation from β-SiO_2 to α-SiO_2 occurs, accompanied by a sharp volumetric change. This specific transformation is technologically significant as it occurs, unlike other phase transformations in quartz, at a relatively low temperature, around 573 °C.

All samples were measured in the same manner to prevent the introduction of errors. Table 4 provides specific results of the final expansion of the samples at a temperature of 1130 °C. The resulting expansion curves for individual sand samples and the specific temperature of the β to α SiO_2 phase transformation are evident from Figure 5. Because the phase transformation temperatures considered during the expansion assessment only account for the temperature of the full course of the volumetric change itself, it was necessary to perform DTA evaluation to determine the onset temperature of this transformation. The results of DTA for individual samples are then mentioned in Figure 6.

Table 4. Linear thermal expansion results at 1130 °C of 6 tested sand samples.

	SH 35	ST 54	GL 21	BG 21	BK 23	SE 22
dL/Lo (%)	2.65	2.21	2.73	2.28	2.78	2.79
B to α SiO_2 (°C)	576.5	574.7	576.3	573.3	579.0	579.1

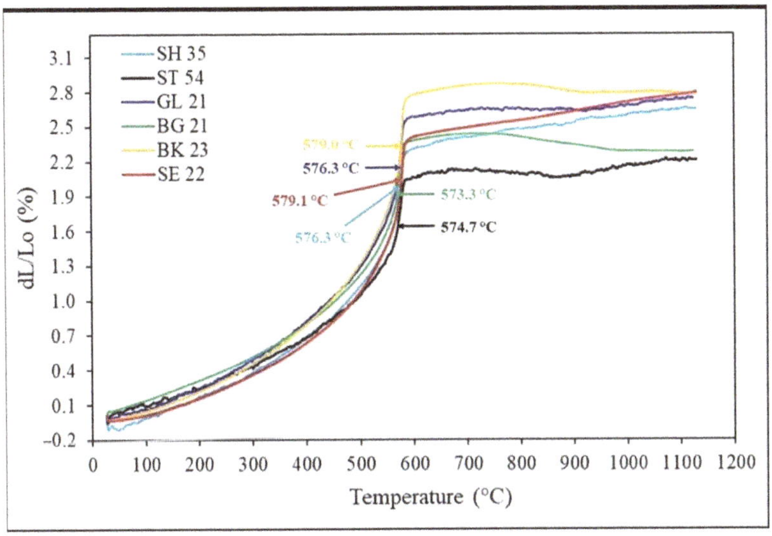

Figure 5. Linear thermal expansion curves for all 6 tested sand samples.

The lowest expansion was achieved in the case of sample ST 54, specifically 2.21%. This sample also exhibited the mildest volumetric change compared to other samples, where the phase transition change was accompanied by a much more abrupt volumetric expansion, as evident from the graph. The actual phase transition change occurred at a temperature of 574.7 °C, the second-lowest temperature compared to other samples. On the other hand, the highest final expansion at 1130 °C was achieved in the case of sample SE 22, specifically 2.79%, which is 26.2% higher compared to ST 54. A related sand sample from a nearby deposit, SH 35, showed a lower final expansion, only 2.65%. Thus, compared to SE 22, it was 5.0% lower and 19.9% higher compared to the lowest measured expansion in ST 54. The most abrupt phase transformation from β to α SiO_2 was recorded in sample BK 23. The highest expansion immediately after the phase transformation achieved with this sample was 2.97%, which is 34.4% more than ST 54, even though the final expansion at 1130 °C was only 2.78%. Sample GL 21 exhibited a final expansion of 2.73%, which is 23.5% higher than ST 54 and 19.7% higher than its related sample BG 21, which, in turn, showed a final expansion of only 2.28%. This was higher than ST 54 by only 3.2%. However, sample BG 21 showed an uneven and somewhat steeper slope of the expansion curve compared to GL 21, exhibiting a significant expansion peak. At the highest point, this peak reached an expansion of 2.49%, and the difference between BG 21 and GL 21 was only 6.8% at this temperature of 798 °C (GL 21 expansion at 798 °C was 2.66%).

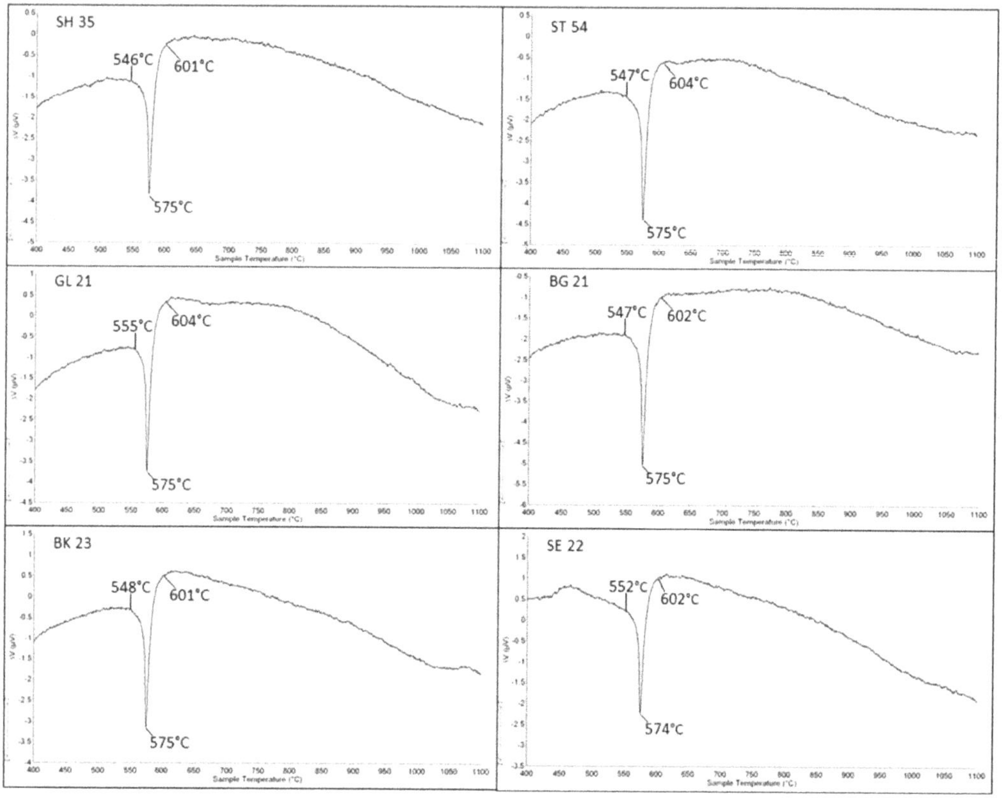

Figure 6. DTA results and the temperature of the beginning of the phase transformation of each tested sand sample.

The most significant factor influencing the overall expansion rate is the grain shape. In the case of samples SH 35, SE 22, and BK 23, they were rounded-grain sands with a smooth grain surface, as evident from Figures 3a and 4e,f. Rounded grains interlock well under load, forming assemblies with minimal pore content and point contact with adjacent grains [32]. The absence of pores between grains and cracks or otherwise irregular surfaces in samples SH 35, SE 22, and BK 23 made these samples more susceptible to higher expansion because there was no space to compensate for grain expansion, thus preventing a certain natural relaxation of this expansion. The resulting samples expanded more than sands with more angular grain shapes (ST 54, as well as GL 21 and BG 21).

In the case of sample ST 54, it was the opposite extreme, i.e., sand with very angular grains and sharp edges. Unlike spheres or rounded-grain sands, where the contact between individual grains could only be point-like, in the case of very angular grains, SEM images (Figure 3b) revealed both edges and entire fracture surfaces of the grain, allowing for several touch variants. These variants included surface-to-surface, surface-to-edge, and edge-to-edge contacts. When assembling angular grains, due to the possibilities of mutual grain contact, there was subsequently an increase in the volume of free spaces between grains, i.e., pores, in contrast to rounded-grain sands. This large number of free pores allowed sample ST 54, unlike rounded-grain sands SH 35, SE 22, and BK 23, to relax thermal grain expansion by allowing the grain to expand into these free spaces.

Samples GL 21 and BG 21, in terms of grain shape, represented an intermediate between rounded-grain sands and angular ST 54. By nature, they were more oval with rounded edges, as evident from SEM Figures 3c and 4d. This shape allowed grains to interlock better while maintaining sufficient porosity and intergranular spaces. Occasional cracks and depressions in the grains also provided additional space for grain expansion compensation. As a result, GL 21 and BG 21, while not achieving as low expansions as ST 54, still exhibited lower expansions compared to rounded-grain sands (SH 35, SE 22, and BK 23).

Another significant factor influencing the extent of dilation is the chemical composition of the sand, specifically the SiO_2 content. In general, a higher SiO_2 content in sands leads to higher dilation, as confirmed by some studies [28,30], but also to various temperature profiles during the phase transformation from β to α SiO_2. From the results of the chemical composition and their relation to the resulting dilation, it was evident that samples with a higher chemical purity exhibited more dilation. This was particularly observed in samples of related character and origin, considering the major influence of grain shape mentioned above. Differential dilation was observed in samples SH 35 and SE 22, as well as GL 21 and BG 21. In the case of rounded-grain sands, a 5.3% greater dilation was observed for the chemically purer SE 22 (91.464% SiO_2) compared to the less pure SH 35 (91.024% SiO_2), as shown in Table 1. The temperature of the phase transition progression, and thus the abrupt volumetric change, increased with increasing SiO_2 content (Figure 5, Table 4), as did the temperature of the beginning of the phase change according to DTA (Figure 6). Chemically purer SE 22 initiated the phase change at 552 °C, which was 6.0 °C higher than the less pure SH 35 (546 °C). The abrupt volumetric change occurred for SE 22 at a temperature of 579.1 °C, which was 2.8 °C higher than for sample SH 35 (576.3 °C). This trend was repeated in the case of oval sands GL 21 and BG 21, where chemically purer GL 21 showed an overall 19.7% greater dilation than the less pure BG 21. The temperature of the onset of the phase transition according to DTA was 555 °C for chemically purer GL 21, which was 8.0 °C higher than for less pure BG 21 (547 °C). The temperature of the abrupt volumetric change for GL 21 was 576.3 °C, a whole 3.0 °C higher than for less pure BG 21 (573.3 °C). The content of impurities, specifically feldspars, reduced dilation and the phase transformation temperature. Feldspars are characterized by lower refractoriness and hardness on the Mohs scale compared to quartz. The reduction in dilation and transformation temperature can be explained by the earlier melting of these minerals, where the melt subsequently filled intergranular spaces, releasing its original space for the dilation of the remaining more refractory grains. Feldspar grains are also significantly softer and likely facilitated the crushing of harder grains (SiO_2) to some extent, thereby releasing space.

In the case of sample BK 23, it was a combination of rounded grains with a smooth surface and higher chemical purity compared to samples SH 35 and SE 22. This particular combination of grain shape, minimal presence of intergranular spaces, and expansiveness based on SiO_2 content, as demonstrated in previous samples, allowed for the sand to exhibit a much more abrupt volumetric change, with a significant dilation reaching higher values (2.97% at a temperature of 932 °C) compared to other examined samples, specifically 34.4% higher than ST 54. Sample ST 54, despite its highest chemical purity, paradoxically exhibited the lowest dilation. The reason was precisely the grain characteristics, where their angularity and structure significantly reduced dilation instead.

4. Conclusions

This experiment investigated the basic characteristics of six types of quartz sands from locations in Central Europe. The examined characteristics primarily included chemical composition and the influence of grain character and shape, which can significantly contribute to the resulting linear thermal dilation of sand and be the cause of many related issues. Based on the conducted tests, the following were confirmed:

- Influence of grain shape: Rounded grains exhibit more dilation than angular grains due to the better interlocking of rounded grains and a higher number of contacts, resulting in a smaller number of intergranular pores for the grain to expand into. On the contrary, angular grains create a larger number of intergranular spaces during stacking, allowing for dilation relaxation.

- It was found that the values of the resulting dilatation for oval grains are on average higher than for square-grained sands but generally lower than for purely round-grained sands with a smooth surface. Influence of chemical purity: Higher SiO_2 content increases susceptibility to dilation and can shift the temperature of the phase transition from β to α SiO_2 to higher temperatures.
- The presence of impurities and feldspars reduces dilation and the temperature of phase transitions by influencing the lower hardness and lower refractoriness of feldspars. The formation of the melt helps relax the expansion of grains.
- The combination of higher chemical purity and rounded grain shape with a smooth surface achieved the highest dilation, namely 2.78%.
- Angular quartz sand with high chemical purity achieved the lowest dilation at 2.21%, despite having the highest SiO_2 content, due to the shape of its grains.

It is not entirely certain how other factors that may affect the resulting dilation of quartz sands, such as grain size and granulometric composition, would impact the final dilatation. This verification could be the subject of further investigation.

Author Contributions: Conceptualization, M.B. and P.L.; methodology, M.B.; validation, P.L. and T.D.; formal analysis, M.B., J.V., S.Z. and M.K.; investigation, M.B., J.V., S.Z. and M.K.; resources, M.B.; writing—original draft preparation, M.B.; visualization, M.B.; supervision, P.L. and T.D.; funding acquisition, P.L. All authors have read and agreed to the published version of the manuscript.

Funding: This research was funded by the project of MEYS (grant number CZ.02.1.01/0.0/0.0/17_049/0008399). This research was carried out with the support of projects of student grant competition (project number SP2023/022). Additionally, the infrastructure used was made available through project No. CZ.02.01.01/00/22_008/0004631, "Materials and Technologies for Sustainable Development", funded by the European Union and the state budget of the Czech Republic within the framework of the Jan Amos Komensky Operational Program.

Data Availability Statement: Data is contained within the article.

Conflicts of Interest: The authors declare no conflicts of interest.

References

1. Mansour, M. Develop a strategic forecast of silica sand based on supply chain decomposition. *Int. J. Eng.* **2015**, *9*, 9–27.
2. Orr, I.; Krumenacher, M. Environmental impacts of industrial silica sand (frac sand) mining. *Heartl. Policy Study* **2015**, *137*, 1–37.
3. Chukwu, P.; Maduabum, A. Silica sand: The architecture material of the 21st century. *Int. J. Innov. Sci. Res. Technol.* **2020**, *5*, 705–709.
4. Kerai, J.V.; Vaniya, S.R. Use of Silica Sand as Fine Material in Concrete. *Int. J. Sci. Res. Dev.* **2015**, *3*, 745–751.
5. Chaudhary, J.L.; Harison, A.; Srivastava, V. Use of silica sand as cement replacement in ppc concrete. *Int. J. Res. Eng. Technol.* **2015**, *4*, 55–58.
6. Malathy, R.; Sentilkumar, S.R.R.; Prakash, A.R.; Das, B.B.; Chung, I.M.; Kim, S.H.; Prabakaran, M. Use of Industrial Silica Sand as a Fine Aggregate in Concrete—An Explorative Study. *Buildings* **2022**, *12*, 1273. [CrossRef]
7. Wang, Y.; Chung, D.D.L. Effects of sand and silica fume on the vibration damping behavior of cement. *Cem. Concr. Res.* **1998**, *28*, 1353–1356. [CrossRef]
8. Tayeh, B.A.; Akeed, M.H.; Qaidi, S.; Bakar, B.A. Influence of sand grain size distribution and supplementary cementitious materials on the compressive strength of ultrahigh-performance concrete. *Case Stud. Constr. Mater.* **2022**, *17*, e01495. [CrossRef]
9. Nakayenga, J.; Cikmit, A.A.; Tsuchida, T.; Hata, T. Influence of stone powder content and particle size on the strength of cement-treated clay. *Constr. Build. Mater.* **2021**, *305*, 124710. [CrossRef]
10. Heeschen, K.U.; Schicks, J.M.; Oeltzschner, G. The promoting effect of natural sand on methane hydrate formation: Grain sizes and mineral composition. *Fuel* **2016**, *181*, 139–147. [CrossRef]
11. Bašistová, M.; Radkovský, F.; Kroupová, I.; Lichý, P. Dilatation of New Progressive Hybrid Sand and Its Effect on Surface Structure, Roughness, and Veining Creation within Grey Cast Iron. *Materials* **2023**, *16*, 2004. [CrossRef] [PubMed]
12. Zhang, M.; Zhu, X.; Shi, J.; Liu, B.; He, Z.; Liang, C. Utilization of desert sand in the production of sustainable cement-based materials: A critical review. *Constr. Build. Mater.* **2022**, *327*, 127014. [CrossRef]
13. Al-Aghbari, M.Y.; Dutta, R.K. Suitability of desert sand cement mixes for base courses in highway pavements. *Electron. J. Geotech. Eng.* **2005**, *10*, 1–17.
14. Ahmad, J.; Zhou, Z.; Martínez-García, R.; Vatin, N.I.; de-Prado-Gil, J.; El-Shorbagy, M.A. Waste foundry sand in concrete production instead of natural river sand: A review. *Materials* **2022**, *15*, 2365. [CrossRef] [PubMed]
15. Bhardwaj, B.; Kumar, P. Waste foundry sand in concrete: A review. *Constr. Build. Mater.* **2017**, *156*, 661–674. [CrossRef]

16. Bostanci, C.; Limbachiya, M.; Kew, H. Portland-composite and composite cement concretes made with coarse recycled and recycled glass sand aggregates: Engineering and durability properties. *Constr. Build. Mater.* **2016**, *128*, 324–340. [CrossRef]
17. Guney, Y.; Aydilek, A.H.; Demirkan, M.M. Geoenvironmental behavior of foundry sand amended mixtures for highway subbases. *Waste Manag.* **2006**, *26*, 932–945. [CrossRef]
18. Ansari, W.S.; Chang, J. Influence of fine cement sand paste in preparation of cementitious materials. *Constr. Build. Mater.* **2020**, *230*, 116928. [CrossRef]
19. Katzer, J. Median diameter as a grading characteristic for fine aggregate cement composite designing. *Constr. Build. Mater.* **2012**, *35*, 884–887. [CrossRef]
20. Al-Harthy, A.S.; Halim, M.A.; Taha, R.; Al-Jabri, K.S. The properties of concrete made with fine dune sand. *Constr. Build. Mater.* **2017**, *21*, 1803–1808. [CrossRef]
21. Ma, Q.; Guo, R.; Zhao, Z.; Lin, Z.; He, K. Mechanical properties of concrete at high temperature—A review. *Constr. Build. Mater.* **2015**, *93*, 371–383. [CrossRef]
22. Xiao, J.; König, G. Study on concrete at high temperature in China—An overview. *Fire Saf. J.* **2004**, *39*, 89–103. [CrossRef]
23. Schneider, U. Concrete at high temperatures—A general review. *Fire Saf. J.* **1988**, *13*, 55–68. [CrossRef]
24. Song, R.; Wang, Y.; Ishutov, S.; Zambrano-Narvaez, G.; Hodder, K.J.; Chalaturnyk, R.J.; Sun, S.; Liu, J.; Gamage, R.P. A comprehensive experimental study on mechanical behavior, microstructure and transport properties of 3D-printed rock analogs. *Rock Mech. Rock Eng.* **2020**, *53*, 5745–5765. [CrossRef]
25. Hodder, J.; Nychka, J.A. Silane treatment of 3D-printed sandstone models for improved spontaneous imbibition of water. *Transp. Porous Media* **2019**, *129*, 583–598. [CrossRef]
26. Song, R.; Wu, M.; Wang, Y.; Liu, J.; Yang, C. In-situ X-CT scanning and numerical modeling on the mechanical behavior of the 3D printing rock. *Powder Technol.* **2023**, *416*, 118240. [CrossRef]
27. Zihms, S.G.; Switzer, C.; Irvine, J.; Karstunen, M. Effects of high temperature processes on physical properties of silica sand. *Eng. Geol.* **2013**, *164*, 139–145. [CrossRef]
28. Svidró, J.; Svidró, J.T.; Diószegi, A. The role of purity level in foundry silica sand on its thermal properties. *J. Phys. Conf. Ser.* **2020**, *1527*, 012039. [CrossRef]
29. Chao, C.H.; Lu, H.Y. Stress-Induced $\beta \rightarrow \alpha$-Cristobalite Phase Transformation in (Na_2O + Al_2O_3)-codoped Silica. *Mater. Sci. Eng. A* **2002**, *328*, 267–276. [CrossRef]
30. Svidró, J.; Diószegi, A.; Svidró, J.T. The origin of thermal expansion differences in various size fractions of silica sand. *Int. J. Cast. Met. Res.* **2020**, *33*, 242–249. [CrossRef]
31. Hrubovčáková, M.; Vasková, I.; Benková, M.; Conev, M. Opening Material as the Possibility of Elimination Veining in Foundries. *Arch. Foundry Eng.* **2016**, *16*, 157–161. [CrossRef]
32. Radkovský, F.; Gawronová, M.; Merta, V.; Lichý, P.; Kroupová, I.; Nguyenová, I.; Kocich, R. Effect of the composition of Hybrid Sands on the change in thermal expansion. *Materials* **2022**, *15*, 6180. [CrossRef] [PubMed]
33. Bašistová, M.; Lichý, P. Differences in characteristic Properties of Silica sands from neighbouring deposits. *Adv. Sci. Technol.* **2023**, *135*, 101–108. [CrossRef]
34. Ndon, U.J.; Bergeson, K.L. Thermal expansion of concretes: Case study in Iowa. *J. Mater. Civ. Eng.* **1995**, *7*, 246–251. [CrossRef]
35. Aleem, S.A.E.; Heikal, M.; Morsi, W.M. Hydration characteristic, thermal expansion and microstructure of cement containing nano-silica. *Constr. Build. Mater.* **2014**, *59*, 151–160. [CrossRef]
36. Liu, H.; Bu, Y.; Zhou, A.; Du, J.; Zhou, L.; Pang, X. Silica sand enhanced cement mortar for cementing steam injection well up to 380 °C. *Constr. Build. Mater.* **2021**, *308*, 125142. [CrossRef]
37. De Souza, W.R.M.; Bouaanani, N.; Martinelli, A.E.; Bezerra, U.T. Numerical simulation of the thermomechanical behavior of cement sheath in wells subjected to steam injection. *J. Pet. Sci. Eng.* **2018**, *167*, 664–673. [CrossRef]
38. Dai, T.; Liu, T.; Jiang, G.; Qu, B.; Zheng, S. Evaluation of Microstructural and Mechanical Properties of Silica-Enriched Oil Well Cement Pastes in Steam Injection Wells, Available at SSRN 4048736. Available online: https://papers.ssrn.com/sol3/papers.cfm?abstract_id=4048736 (accessed on 21 November 2023).
39. Bamonte, P.; Gambarova, P. Properties of concrete subjected to extreme thermal conditions. *J. Struct. Fire Eng.* **2014**, *5*, 47–62. [CrossRef]
40. Thiel, J. *Thermal Expansion of Chemically Bonded Silica Sand*; American Foundry Society: Schaumburg, IL, USA, 2011; pp. 1–10.
41. Wang, H.; Lu, Y.; Ripplinger, K.; Detwiler, D.; Luo, A.A. A statistics-based cracking criterion of resin-bonded silica sand for casting process simulation. *Metall. Mater. Trans. B.* **2017**, *48*, 260–267. [CrossRef]

buildings

MDPI

Article

Damage Model of Basalt-Fiber-Reinforced Cemented Soil Based on the Weibull Distribution

Lina Xu [1], Runze Zhang [2], Lei Niu [2,*] and Chenhui Qi [2]

[1] School of Transportation Science and Engineering, Jilin Jianzhu University, Changchun 130118, China
[2] School of Civil Engineering, Jilin Jianzhu University, Changchun 130118, China
* Correspondence: niulei@jlju.edu.cn

Abstract: This study investigates the mechanical performance and a constitutive model of basalt-fiber-reinforced cemented soil (BFRCS) containing 0%, 0.1%, 0.3%, 0.5%, and 0.7% basalt fibers with lengths of 3, 6, 12, 20, and 35 mm, respectively. Unconfined compressive strength tests were used to examine the mechanical performance of BFRCS with varying basalt fiber contents and lengths. The test results demonstrate that the basalt fiber content of optimal quality is 0.1%, and that the fiber distribution uniformity and density have a significant impact on the strength of BFRCS. Based on the Weibull distribution of BFRCS for the degree of damage, a damage model for BFRCS, accounting for the fiber length and fiber content, is proposed here. Moreover, in this study we explored the relationship between the scale parameter as well as shape parameter of the Weibull distribution and fiber content as well as length. Furthermore, the evaluation methods for the mechanical properties of BFRCS according to the scale and shape parameters of the Weibull distribution are discussed. The results suggest that the proposed constitutive model captures the compressive stress–strain relationship of BFRCS; the theoretical results are in strong agreement with the data obtained.

Keywords: basalt-fiber-reinforced cemented soil; damage model; Weibull distribution; unconfined compressive strength

Citation: Xu, L.; Zhang, R.; Niu, L.; Qi, C. Damage Model of Basalt-Fiber-Reinforced Cemented Soil Based on the Weibull Distribution. *Buildings* **2023**, *13*, 460. https://doi.org/10.3390/buildings13020460

Academic Editor: Tomáš Dvorský

Received: 5 January 2023
Revised: 2 February 2023
Accepted: 6 February 2023
Published: 7 February 2023

1. Introduction

Currently, cemented soil is popular in various areas of soil reinforcement engineering [1,2], but its high brittleness and low tensile properties [3,4] render it susceptible to cracking during use, thereby decreasing its durability. To improve its mechanical properties, researchers have added various types of fibers to cemented soil to transform it into fiber-reinforced cemented soil. Studies have demonstrated that incorporating fibers can improve the strength and toughness of cemented soil [5–7].

Basalt fiber is a new environmental fiber material with a high tensile strength and erosion resistance [8–10], rendering it a common soil reinforcement material. Yang et al. [11] reported that when the fiber content of soil increased, its strength and stability increased. Furthermore, Otieno et al. [12] reported the reinforcing effect of basalt fiber and rice husk ash (RHA) co-cured with cement on the compressibility and swelling of soil. They reported that the addition of 12 mm of basalt fibers, 5% of RHA, and 3% of cement could effectively reinforce soil and enhance its ultimate strength. Ma et al. [13] examined the dynamic mechanical properties of basalt-fiber-reinforced cemented soil (BFRCS) and reported that the basalt fiber content affected the energy absorption capacity and improved the dynamic strength of BFRCS. Moreover, Cao [14] observed that adding basalt fibers to cement–fly-ash-stabilized soil considerably enhanced the static–dynamic mechanical properties of stabilized soil. With a basalt fiber content of 0.6%, a stable internal space structure was produced in stabilized soil; however, multiple weak interfaces between the fibers appeared in stabilized soil with a basalt fiber content of 1.2%. Shen et al. [15] examined the strength properties of fiber-reinforced clayey soil with a lower content of lime or cement (5% by weight); however, for the lime-treated soil specimens the peak values were observed when the fiber content was 0.1%. Furthermore, Zhu et al. [16] considered that the interfacial peak/residual shear resistance and shear stiffness of fiber-reinforced soil considerably depend on the soil compaction conditions. The addition of cement to soil can improve the soil compaction conditions, thus improving the effect of fiber-reinforced soil [17]. Wang et al. [18] reported that the inclusion of basalt fibers improves the ductility and weakens the brittleness of cemented kaolinite. The compressive strength increases with the basalt fiber content and curing time, peaks at a 0.2% fiber content, and then decreases because of the formation of a weak zone at higher fiber contents. Xiao et al. [19] discovered that the unconfined compressive strength (UCS) and splitting tensile strength of basalt-fiber-reinforced biocement with a given basalt fiber content significantly increased with an increase in the calcite content. Moreover, Lv et al. [20] reported that, as the fiber content increased, the cohesion and residual cohesion first increased and then decreased, and that the brittleness index decreased. Furthermore, the internal friction angle and residual internal friction angle exhibited no clear variation patterns. Dasaka et al. [21] reported that the UCS of soil significantly increased with the addition of coconut shell fibers. Divya et al. [22] used a digital image cross-correlation analysis of the top view of the soil captured during tensile tests to obtain the displacement vectors and strain field distribution. The fiber content and fiber length increased, such that the other parameters remained constant. Thus, the strain upon crack initiation and the energy absorption capacity of the fiber-reinforced soil increased, and the post-cracking behavior improved.

The aforementioned study showed a synergistic effect between fibers and cement, and that their joint addition can have a coupled reinforcing effect. Moreover, the fibers can effectively improve the mechanical properties of the reinforced material as long as they are present in the appropriate content and are of a suitable length.

Gao et al. [23] proposed a critical-state fiber-reinforced soil (FRS) model in the triaxial stress space by using the concept of effective skeleton stress and the void ratio. Sivakumar Babu et al. [24] reported an analytical model for the analysis of FRS in the framework of a modified cam clay model. Furthermore, Salih et al. [25] designed a stress–strain model describing the soil brick response to compressive loading for each fiber type, obtained via a regression analysis. Kanchi et al. [26] formulated an analytical model for the FRS based on the modified cam clay model concept. The results demonstrated that as the inclination of the fibers in the horizontal plane increased, the contribution of the fibers in improving the strength of FRS decreased. Wu et al. [27] developed a uniaxial strain-softening constitutive model for FRS based on the disturbed state concept. This study on the intrinsic structural model of reinforced soil primarily focused on fiber-reinforced soil or cement-reinforced soil; however, only a few studies are related to the constitutive model of fiber-reinforced cemented soil, and the results have not enabled researchers to form a unified theory.

Based on UCS tests, this study proposes a Weibull-distribution-based damage model. Moreover, the effects of fiber content and fiber length on the mechanical properties of BFRCS were analyzed. The model parameters were determined, and the stress–strain relationship of BFRCS under uniaxial pressure was fitted and analyzed for damage evolution. This study aids researchers in comprehending the evolution of the damage properties of BFRCS and provides a theoretical foundation for their accurate analysis and evaluation.

2. Material Properties and Test Procedure

2.1. Test Materials

2.1.1. Soil Properties

The test soil was collected from a side slope in Changchun City, China, and its fundamental properties are listed in Table 1. Figure 1 shows the appearance of the test soil, and Figure 2 shows the original spectrum obtained in the XRD test. The primary components are quartz, sodium feldspar, and potassium feldspar. The grain particles are shown in Figure 3.

Table 1. Physical indices of the soil.

Physical and Mechanical Index	Value
Natural water content (%)	33.2
Plastic limit (%)	41
Plasticity index	16
Natural density (g/cm^3)	2.71
Void ratio	1.03
Liquid limit (%)	25

Figure 1. Image of the experimental soil.

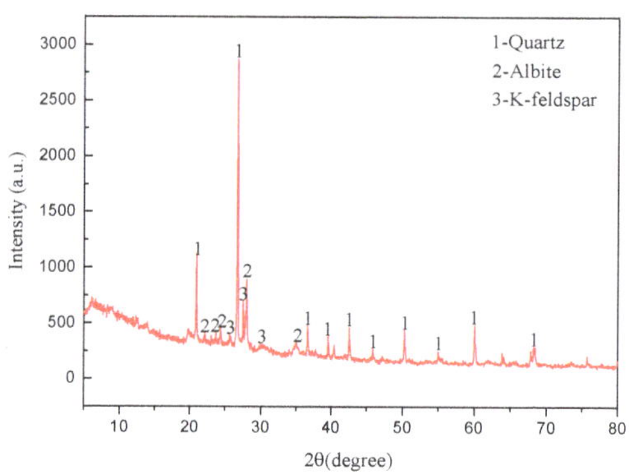

Figure 2. Original XRD spectrum of the tested soil.

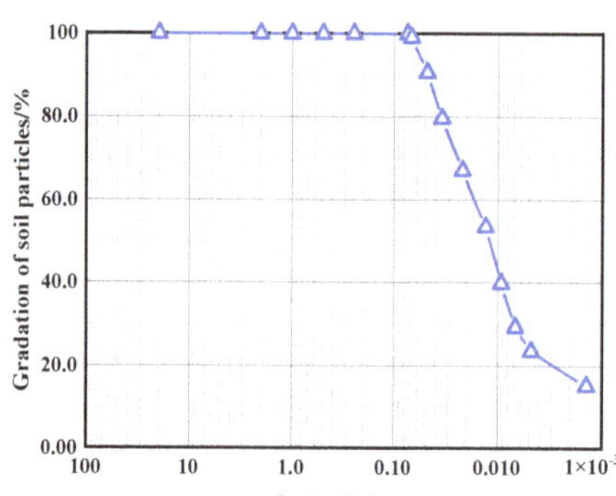

Figure 3. Grain size distribution curve.

2.1.2. Basalt Fibers

The basic properties and composition of the basalt fibers are shown in Tables 2 and 3. Figure 4 shows the appearance of the basalt fibers.

Table 2. Basic properties of the basalt fibers.

Main Component	Si	Ca	Al	Mg	O	K	Na	Ti	Fe
Atomic fraction/%	26.6	18.93	7.89	6.90	31.81	1.18	1.63	1.26	4.04

Table 3. Basalt fiber composition.

Diameter/μm	Tensile Strength/MPa	Modulus of Elasticity/GPa	Density/(g/cm^3)	Thermal Conductivity/(W-m/k)	Operating Temperature/°C
7–15	3000–4800	91–110	2.63–2.65	0.030–0.038	−269–650

Figure 4. Basalt fibers.

2.1.3. Cement

Ordinary Portland cement (grade 42.5), produced by the Changchun Yatai Group of China, was used as the cementitious material, and its basic properties are shown in Table 4.

Table 4. Basic properties of the cement.

Strength Grade	Specific Surface Area/(m^2/kg)	Initial Setting Time/h	Final Set Time/h	Cementitious Sand Flexural Strength/MPa		Cementitious Sand Compressive Strength/MPa	
				3 d	28 d	3 d	28 d
P-O 42.5	345	2	2.5	5	8	24	49

2.2. Specimen Preparation

Cubic specimens were prepared based on the JGJ/T233-2011 Specification for Mix Proportion Design of Cement Soil [28], with dimensions of 70.7 mm × 70.7 mm × 70.7 mm. The cement content was 10%. To produce the BFRCS specimens, basalt fibers, cement, and soil were mixed for 3 min. Water was then added, and the mixture was stirred for 6 min. Finally, the mixture was molded by 70.7 mm^2 cubic molds and shaken for 3 min using a shaking table. The specimen preparation flowchart is shown in Figure 5.

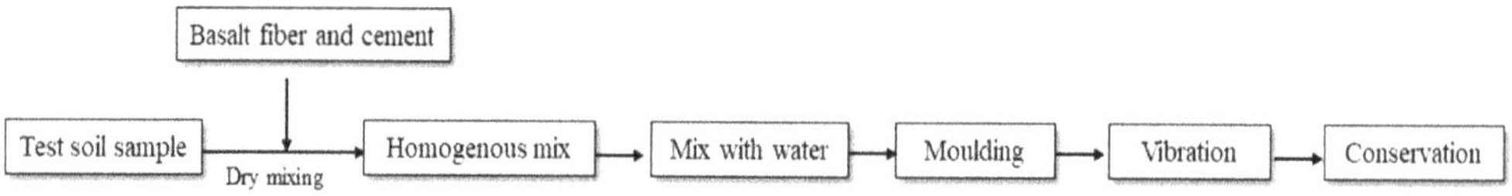

Figure 5. Specimen preparation flowchart.

2.3. Test Design

The mix designs of 21 groups of BFRCS are listed in Table 5, where C indicates the fiber content and L indicates the fiber length. For instance, C0.1-L20 indicates that the fiber content is 0.1% and the fiber length is 20 mm. Each group had three duplicate specimens to reduce the test errors.

Table 5. Mixture designs of the test specimens.

No.	Fiber Content/%	Fiber Length/mm	No.	Fiber Content/%	Fiber Length/mm
C0-L0	0	0	C0.5-L3	0.5	3
C0.1-L3	0.1	3	C0.5-L6	0.5	6
C0.1-L6	0.1	6	C0.5-L12	0.5	12
C0.1-L12	0.1	12	C0.5-L20	0.5	20
C0.1-L20	0.1	20	C0.5-L35	0.5	35
C0.1-L35	0.1	35	C0.7-L3	0.7	3
C0.3-L3	0.3	3	C0.7-L6	0.7	6
C0.3-L6	0.3	6	C0.7-L12	0.7	12
C0.3-L12	0.3	12	C0.7-L20	0.7	20
C0.3-L20	0.3	20	C0.7-L35	0.7	35
C0.3-35	0.3	35			

2.4. Testing Apparatus and Procedure

The UCS test was performed with a WAW-600 microcomputer-controlled servo universal testing machine, produced by Changchun Kexin Testing Instruments in China, with displacement uniformity control and a loading speed of 0.1 mm/s. The acquisition of the UCS data was performed by using an automatic acquisition system matched with the WAW-600 universal testing machine, including the loads and displacements.

3. Experimental Results and Discussion

3.1. Effect of the Fiber Length on the Mechanical Properties

Damage strain is defined as the strain corresponding to the peak stress in the stress–strain curve (ε_1). Figure 6a–d shows the variations in the damage strain versus the fiber length for the specimens with varying fiber contents, with the dashed line representing the value of the damage strain of the cemented soil without fibers. Figure 6 shows that the damage strain values of the specimens under different reinforcement conditions are higher than those of the specimens without fibers, indicating that the admixture of basalt fibers can improve the deformation resistance and toughness of the cemented soil. The damage strain curves tended to increase first and then decrease with an increase in the fiber length. The maximum value of the damage strain occurred in the case of the 6 mm fiber length, when the fiber contents were 0.1% and 0.3%, as well as the 20 and 12 mm fiber lengths, when the fiber contents were 0.5% and 0.7%. In conclusion, the damaged strain reached its maximum at a fiber content of 0.1% and a fiber length of 6 mm.

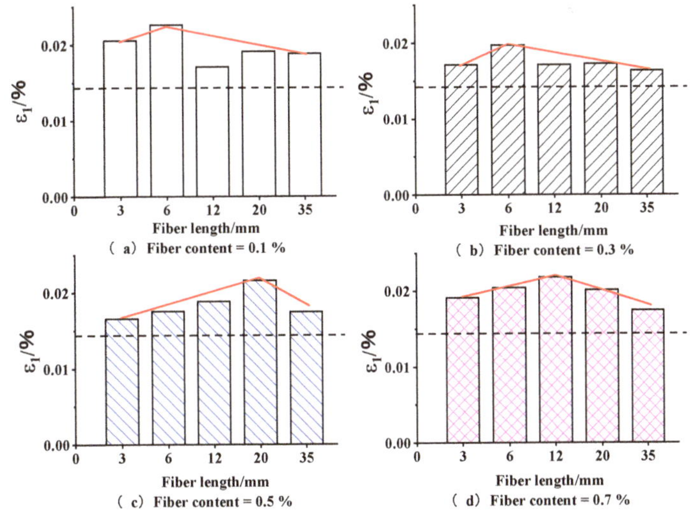

Figure 6. Variations in the damage strain with fiber length and different fiber contents.

Considering the peak stress in the stress-strain curve as the peak strength (σ_{max}) for analysis, Figure 7a–d illustrates the variations in the peak strength of the specimens with different fiber lengths and fiber contents. The dashed lines represent the cemented soil's peak strengths without the addition of fibers. Figure 7 shows that the incorporation of basalt fibers can effectively increase the peak strengths of the specimens, and that the peak strengths of all of the specimens with fibers incorporated are greater than those of the specimens without fibers incorporated. The peak strength curves showed the tendency to decrease with an increase in the fiber length, except for a fiber content of 0.1%. The maximum peak strength values occurred in the case of the 3 mm fiber length, when the fiber contents were 0.3%, 0.5%, and 0.7%, and the 12 mm fiber length, when the fiber content was 0.1%. To sum up, the peak strength reached its maximum at a fiber content of 0.1% and a fiber length of 12 mm.

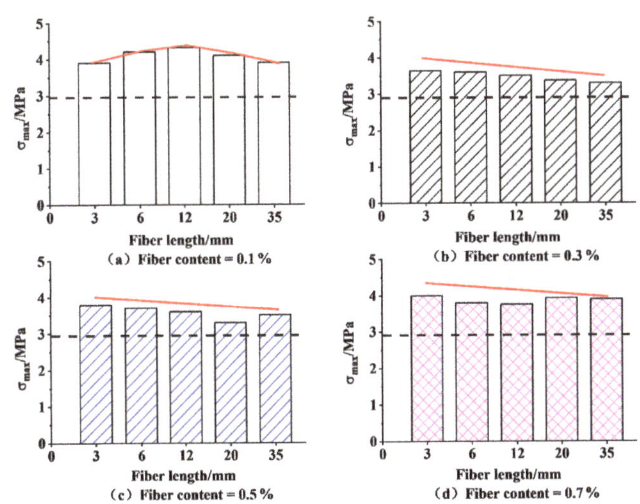

Figure 7. Variations in the peak strength with fiber length and different fiber contents.

3.2. Effect of the Fiber Content on Mechanical Properties

Figure 8a–e shows the variations in the damage strains of the specimens with different fiber contents and fiber lengths. The dashed lines in the figure show the damage strain values of the cemented soil without the addition of fibers. From Figure 8, we can observe that the damage strain curves tended to decrease first and then increase with an increase in the fiber content when the fiber lengths were 3, 6, and 12 mm. Furthermore, when the fiber lengths were 12 and 35 mm the damage strain curves tended to linearly increase and linearly reduce. The tendencies of the damage strain curves for the individual fiber lengths are not identical.

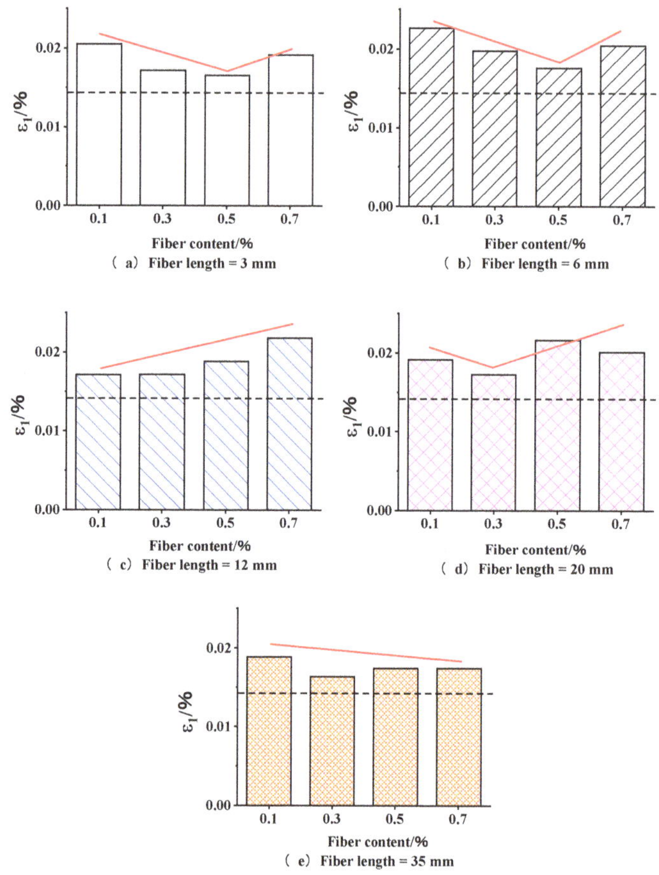

Figure 8. Variations in the damage strain with fiber content and different fiber lengths.

Figure 9a–e shows the variations in the peak strength of the specimens with different fiber contents and fiber lengths. Figure 9 shows that the peak strengths tended to increase and then decrease with an increase in the fiber content for all the fiber lengths, while for all of the fiber lengths a fiber content of 0.1% resulted in the most noticeable improvement in the peak strength of BFRCS.

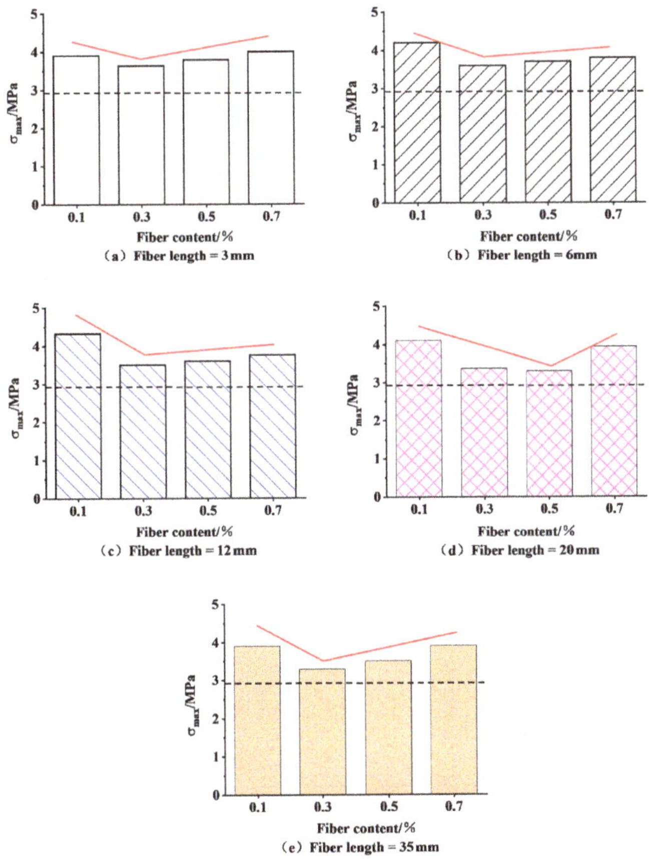

Figure 9. Variations in the peak strength with fiber content and different fiber lengths.

3.3. Reinforcement Mechanism of Fibers in Cemented Soil

The effect of the fiber material on cemented soil reinforcement primarily lies in the adhesion between fibers and cemented soil in addition to the distribution law of fibers, where the adhesion is related to the soil quality, particle gradation, cement content, conservation age, and other factors. Moreover, the distribution law of fibers is closely related to fiber length, content, and other factors. Furthermore, the observation of a loaded specimen reveals that the fibers in the specimen appear to be clustered. The distribution cannot be fully dispersed, and even fibers dispersed in filaments have an uneven distribution, as shown in Figure 10. From a microscopic perspective, constructing a BFRCS constitutive model becomes relatively complex and challenging in light of the complex fiber arrangement. Figure 10 shows that when the fiber density is too high the reinforcing effect is diminished by a mechanism similar to the anchor group effect, whereas it is difficult to exert a reinforcing effect when the density is too low. Therefore, a suitable fiber content, length, and arrangement can enhance the strength of BFRCS.

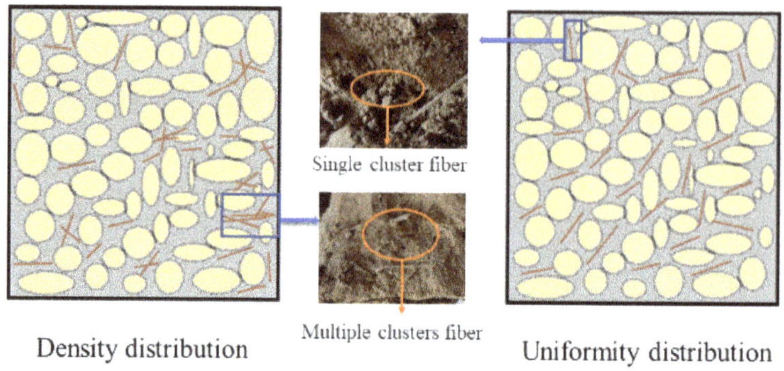

Figure 10. Dense and uniform fiber distributions in the matrices.

4. Damage-Theory-Based Constitutive Model

Damage mechanics is based on continuum mechanics, used to examine the laws of solid materials under certain loads and in erosive environments, according to which materials that develop damage will eventually undergo the destruction process. Statistical damage mechanics assumes that the degree of microelement damage within a material follows a probability density distribution, based on which the damage of the material is described. Because the integral distribution function of the Weibull distribution is easy to integrate, its mean value is greater than 0, etc. [29,30], the Weibull distribution is widely used in the description of rock and concrete materials [31–34]. This paper's statistical damage model of BFRCS is based on the following fundamental postulates [35]:

1. The number of damaged microelements in relation to the pore area and effective bearing area of BFRCS obeys the Weibull distribution.
2. Under the action of an external load, the material comprises damaged and undamaged parts; the undamaged part bears the load on any microelements of the material.
3. The damage evolution of microelements during loading is a continuously changing process.
4. The material is an isotropic damage material, and the stress-strain relationship of undamaged microelements is in accordance with Hooke's law.

4.1. Establishment of the Constitutive Model

Under the load, the damage in BFRCS gradually increases and expands, leading to changes in the effective bearing area and internal damage area of BFRCS. According to the pertinent theory of damage mechanics [36], the nominal stress and effective stress of BFRCS satisfy the following relationship:

$$\sigma = \frac{F}{A} \tag{1}$$

$$\sigma^* = \frac{F}{A^*} \tag{2}$$

where σ is the nominal stress of BFRCS (MPa), σ^* is the effective stress of BFRCS (MPa), A is the nominal bearing area (mm^2), A^* is the effective bearing area (mm^2), and F is the load on BFRCS (N).

The relationship between the two kinds of stress can be determined from the above two equations as follows:

$$\frac{\sigma}{\sigma^*} = \frac{A^*}{A} \tag{3}$$

For BFRCS, the concept of a damage variable can be used to express the degree of internal damage, to which the damage variable D can be introduced to express the ratio of the internal damage area to the total load-bearing area:

$$D = \frac{A - A^*}{A} = 1 - \frac{A^*}{A} \tag{4}$$

The above equation contains D as its damage variable, $0 \leq D \leq 1$, and its core is the ratio of the damaged area to the total area, which provides a parameter for the subsequent damage level.

Substituting Equation (4) into Equation (3) yields Equation (5), as follows:

$$\sigma = (1 - \mathrm{D})\sigma^* \tag{5}$$

According to the Lemaitre strain equivalence hypothesis [37], the intrinsic relationship of the damaged material can be expressed in terms of the stresses borne by the undamaged part of the material. Therefore,

$$\sigma = (1 - \mathrm{D})E\varepsilon \tag{6}$$

where σ is the stress to which BFRCS is subjected (MPa), D is the damage variable of BFRCS ($0 \leq D \leq 1$), E is the modulus of elasticity of BFRCS, and ε is the strain of BFRCS.

4.2. Calculation of the Damage Variables at Different Stages

As shown in Figure 11, according to the stress–strain curve of BFRCS, the stress damage process is divided into two stages: the compression stage and damage stage. In the compression stage, with an increase in the strain, the main damage that affects the material is destruction of the pore structure by compression, and the pore unit of BFRCS decreases with an increase in the strain. In the damage phase, the main damage in the soil is the compression of the structural units, and the structural units of BFRCS decrease with an increase in the strain.

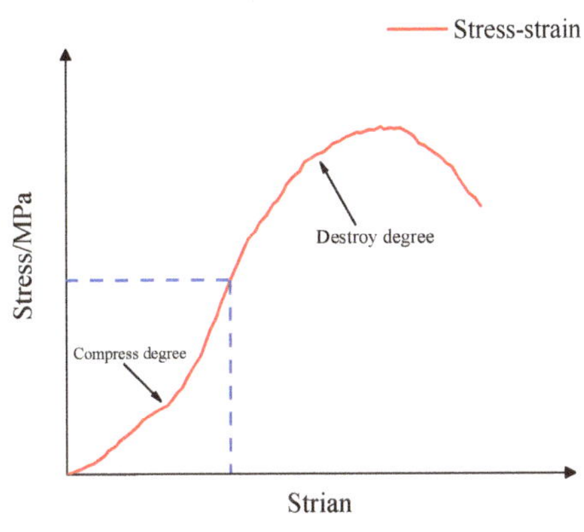

Figure 11. Different degrees of BFRCS in the loading process.

4.3. Calculation of the Damage Variables in the Compression Phase

According to statistical damage mechanics, BFRCS materials can be approximated as structural systems composed of a large number of microelements. Furthermore, in the present study [38], where the degree of damage of BFRCS can be expressed as the ratio of the number of pore microelements to the total number of microelements, it was assumed that, with an increase in strain, a proportion of the initial pores in BFRCS would be destroyed during the compression, and that the damage phases would follow the Weibull distribution as the displacement increased. Thus, Equation (7) can be derived as follows:

$$p(\varepsilon) = \frac{n}{m}\left(\frac{\varepsilon}{m}\right)^{n-1} e^{-\left(\frac{\varepsilon}{m}\right)^{n}} \tag{7}$$

where $p(\varepsilon)$ is the Weibull probability density function, m and n are the scale parameter and shape parameter of the Weibull distribution, respectively, and ε is the strain variable.

With an increase in displacement, N is the number of damaged microelements in the pore microelements. The number of damaged microelements in BFRCS is shown as follows:

$$N = \int_0^{\varepsilon_1} N_t p(\varepsilon) d\varepsilon = N_t\left(1 - e^{-\left(\frac{\varepsilon}{m}\right)^{n}}\right) \tag{8}$$

In this case, by defining the total number of microelements as M, the number of pore microelements is N_t. The damage effect, at this time, is believed to be much less than the effect of compression; therefore, the effect of compression is primarily considered. Since the compression stage basically occurs in the early low-strain stage, it is believed that its effect on the total number of units is negligible.

Let the initial damage factor be D_0. Then:

$$D_0 = \frac{N_t}{M} \tag{9}$$

Then, from the definition of the damage variable of statistical damage mechanics, it is clear that the damage variable of BFRCS is the ratio of the number of damaged microelements to the total number of microelements. The damage coefficient in the compression phase is as follows:

$$D = \frac{(N_t - N)}{M} = e^{-\left(\frac{\varepsilon}{m}\right)^{n}} \frac{N_t}{M} = e^{-\left(\frac{\varepsilon}{m}\right)^{n}} D_0 \tag{10}$$

From Equations (6) and (10), we yield the following:

$$\sigma = E\varepsilon\left(1 - D_0 e^{-\left(\frac{\varepsilon}{m}\right)^{n}}\right) \tag{11}$$

4.4. Calculation of the Damage Variables in the Damage Phase

With an increase in deformation, the proportion of the solid unit damage in BFRCS follows the Weibull distribution as the displacement increases during the compression and elastic phases, and the following equation can be obtained:

$$p_1(\varepsilon) = \frac{n_1}{m_1}\left(\frac{\varepsilon}{m_1}\right)^{n_1-1} e^{-\left(\frac{\varepsilon}{m_1}\right)^{n_1}} \tag{12}$$

where $p_1(\varepsilon)$ is the Weibull probability density function, m_1 and n_1 are the scale and shape parameters of the Weibull distribution, respectively, and ε is the strain variable.

Let M be the total number of microelements and N_1 be the number of microelements that are damaged. It is believed that the damage phase is present from the start, but in the early stage its influence is negligible and in the late stage it is pronounced. Thus, the integration based on the deformation of 0 yields the following equation:

$$N_1 = \int_0^{\varepsilon} M p(\varepsilon) d\varepsilon = M\left\{\left(1 - e^{-\left(\frac{\varepsilon}{m_1}\right)^{n_1}}\right)\right\} \tag{13}$$

Then, from the definition of the damage variable of statistical damage mechanics, it is clear that the damage variable of BFRCS is, at the same time, the ratio of the number of damaged microelements to the total number of microelements. Thus, the damage coefficient of the damage phase is as follows:

$$D = \frac{N_1}{M} = 1 - e^{-(\frac{\varepsilon}{m_1})^{n_1}} \tag{14}$$

From Equations (6) and (14), we yield the following:

$$\sigma = E\varepsilon\left(e^{-(\frac{\varepsilon}{m_1})^{n_1}}\right) \tag{15}$$

Combining Equations (10) and (15), the following Equation (16) can be obtained:

$$\begin{cases} \sigma = E\varepsilon\left(1 - D_0 e^{-(\frac{\varepsilon}{m})^{n}}\right) & 0 \leq \varepsilon < \varepsilon_0 \\ \sigma = E\varepsilon\left(e^{-(\frac{\varepsilon}{m_1})^{n_1}}\right) & \varepsilon_0 \leq \varepsilon \end{cases} \tag{16}$$

Equation (16) is the damage model for BFRCS, where E is the modulus of elasticity of BFRCS, D_0 is the initial damage coefficient of BFRCS, ε_0 is the strain value for the transition from the compression phase to the damage phase, ε is the strain of BFRCS, and σ is the UCS of BFRCS.

The values of the parameters are m = 0.0082, m_1 = 0.0233, n = 1.92, and n_1 = 1.49 when the fiber content is 0. Based on Equation (16), the regression calculation of the stress–strain curve of BFRCS can be obtained by using the constitution model, as the parameters m, n, m_1, and n_1 vary with the fiber length, L, and the fiber content, C, as shown in Table 6 below.

Table 6. Calculation results of the parameters m, n, m_1, and n_1.

Parameter Expressions in Different Degrees	
Compress Degree	**Destroy Degree**
m = 9.52 × 10^{-3}–6.05 × 10^{-3} C–1.08469 × $10^{-4.}$ * L–7.49 × 10^{-3} * C^2 + 3.20978 × 10^{-6} * L^2 –3.94778 × 10^{-5} * C* L	m_1 = 2.918 × 10^{-2}–1.791 × 10^{-2} * C–2.94713 × 10^{-4} * L + 2.495 × 10^{-2} * C^2 + 6.91643 × 10^{-6} * L^2 + 1.03341 × 10^{-4} * C * L
n = 5.2595–7.80631 * C–7.979 × 10^{-2} * L + 7.16081 * C^2 + 1.63 × 10^{-3} * L^2 + 4.1 × 10^{-2} * C * L	n_1 = 4.888–7.762 * C–4.83 × 10^{-3} * L – 5.243 * C^2 + 1.248 × 10^{-4} * L^2 –4.93 × 10^{-3} * C * L

5. Results and Discussion

5.1. Weibull Parameters

Figure 12 shows the results of the scale parameters and shape parameters, m, n, m_1, and n_1, at different stages according to the fiber content and fiber length. When the specimen is in the compression stage, m and n both increase and then decrease with an increase in the fiber content, while they first decrease and then increase with an increase in the fiber length. When the specimen is in the damage stage, m_1 decreases and then increases with an increase in the fiber content and fiber length, and n_1 decreases with the increase in the fiber content, while the effect of the fiber length is insignificant.

Early research viewed the scale and shape parameters primarily as parameters that influence and determine the variations in their model curves, but their physical significance was not fully understood. Some studies reported that the shape parameters n and n_1 reflect the degree of concentration of the internal microelement strength [39]. According to the analysis reported in this paper, n and n_1 are characterized by the brittleness of BFRCS. In the process of compressive deformation, the smaller the values of n and n_1 are the more concentrated the microelemental strength of the material will be. The material will be more brittle, and the arc of the curve between the compression and destruction stages of the loading process, on the one hand, and between the obvious compression and destruction stages, on the other hand, will be larger. The parameters m and m_1 represent the material's strength at a particular level.

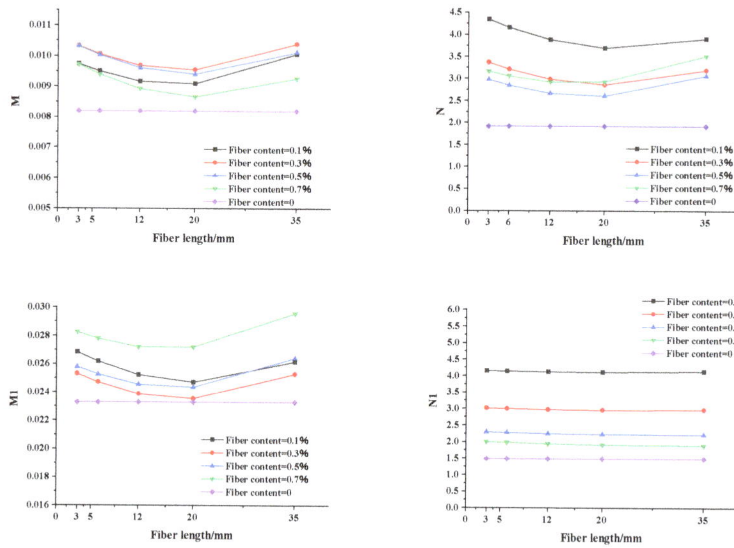

Figure 12. Varying trends in the parameters m, n, m_1, and n_1.

The results of the analysis indicate that the addition of basalt fibers can effectively reduce the stress concentration of the material, which improves the material's brittleness and enhances its average strength, its plastic deformation capacity under an external load, and its bearing capacity after the peak stress. These improvements are influenced by the coupling of the fiber length and fiber content. It can be seen from the relationship between Weibull's shape parameters, n and n_1, and the fiber content as well as length that the main factors affecting the brittleness of BFRCS in the compression stage are the fiber content and fiber length. Moreover, the main factor affecting the brittleness of BFRCS in the damage stage is the fiber content, while the influence of the fiber length is not evident; the size parameters m and m_1 are jointly influenced by the fiber content and fiber length. As per the results of the overall analysis, the dimensional and shape parameters are excellent at a content of 0.1%, indicating the superior overall performance and confirming the conclusion of the previous analysis that 0.1% is the optimal content.

5.2. Stress–Strain Relationship Based on Statistical Damage

By substituting C and L into Table 6 and inserting m, n, m_1, and n_1 into Equation (16), the simulated constitutive relationships between different types of BFRCS with various fiber contents and lengths can be obtained (Figure 13). In Figure 13 the black lines represent the measured compressive stress-strain curves, while the red lines represent their respective theoretical curves for the same fiber content and length. The theoretical curve based on the presented damage model is similar to the test results, indicating that the Weibull distribution is a sound indicator of BFRCS strength and that the presented damage model is applicable to BFRCS.

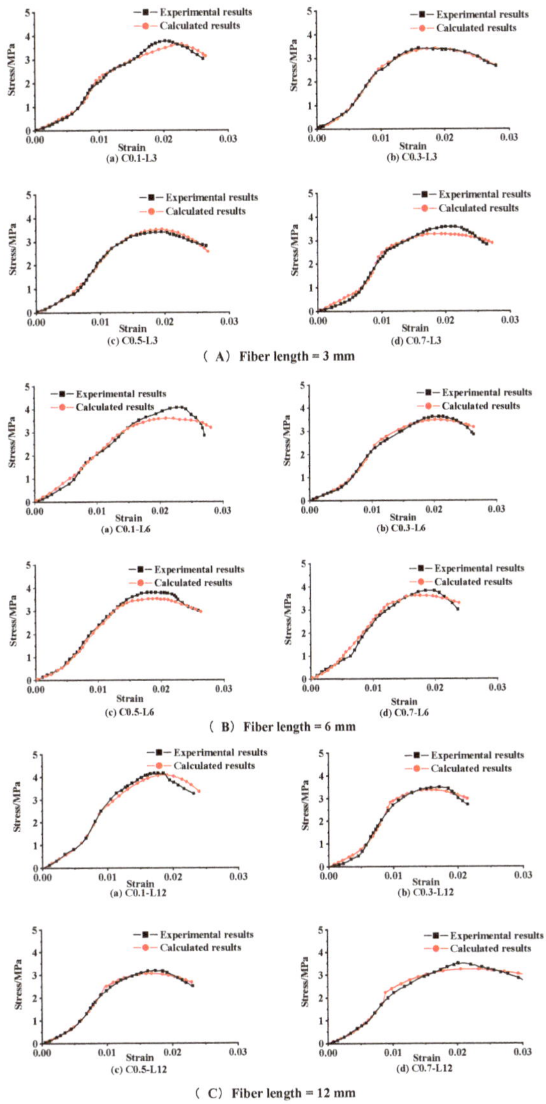

Figure 13. *Cont.*

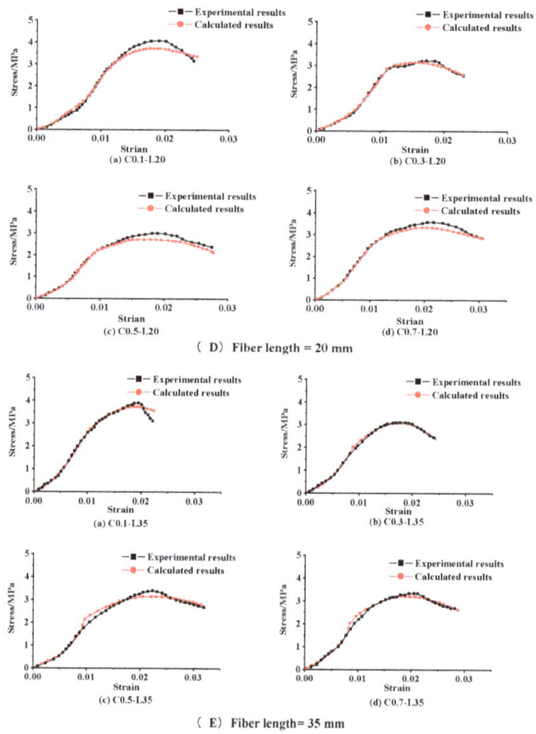

Figure 13. Comparison of the calculation results of the proposed constitutive models with the experimental results of stress–strain curves of the fiber-reinforced cemented soil.

5.3. Constitutive Model Validation

In order to verify the rationality of the constitutive model, we performed BFRCS curing for 14 days, where the fiber content was 0.1%, the fiber lengths were 3, 6, 12, and 20 mm, and the cement content of BFRCS was 10%. For instance, C0.1-L3-14d indicates that the fiber content is 0.1%, the fiber length is 3 mm, and the curing time is 14 d. The related parameters are shown in Table 7.

Table 7. Related parameters of the BFRCS subjected to curing for 14 d.

Related Parameters	E	D_0	n	m	m_1	n_1
C0.1-L3-14d	202	0.61	4.3380771	0.008857147	0.026847	4.036524
C0.1-L6-14d	210	0.57	4.1550171	0.008606567	0.026181	4.023925
C0.1-L12-14d	214	0.6	3.8769171	0.008604154	0.025221	4.005465
C0.1-L20-14d	220	0.59	3.6886771	0.008743518	0.024714	3.99483

The curing of BFRCS for 14 days was tested in the same way as the curing of BFRCS for 28 days. By substituting C and L into Table 6 and inserting m, n, m_1, and n_1 into Equation (16), we can obtain the simulated constitutive relationships between different types of BFRCS with various fiber contents and lengths subjected to curing for 14 days (Figure 14). In Figure 14 the black lines represent the measured compressive stress-strain curves and the red lines represent the stress-strain curves calculated by the constitutive model.

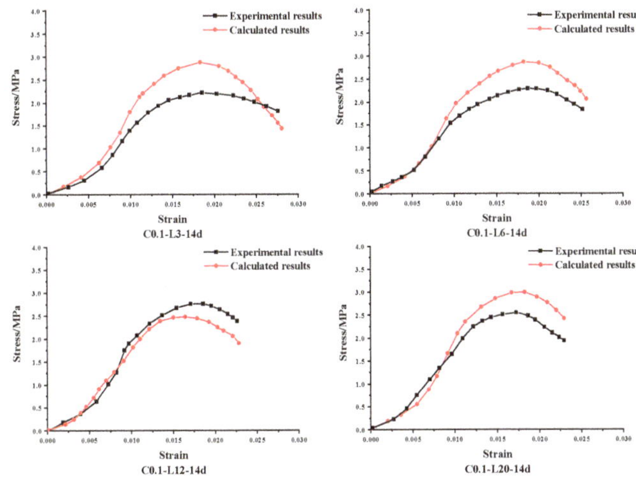

Figure 14. Comparison of the calculation results obtained by the constitutive model with the experimental results of stress-strain curves of BFRCS subjected to curing for 14 days.

According to Figure 14, it can be seen that the theoretical stress-strain curves calculated by the constitutive model are similar to those of the stress-strain curves of BFRCS. The theoretical stress-strain curves calculated by the constitutive model reflect the trend in the stress-strain curves of BFRCS subjected to curing for 14 days. They also reflect the strain corresponding to the peak stress, but the stress-strain curves calculated by the constitutive model are not very effective in predicting the magnitude of the peak stress.

6. Conclusions

The addition of basalt fibers can effectively improve the peak strength and damage strain of cemented soil. When the fiber content is 0.1% the peak strength and breaking strain of BFRCS are excellent. In this study, the peak strength reached its maximum at a fiber content of 0.1% and a fiber length of 12 mm, while the damaged strain reached its maximum at a fiber content of 0.1% and a fiber length of 6 mm.

The variation in the basalt fiber content of fiber-reinforced cemented soil has a greater influence on the UCS than fiber length. This is primarily due to the greater influence of fiber content on the grip force between the fibers and cement in the soil. Furthermore, in terms of damage strain, the influence of fiber content and length is not evident when considered as a single factor, primarily due to the toughness of fiber-reinforced cemented soil. This, in turn, is mainly because the toughness of fiber-reinforced cemented soil results from the coupling of these two factors.

A constitutive model based on the Weibull distribution, which considers the fiber length and fiber content, was established, and the determination method of the model parameters was provided. Through example verification, we determined that the model calculation results presented in this paper are in strong agreement with the experimental results and accurately reflect the influences of the fiber length and fiber content on the inherent behavior of fiber-reinforced cemented soil. The fiber content has a greater effect on the material's brittleness than the fiber length, but the effects of the fiber content and fiber length on the material's strength are more significant.

Author Contributions: Formal analysis, R.Z.; investigation, R.Z. and C.Q.; data curation, L.N.; writing—original draft preparation, R.Z.; writing—review and editing, L.X.; supervision, L.X. All authors have read and agreed to the published version of the manuscript.

Funding: This research was funded by the National Natural Science Foundation of China (grant number: 52008185) and Jilin Provincial Science and Technology Department (grant number: 20220203063SF).

Institutional Review Board Statement: Not applicable.

Informed Consent Statement: Not applicable.

Data Availability Statement: The data are contained within the article.

Conflicts of Interest: The authors declare no conflict of interest.

References

1. Ayeldeen, M.; Kitazume, M. Using fiber and liquid polymer to improve the behaviour of cement-stabilized soft clay. *Geotext. Geomembr.* **2017**, *45*, 592–602. [CrossRef]
2. Boz, A.; Sezer, A.; Özdemir, T.; Hızal, G.E.; Azdeniz Dolmacı, Ö. Mechanical properties of lime-treated clay reinforced with different types of randomly distributed fibers. *Arabian J. Geosci.* **2018**, *11*, 122. [CrossRef]
3. Baldovino, J.d.J.A.; Moreira, E.B.; Carazzai, É.; Rocha, E.V.d.G.; dos Santos Izzo, R.; Mazer, W.; Rose, J.L. Equations controlling the strength of sedimentary silty soil-cement blends: Influence of voids/cement ratio and types of cement. *Int. J. Geotech. Eng.* **2019**, *15*, 359–372. [CrossRef]
4. Chen, D.; Liao, Y.; Jiang, C.; Feng, X. The mechanical properties of coastal soil treated with cement. *J. Wuhan Univ. Technol.* **2013**, *28*, 1155–1160. [CrossRef]
5. Xu, L.; Deng, H.; Niu, L.; Zheng, J.; Qian, Y. Experimental study on mechanical properties of fiber-reinforced cemented soil with the low temperature curing condition subjected to effect of salt and drying-wetting cycles. *J. Civil EE.* **2022**, *44*, 10–19. [CrossRef]
6. Xu, L.; Niu, L. Effect of polypropylene fiber on frost resistance of cemented soil. *Mater. Plast.* **2019**, *57*, 78–86. [CrossRef]
7. Liu, J.; Bai, Y.; Song, Z.; Wang, Y.; Chen, Z.; Wang, Q.; Kanungo, D.P.; Qian, W. Debi Prasanna Kanungo, and Wei Qian, Effect of basalt fiber on the strength properties of polymer reinforced sand. *Fibers Polym.* **2018**, *19*, 2372–2387. [CrossRef]
8. Zhang, X.; Gu, X.; Lv, J.; Zou, X. 3D numerical model to investigate the rheological properties of basalt fiber reinforced asphalt-like materials. *Constr. Build. Mater.* **2017**, *138*, 185–194. [CrossRef]
9. Zhang, X.; Gu, X.; Lv, J.; Zhu, Z.; Zou, X. Numerical analysis of the rheological behaviors of basalt fiber reinforced asphalt mortar using ABAQUS. *Constr. Build. Mater.* **2017**, *157*, 392–401. [CrossRef]
10. Shi, F.J. A study on structure and properties of basalt fiber. *Appl. Mech. Mater.* **2012**, *238*, 17–21. [CrossRef]
11. Yang, K.; Wei, S.; Adilehou, W.M.; Ho, H. Fiber-reinforced internally unstable soil against suffusion failure. *Constr. Build. Mater.* **2019**, *222*, 458–473. [CrossRef]
12. Otieno Owino, A.; Nahar, N.; Hossain, Z.; Tamaki, N. Dimensional influence of basalt fiber reinforcements on the consolidation behaviour of rice husk ash stabilized soils. *Constr. Build. Mater.* **2022**, *339*, 127686. [CrossRef]
13. Ma, Q.; Gao, C. Effect of basalt fiber on the dynamic mechanical properties of cement-soil in SHPB test. *J. Mater. Civ. Eng.* **2018**, *30*, 04018185. [CrossRef]
14. Cao, Z.; Ma, Q.; Wang, H. Effect of basalt fiber addition on static-dynamic mechanical behaviors and microstructure of stabilized soil compositing cement and fly ash. *Adv. Civ. Eng.* **2019**, *0401815*, 821453. [CrossRef]
15. Shen, Y.; Tang, Y.; Yin, J.; Li, M.; Wen, T. An experimental investigation on strength characteristics of fiber-reinforced clayey soil treated with lime or cement. *Constr. Build. Mater.* **2021**, *294*, 12357. [CrossRef]
16. Zhu, H.-H.; Zhang, C.-C.; Tang, C.-S.; Shi, B.; Wang, B.-J. Modeling the pullout behavior of short fibers in reinforced soil. *Geotext. Geomembr.* **2014**, *42*, 329–338. [CrossRef]
17. Zhu, C.; Wang, Z.; Ju, J.; Liu, J.; Wang, X. The research on compaction characteristics of cement soil. In Proceedings of the Fourth International Conference on Digital Manufacturing Automation, Qindao, China, 29–30 June 2013; Volume 2, pp. 1521–1524. [CrossRef]
18. Wang, D.; Wang, H.; Larsson, S.; Benzerzour, M.; Maherzi, W.; Amar, M. Effect of basalt fiber inclusion on the mechanical properties and microstructure of cement-solidified kaolinite. *Constr. Build. Mater.* **2020**, *241*, 118085. [CrossRef]
19. Xiao, Y.; He, X.; Evans, T.M.; Stuedlein, A.W.; Liu, H. Unconfined compressive and splitting tensile strength of basalt fiber-reinforced biocemented sand. *J. Geotech. Geoenviron. Eng.* **2019**, *145*, 04019048. [CrossRef]
20. Lv, X.; Zhou, H.; Liu, X.; Song, Y. Experimental study on the effect of basalt fiber on the shear behavior of cemented sand. *Environ. Earth Sci.* **2019**, *78*, 688. [CrossRef]
21. Dasaka, S.M.; Sumesh, K.S. Effect of coir fiber on the stress-strain behavior of a reconstituted fine-grained soil. *J. Nat. Fibers* **2011**, *8*, 189–204. [CrossRef]
22. Divya, P.V.; Viswanadham, B.V.S.; Gourc, J.P. Evaluation of tensile strength-strain characteristics of fiber-reinforced soil through laboratory tests. *J. Mater. Civ. Eng.* **2014**, *26*, 14–23. [CrossRef]
23. Gao, Z.; Lu, D.; Huang, M. Effective skeleton stress and void ratio for constitutive modeling of fiber-reinforced sand. *Acta Geotech.* **2020**, *15*, 2797–2811. [CrossRef]
24. Sivakumar Babu, G.L.; Chouksey, S.K. Model for analysis of fiber-reinforced clayey soil. *Geomech. Geoeng.* **2010**, *5*, 277–285. [CrossRef]

25. Salih, M.M.; Osofero, A.I.; Imbabi, M.S. Constitutive models for fibre reinforced soil bricks. *Constr. Build. Mater.* **2020**, *240*, 117806. [CrossRef]
26. Kanchi, G.M.; Neeraja, V.S.; Sivakumar Babu, G.L. Effect of anisotropy of fibers on the stress-strain response of fiber-reinforced soil. *Int. J. Geomech.* **2015**, *15*, 06014016. [CrossRef]
27. Wu, Z.; Xu, J.; Li, Y.; Wang, S. Disturbed state concept-based model for the uniaxial strain-softening behavior of fiber-reinforced soil. *Int. J. Geomech.* **2022**, *22*, 4022092. [CrossRef]
28. Specification for mix Proportion Design of Cement Soil. In *Chinese Standard JGJ-T233-2011*; Fujian Academy of Building Research: Fuzhou, China, 2011. (In Chinese)
29. Lemaitre, J. A continuous damage mechanics model for ductile fracture. *Trans. ASME, Ser. H.* **1985**, *107*, 83–89. [CrossRef]
30. Tian, H.; Ziegler, M.; Kempka, T. Physical and mechanical behavior of claystone exposed to temperatures up to 1000 °C. *Int. J. Rock Mech Min.* **2014**, *70*, 144–153. [CrossRef]
31. Dusan, K.; Silva, M.A.G. Statistical aspects of the continuous damage theory. *Int. J. Solids Struct.* **1982**, *18*, 551–562. [CrossRef]
32. Mcdowell, G.R.; Amon, A. The application of Weibull statistics to the fracture of soil particles. *J. Jpn. Geotech Soc.* **2000**, *40*, 133–141. [CrossRef]
33. Munkholm, L.; Perfect, E. Brittle fracture of soil aggregates. *Soil Sci. Soc. Am. J.* **2005**, *69*, 1565–1571. [CrossRef]
34. Frandsen, H.L. Weibull statistics effective area and volume in the ball-on-ring testing method. *Mech. Mater.* **2014**, *73*, 28–37. [CrossRef]
35. Ding, Y.; Wei, W.; Pan, B.; Huang, Q.; Liu, Z.; Liu, D. Statistical damage model for fiber-reinforced vegetation concrete substrate. *Chin. J. Geotech. Eng.* **2022**, *044*, 652–659.
36. Kachanov, M.; Montagut, E. Interaction of a crack with certain microcrack arrays. *Eng. Fract. Mech.* **1986**, *25*, 625–636. [CrossRef]
37. Lemaite, J. How to use damage mechanics. *Nucl. Eng. Des.* **1984**, *80*, 233–245. [CrossRef]
38. Tao, C.; He, H.; Yan, W.; Zhou, D. Compression damage constitutive model of hybrid fiber reinforced concrete and its experimental verification. *Constr. Build. Mater.* **2020**, *264*, 120026. [CrossRef]
39. Xie, S.; Lin, H.; Wang, Y.; Xiong, W.; Zhao, Y.; Du, S. A statistical damage constitutive model considering whole joint shear deformation. *Int. J. Damage Mech.* **2020**, *29*, 988–1008. [CrossRef]

buildings

MDPI

Article

Development of Non-Proprietary Ultra-High Performance Concrete Mixtures

Tawsif Mohammad Hasan [1], Levi Gilbert [2], Srinivas Allena [1,*], Josiah Owusu-Danquah [1] and Anthony Torres [3]

1 Department of Civil and Environmental Engineering, Cleveland State University, Cleveland, OH 44115, USA
2 Department of Civil and Environmental Engineering, Washington State University, Richland, WA 99164, USA
3 Department of Engineering Technology, Texas State University, San Marcos, TX 78666, USA
* Correspondence: s.allena@csuohio.edu

Abstract: The development of non-proprietary Ultra-High Performance Concrete (UHPC) is one way to reduce the initial cost of construction. However, workability is a major issue for which such mixtures are not practical in field conditions. Ultra-high performance cannot be achieved in field conditions if the concrete is not placed, finished, and compacted properly during placement. In this research, six UHPC mixtures were developed (three with steel fibers and three without fibers) using materials which are readily available on the local marketplace with water-to-cementitious materials ratios ranging between 0.17 to 0.30. The workability was determined using standard ASTM flow table apparatus, and specimens were prepared to determine compressive strength, splitting tensile strength, and permeable porosity. Flow table test exhibited flow values greater than 250 mm. Such high workability of the mixtures was achieved by optimizing the silica fume content and water reducing admixture dosage. These mixtures exhibited compressive strengths greater than 120 MPa and splitting tensile strengths greater than 5.10 MPa in both ambient and elevated curing temperatures. Results indicated that UHPC can be produced with a water-to-cementitious materials ratio as high as 0.30. Steel fibers helped to increase splitting tensile strength due to fiber-matrix interactions. Very low permeable porosity (1.7–16.7%) was observed which indicates superior durability due to the significant reduction of ingress of deleterious ions.

Keywords: UHPC; local materials; curing regimens; workability; mechanical strengths; permeable porosity

check for updates

Citation: Hasan, T.M.; Gilbert, L.; Allena, S.; Owusu-Danquah, J.; Torres, A. Development of Non-Proprietary Ultra-High Performance Concrete Mixtures. *Buildings* **2022**, *12*, 1865. https://doi.org/10.3390/buildings12111865

Academic Editor: Tomáš Dvorský

Received: 6 October 2022
Accepted: 1 November 2022
Published: 3 November 2022

1. Introduction

Ultra-high-performance concrete (UHPC) is a relatively new class of cementitious composite material that is attracting interest globally due to its superior mechanical properties (very high strength and ductility) and excellent durability properties. UHPC was first developed in the early 1990s [1], with compressive strengths greater than 150 MPa. Apart from high strength, discrete steel fibers in UHPC help to improve tensile ductility. Due to its dense microstructure, UHPC possesses superior durability properties including resistance to alkali silica reactivity (ASR), rapid freeze–thaw cycles, negligible to very low chloride penetration, and significantly lower permeable porosity. Therefore, this innovative material can be used in a wider variety of highway infrastructure applications including closure and connection pours, thin-bonded overlays on deteriorated bridge decks, bridge shear keys (as a grout material), and connections between deck panels, and seismic retrofit of bridge substructures. Although UHPC offers several benefits to performance and design, like any other advanced material, the issue of the initial cost is a concern that needs to be addressed in order to make this innovative material available for a wide variety of applications. Several pre-packaged UHPC products are available in the commercial market [2–4], which are expensive, consequently increasing the initial cost of construction. This is the main reason for not being able to fully utilize the benefits of UHPC. Therefore, the use of local materials to produce non-proprietary UHPC products should be explored.

UHPC is typically produced with portland cement, silica fume, fine sand (particle size range of 75–600 μm), quartz powder, high range water reducing admixture (HRWRA), accelerators, discrete steel fibers, and water. UHPC mixtures are developed using a very low water-cementitious materials (w/cm) ratio, typically less than 0.20. Silica fume help enhance the microstructure by filling up the voids with the secondary hydrates produced due to its reaction with calcium hydroxide to form secondary C-S-H. Quartz powder is used to increase the pozzolanic reaction. The HRWRA is added to improve the workability of the mixture, while the steel fibers are used to improve the tensile ductility. Each of these materials' selection plays an important role in developing UHPC that exhibits superior mechanical and durability properties.

UHPC's superior properties are mainly due to its dense microstructure. Dense microstructure of UHPC can be achieved by optimizing the particle packing density at a very low w/cm ratio (less than 0.20). However, achieving optimum packing density in field conditions depends on proper pumping, casting, molding, and consolidation [5,6]. Otherwise, entrapped air, fiber segregation, and clustering will occur, leading to continued porous structure [7,8]. As a result, mechanical and durability properties will be negatively affected.

Developing workable UHPC mixtures is very important to achieve better consolidation and pumpability. Most studies on UHPC have focused on achieving and assessing mechanical properties and durability, and limited studies have elucidated the workability aspects of UHPC. Difficulties in the handling UHPC have been recognized as a major problem during construction because a significant workability reduction occurs due to a very low w/cm ratio and very high dosages of fine powders. Ultra-high performance could not be achieved if the required workability of UHPC does not meet. Therefore, in this research, an attempt has been made to develop a highly workable UHPC without compromising its mechanical strength.

The workability of UHPC depends on particle packing density, the specific surface area of constituent materials, chemical admixtures, absorption or release of water during the cement hydration process and temperature of the mixture [9]. So, the choice of material constituents can influence many parameters of UHPC, including strength, shrinkage, durability, etc. Choice of materials includes the cement type, supplementary cementitious material (SCM), aggregate type and size, type of fibers, and type of admixtures. Each of these factors can significantly change the fresh and hardened properties of UHPC. For example, short straight fibers produce a higher flow value than long straight and hooked fibers [10]. Again, an increase in fiber content decreases the workability while increasing the air content and permeable porosity [11]. At the same time, both compressive and flexural strength increases with the increase in fiber content in the case of straight fibers at 1.5% by volume fiber content [11]. However, hooked fibers showed 17–50% improved bending strength for the same fiber content compared to straight fiber counterparts [10,12].

HRWRAs are polymers or copolymers with long chains [13] and help to reduce the w/cm ratio as low as 0.11 [1]. Cementitious particles absorb the HRWRA molecules and create steric and electrostatic forces [14,15]. Therefore, they repel each other, and water is released due to dispersion. As a result, flowability increases with increasing HRWRA dosage [13,16]. However, the type of HRWRA can affect the workability as their dispersion capacity varies with the chemical composition [17]. Therefore, the choice of HRWRA can depend on SCM used in the UHPC mixture.

Researchers use different types of SCMs to identify their suitability to produce UHPC. Since the specific surface area of different SCMs varies, their water absorption capacity and the resultant workability change. Nano-metakaolin reduces flowability due to higher specific surface area, for which the water absorption rate is high [18,19]. For a similar reason, rice husk ash and blast furnace slag should not be used at more than 40% replacement level [20,21]. Nevertheless, fly ash was found to increase workability due to its spherical shape, which creates a ball-bearing effect between particles [22]. However, the effect of silica fume (SF) on workability is not straight forward as other SCMs. The workability of the mixtures was improved due to the addition of silica fume [1,23]. However, other researchers reported that workability was reduced due to the addition of silica fume. Due to the large specific surface area of SF particles, water demand increases in UHPC mixtures [24]. Therefore, high SF content increases viscosity and requires a high HRWRA dosage to attain target workability [25]. Replacing 10–20% cement with SF was an optimum replacement level without negatively affecting the flowability [20,26]. However, Duval and Kadri [26] proposed a 10% optimum replacement level to maximize workability and strength. Type of SF used by these researchers may be one reason for this variation. Abdulkareem et al. [17] demonstrated that white SF outperformed grey SF in terms of workability. However, white SF underperformed in terms of compressive strength due to lower specific surface area, which reduces the rate of pozzolanic reaction. Carbon content in SF may also affect workability [27]. Nevertheless, the effect of SF on workability is still unclear [12].

The combined action of HRWRA and SF helped disperse flocculated cement particles, making more water available for lubrication [26]. At a 10–20% replacement level, spherical fine SF particles can provide a lubricating effect for which less HRWRA dosage is required [26]. After 20%, a higher specific surface area may subdue the lubricant effect of spherical particles, which increases the HRWRA requirement [20]. However, Duval and Kadri [26] showed that target workability could be achieved by minimum HRWRA dosage at a 10% SF replacement level. A significant amount of superplasticizer was adsorbed on the SF particles if the SF replacement level is high [28]. Liu et al. [13] found that at a constant w/cm ratio, HRWRA has a maximum adsorption capacity. When maximum capacity is reached, additional HRWRA remains unadsorbed which may agglomerate and increase the viscosity. Therefore, dispersive capacity and water reducing effect diminish with increased HRWRA dosage. This effect is more pronounced at a very low w/cm ratio (less than 0.20) with a high HRWRA dosage. However, the research was focused on attaining target workability by changing HRWRA dosage at a fixed SF content. Therefore, HRWRA dosage as high as 2% and more by mass of total cementitious material content were used by Richard & Cheyrezy [1], Shi et al. [21], Duval & Kadri [26], Aldahdooh et al. [27] at a very low w/cm ratio.

In this research, cost-effective, sustainable, and workable UHPC mixtures were developed using constituent materials that are local to northeast Ohio. This work demonstrated that a very low w/cm ratio (less than 0.20) is not needed to develop UHPC with comparable mechanical performance but can be developed using a w/cm as high as 0.30. The materials used were Type I/II portland cement, silica fume, fine sand, steel fibers that were 13 mm long (aspect ratio of 65), and HRWRA. Apart from the traditional UHPC containing steel fibers, mixtures without fibers were also studied to produce a cost-effective option for applications that do not require high tensile ductility. Flow test was performed to identify the workability of UHPC in fresh state while compressive strength and splitting tensile strength tests were performed in the hardened state to determine the mechanical properties. Permeable porosity was also determined to evaluate physical properties. The test results showed that the presented mixtures conformed to the UHPC class concrete as per ASTM C1856 [29].

2. Materials and Experimental Methods

2.1. Materials

Type I/II ordinary portland cement with a specific gravity of 3.15 was used in this study. Commercially available silica fume with a specific gravity of 2.24 was used as a mineral admixture. The chemical and physical properties of Type I/II cement and silica fume used in this study are presented in Table 1.

Table 1. Chemical compositions and physical properties of cementitious materials.

Properties	Chemical Properties (%)								Physical Properties	
	SiO_2	Al_2O_3	Fe_2O_3	CaO	MgO	SO_3	Na_2O	K_2O	Relative Density	Moisture Content (%)
Cement	22.3	3.40	3.10	64.30	2.30	2.50	0.22	0.47	3.15	-
Silica fume	96.90	0.20	0.20	0.30	0.20	0.10	0.20	0.30	2.24	0.04

Local sand conforming to ASTM C33 [30] was used in this work. The sand was first sieved through ASTM No. 30 sieve and then washed over ASTM No. 200 sieve to achieve particle size between 75–600 μm. After washing, the sand was kept in an oven at 105 °C to achieve 0% moisture content. Specific gravity and absorption values of the sand were determined according to ASTM C128 [31] and were found to be 2.78 and 2.35%, respectively. The particle size distribution of fine sand is shown in Figure 1, which is determined as per ASTM C136 [32].

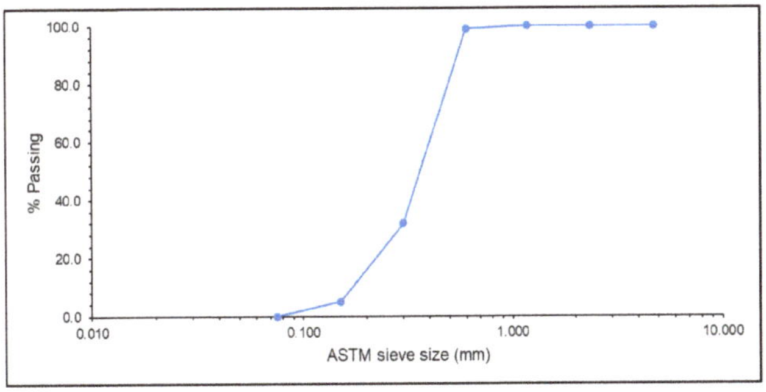

Figure 1. Particle size distribution of fine aggregate.

Straight steel fibers (1.5% by volume) with a length of 13 mm and an aspect ratio of 65 were used to provide ductility. Polycarboxylate-based HRWRA (MasterGlenium 3030 from Master Builders Solutions) with an ability to reduce water demand by 25–30% was used to achieve workability".

2.2. Mixture Proportioning of UHPC

Table 2 shows the mixture proportions, w/cm ratio, water-to-cement ratio (w/c), silica fume-to total cementitious materials (sf/cm) ratio, sand to-cementitious materials (s/cm) ratio, and HRWRA/cementitious materials ratio for the final six mixtures that were obtained by modifying the mixture proportions proposed by Allena and Newtson [33]. These mixtures were categorized into three groups: A, B, and C, based on silica fume content and w/cm ratio.

Category A mixtures had the greatest silica fume content (20% by mass of total cementitious materials) with a 0.30 w/cm ratio. Mixture categories B and C were developed by modifying silica fume content and w/cm ratio used in the category A mixture. The w/cm ratio was reduced to 0.20 for category B mixtures which was further reduced to 0.17 for Category C mixtures. Silica fume content was also reduced to 16.7% and to 16.0% by mass of total cementitious material content for the corresponding mixtures to maintain workability.

2.3. Mixing Procedure, and Workability

A vertical shaft mixer with a 38 rpm paddle speed was used to mix the constituents of UHPC. Initially, sand and cementitious materials were dry mixed. After dry mixing for 2 min, 75% of the total water was added to the mixer. After thorough mixing, HRWRA, followed by the remaining 25% of water was added, and mixing was continued. Steel fibers were added to the mix at the end. The total mixing time varied from 15–20 min. The mixing sequence is illustrated in Figure 2. After mixing, the workability of freshly mixed UHPC was measured by conducting a flow table test as per ASTM C1437 [34].

Table 2. Mixture proportions of formulated UHPC mixtures.

Mixture Category	A		B		C	
Mixture Type	**Plain**	**Fiber Reinforced**	**Plain**	**Fiber Reinforced**	**Plain**	**Fiber Reinforced**
Mixture Designation	AP	AF	BP	BF	CP	CF
Cement (kg/m^3)	890	890	890	890	890	890
Silica fume (kg/m^3)	223	223	178	178	169	169
Fine sand (kg/m^3)	592	549	982	939	1059	1016
Steel fiber (kg/m^3)	0	119	0	119	0	119
HRWRA (L/m^3)	50	50	50	50	59	59
Water (kg/m^3)	334	334	214	214	180	180
w/c	0.37	0.37	0.24	0.24	0.20	0.20
w/cm	0.30	0.30	0.20	0.20	0.17	0.17
sf/cm	0.20	0.20	0.17	0.17	0.16	0.16
s/cm	0.53	0.49	0.92	0.88	1.00	0.96
HRWRA/cm	0.047	0.047	0.048	0.048	0.059	0.059
Flow diameter (mm) Static	>250	230	>250	230	>250	230
Flow diameter (mm) Dynamic	>250	>250	>250	>250	>250	>250

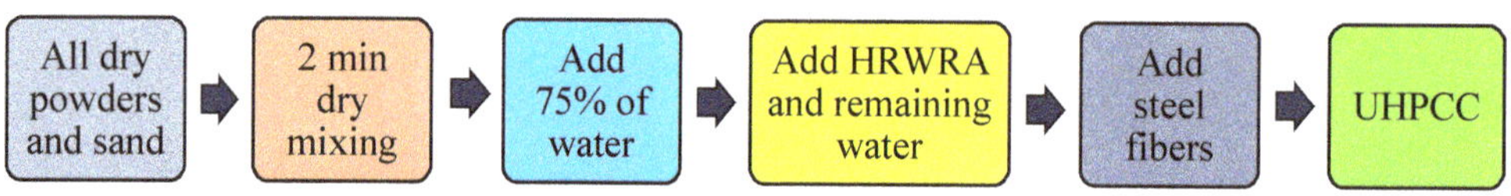

Figure 2. Mixing procedure for UHPC.

2.4. Specimen Preparation

50 mm cubes, 100 mm cubes, 100 by 200 mm cylinders, and discs of 50 mm thick and 100 mm diameter were produced from each batch. A high frequency vibration table was used for consolidation. The time of vibration was selected based on the consistency of the individual mixture that ranged between 30–90 s. Specimens were cast into molds and were covered with plastic sheets to avoid loss of moisture and kept at room temperature for 24 h. After 24 h, specimens were demolded and placed under specified curing regimens.

2.5. Curing Regimes

Four different curing regimens were investigated in this research to study the effect of curing conditions on UHPC properties. The details of these four curing regimens are shown in Table 3.

Table 3. Curing regimens.

Type	Designation	Specification
Standard curing	SC	Air cured in the molds for 24 h. After demolding, specimens were immersed under water until the day of testing.
Moist curing	MC	Air cured in the molds for 24 h. After demolding, the specimens were placed in a closed curing box at room temperature to maintain ~100% relative humidity until the day of testing. Water was filled to cover the bottom 50 mm depth of the curing box. A screen was used to place the specimens so that the specimens were not in direct contact with water.
Warm bath curing	WB	Air cured in the molds for 24 h. After demolding, specimens were cured in a water bath at 90 °C (194 °F) until the time of testing.
Heat curing	HC	Air cured in the molds for 24 h. After demolding, specimens were cured in a water bath at 90 °C (194 °F) until two days prior to testing and were then removed from water bath, and dry-cured at 90 °C (194 °F) in an oven for the last two days. This curing method is used to accelerate pozzolanic reaction of silica fume and to enhance the microstructure of UHPC.

2.6. Permeable Porosity

Water permeable porosity was determined by following the procedure reported in Yu et al. [35]. Disk specimens (100 mm diameter and 50 mm thickness) were prepared and cured for 7 and 28 days. Two disk specimens were tested on each day. Specimens were water saturated by submerging specimens under the water for 24 h. Water temperature was maintained at 23 ± 2 °C (73.5 ± 3.5 °F). After 24 h of saturation, the mass of the samples was measured underwater (m_w). Samples were then air-dried to achieve saturated surface dry (SSD) condition, and the mass of SSD samples (m_s) was recorded. Air dried samples were then kept in an oven at 90 °C (194 °F) temperature for 24 h to achieve the oven-dry condition. After 24 h of oven drying, the oven-dry mass (m_d) of the samples was measured. The water permeable porosity of each sample was then calculated using the following equation.

$$\varnothing = \frac{m_s - m_d}{m_s - m_w} \times 100 \tag{1}$$

2.7. Compressive Strength

A total of 144 cubes of 50 mm size and 96 cubes of 100 mm size were tested to evaluate the compressive strength of UHPC as per the procedure outlined in ASTM C109 [36] at 7 and 28 days.

2.8. Splitting Tensile Strength

A total of 96 cylinder specimens of size 100 by 200 mm were tested to evaluate the split tensile strength of UHPC as per ASTM C496 [37] at 7 and 28 days.

3. Results and Discussion

3.1. Workability

All the mixtures exhibited flow diameters larger than 200 mm as specified in ASTM C1856 [29] (Table 2). The plain mixtures exhibited flows greater than 250 mm (overflowing the flow table, Figure 3a) in both static (without jolting) and dynamic (with jolting) conditions. However, fiber-reinforced mixtures overflowed the flow table in dynamic conditions only. The shape of fibers increases the friction to achieve improved anchorage between fiber and matrix [38], resulting in reduced flow of the mixture.

(a)

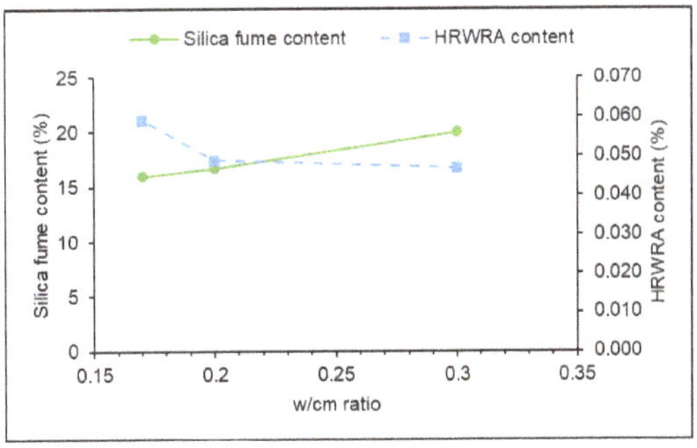

(b)

Figure 3. Workability of UHPC mixtures; (**a**) flow table test, and (**b**) effect of silica fume content and HRWRA dosage on workability.

The workability of UHPC is affected by silica fume content and HRWRA dosage, as can be seen from Figure 3b. Figure 3b showed that to produce the same workability in all mixtures, silica fume content needs to be reduced with increased HRWRA dosage for low w/cm ratio mixtures.

3.2. Permeable Porosity

Permeable porosity values for all UHPC mixtures are presented in Table 4 and Figure 4. UHPC mixtures exhibited very low porosity values. These values ranged from 1.7 to 21.8% after seven days of curing, while this range was between 1.7 and 16.7% after 28 days of curing. Here, porosity increased with the w/cm ratio. These values are consistent with the literature [39,40]. It has been observed that the curing period, curing temperature, fiber dosage, and silica fume content affect permeable porosity.

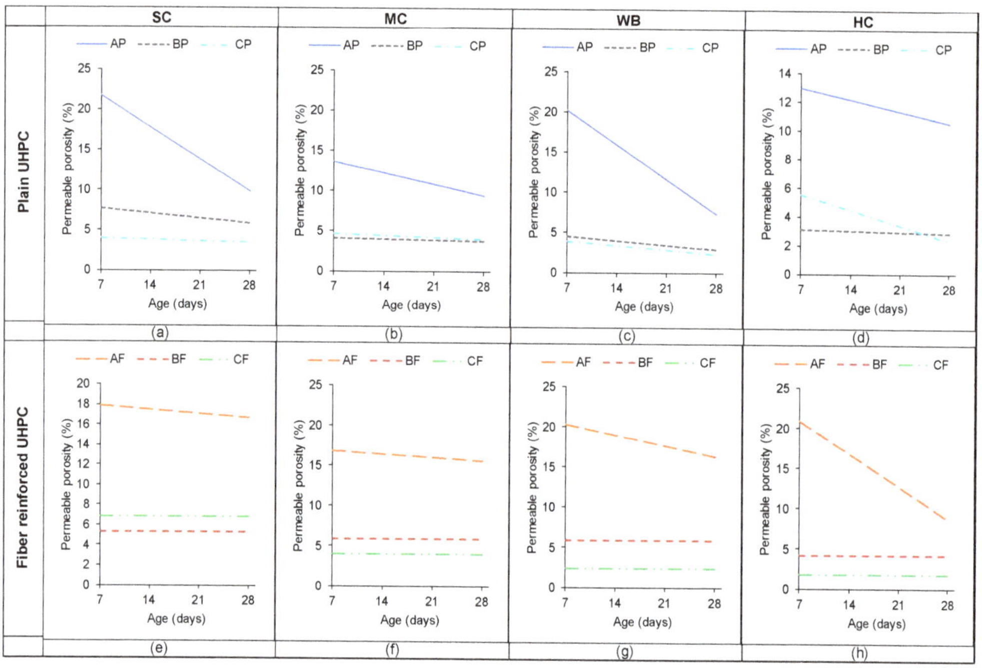

Figure 4. Permeable porosity of plain UHPC mixtures cured under (**a**) SC, (**b**) MC, (**c**) WB, and (**d**) HC regimens and permeable porosity of fiber reinforced mixtures cured under (**e**) SC, (**f**) MC, (**g**) WB, and (**h**) HC regimens.

Figure 4 shows the effectiveness of a longer curing period in reducing permeable porosity. After 28 days of curing, the permeable porosity of AP and AF mixtures was reduced by 45.9% and 24.4%, respectively, compared to permeable porosity after seven days of curing. The reduction was 22.1% and 33.1% for BP and CP mixtures, respectively. However, the permeable porosity of BF and CF mixtures did not change with an extended curing period.

Table 4. Mechanical and Physical Properties of UHPC Mixtures.

Mixture	w/cm Ratio	Curing		Permeable Porosity (%)		Compressive Strength (MPa)				Split Tensile Strength (MPa)	
						50 mm Cube		100 mm Cube			
				7 Day	28 Day	7 Day	28 Day	7 Day	28 Day	7 Day	28 Day
AP	0.3	SC	mean	21.8	9.9	79.7	88.3	43.0	54.3	3.85	7.60
			std.dev	2.2	1.3	0.7	0.9	3.5	2.5	0.70	0.30
		MC	mean	13.7	9.4	91.9	99.3	51.4	54.3	5.15	6.95
			std.dev	0.7	1.1	3.7	1.0	3.7	1.1	0.80	0.80
		WB	mean	20.3	7.4	108.3	113.0	91.9	103.8	4.80	5.10
			std.dev	1.4	0.4	2.6	2.5	3.8	0.4	1.40	0.50
		HC	mean	13.0	10.5	98.0	127.0	83.5	96.8	5.55	3.80
			std.dev	1.1	0.9	0.2	5.9	5.3	4.8	0.20	0.30
AF	0.3	SC	mean	17.9	16.7	100.5	113.9	50.9	83.4	6.65	8.85
			std.dev	1.6	0.4	3.5	1.5	5.5	1.7	1.60	0.20
		MC	mean	16.8	15.6	63.7	94.3	55.9	65.2	7.85	8.60
			std.dev	0.2	0.2	9.7	1.7	0.1	0.1	1.20	1.70
		WB	mean	20.3	16.4	43.7	138.0	69.4	73.9	7.60	9.75
			std.dev	0.2	0.6	2.1	0.1	0.9	2.1	0.20	2.50
		HC	mean	20.9	8.7	112.6	123.9	82.8	88.1	9.20	10.45
			std.dev	0.6	0.1	0.4	2.8	2.1	1.9	0.60	4.30
BP	0.2	SC	mean	7.7	5.9	96.1	103.7	41.5	47.7	7.15	8.25
			std.dev	0.3	0.1	1.7	2.0	0.6	0.6	0.20	0.20
		MC	mean	4.1	3.7	100.5	128.2	49.2	90.8	7.35	8.35
			std.dev	0.2	0.2	2.0	2.5	8.5	0.4	0.50	0.20
		WB	mean	4.5	2.9	125.1	138.6	122.1	129.4	8.50	8.75
			std.dev	0.4	0.2	4.4	1.7	3.2	1.8	0.80	0.30
		HC	mean	3.1	2.8	124.2	126.5	89.5	103.8	3.70	6.75
			std.dev	0.6	0.1	0.3	1.3	22.7	2.5	1.10	1.20
BF	0.2	SC	mean	5.3	5.3	116.6	128.9	75.4	105.9	12.15	14.35
			std.dev	0.2	0.3	3.3	2.9	1.7	2.1	0.90	0.40
		MC	mean	5.9	5.9	119.4	131.7	81.4	98.1	12.10	14.70
			std.dev	0.2	0.2	2.6	1.0	0.9	0.7	0.20	0.70
		WB	mean	5.9	5.9	127.0	138.0	106.2	115.9	11.35	11.40
			std.dev	0.3	0.1	2.6	0.7	4.1	0.7	0.10	0.70
		HC	mean	4.2	4.2	128.6	136.9	100.1	115.7	13.60	13.40
			std.dev	0.1	0.1	2.6	2.6	5.4	8.8	0.70	1.30
CP	0.17	SC	mean	4.0	3.5	84.7	92.7	78.1	95.0	8.05	9.30
			std.dev	0.4	0.5	0.3	2.5	2.7	0.5	0.70	0.30
		MC	mean	4.7	4.0	121.2	131.4	79.8	89.5	6.45	8.75
			std.dev	0.2	0.1	0.8	3.0	0.6	1.3	0.40	0.30
		WB	mean	3.9	2.3	114.0	130.6	58.0	127.8	7.70	7.95
			std.dev	0.2	0.3	4.6	7.8	0.7	13.4	0.70	0.20
		HC	mean	5.5	2.3	115.4	134.0	87.3	100.5	7.10	7.65
			std.dev	0.1	0.5	2.3	5.0	0.4	0.9	0.70	0.10
CF	0.17	SC	mean	6.8	6.8	121.7	136.9	81.7	103.2	11.70	13.25
			std.dev	0.5	0.2	1.7	0.2	1.0	4.4	0.60	0.80
		MC	mean	4.0	4.0	116.9	118.3	68.9	88.6	12.60	13.55
			std.dev	0.2	0.2	3.4	3.3	1.6	2.5	0.90	1.90
		WB	mean	2.4	2.4	137.7	138.1	125.2	131.9	13.95	14.20
			std.dev	0.1	0.2	3.6	3.8	2.1	3.1	2.00	0.10
		HC	mean	1.7	1.7	142.2	147.2	138.0	137.7	13.15	14.60
			std.dev	0.1	0.2	2.2	4.4	5.9	5.8	3.50	0.70

Std.dev = standard deviation.

3.2.1. Effect of Curing Regimen on Permeable Porosity

The specimens cured under MC, WB, and HC curing regimens exhibited similar permeable porosity values and were lower than those of the SC curing regimen (Figure 5). However, the permeable porosity values exhibited by specimens cured at elevated temperatures (WB and HC) were the lowest compared to those cured under SC and MC curing regimens. One possible reason can be the greater amount of secondary calcium-silicate-hydrate (C-S-H) gel produced due to the accelerated pozzolanic reaction of silica fume at elevated temperature [41]. However, the MC curing regimen performed better than SC and produced almost similar permeable porosity as that of WB and HC. This can be attributed to the heat entrapped within the closed curing box (Table 3), and this heat might have increased the air temperature which might have eventually increased the hydration rate in specimens cured under MC curing regimen.

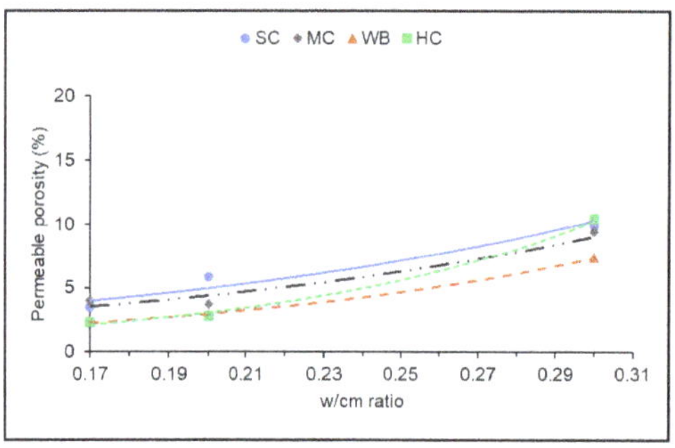

(**a**)

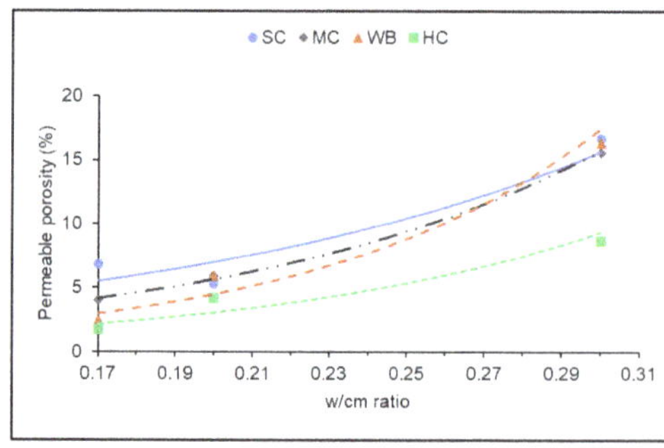

(**b**)

Figure 5. Effect of curing regimens on permeable porosity of (**a**) plain UHPC and (**b**) fiber reinforced UHPC.

3.2.2. Effect of Steel Fibers on Permeable Porosity

Figure 6 shows that the permeable porosity of fiber reinforced mixtures is greater than the corresponding plain mixtures. Yu et al. [11] reported a similar trend for UHPCs with different fiber content. The Addition of steel fibers in a matrix of uniformly graded fine aggregate can reduce the packing density, which, in return might have increased the permeable porosity of the fiber-reinforced mixtures compared to corresponding plain mixtures [42,43].

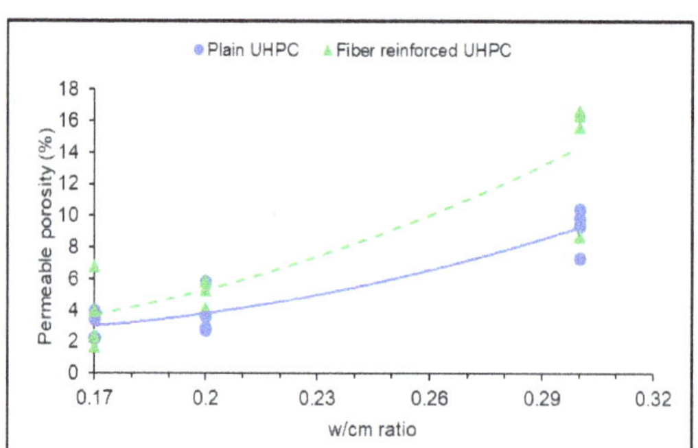

Figure 6. Permeable porosity vs. w/cm ratio for formulated UHPCs.

3.2.3. Effect of Silica Fume Content on Permeable Porosity

The extreme fineness of silica fume (21.82 m^2/gm) should help reduce the permeable porosity of concrete. However, in this study, a decrease in permeable porosity was observed with a decrease in silica fume content. Permeable porosities of plain and fiber reinforced mixtures were reduced by 67.5% and 74.0%, respectively, as the silica fume to cementitious materials (sf/cm) ratio decreased from 0.20 to 0.16 (Figure 7).

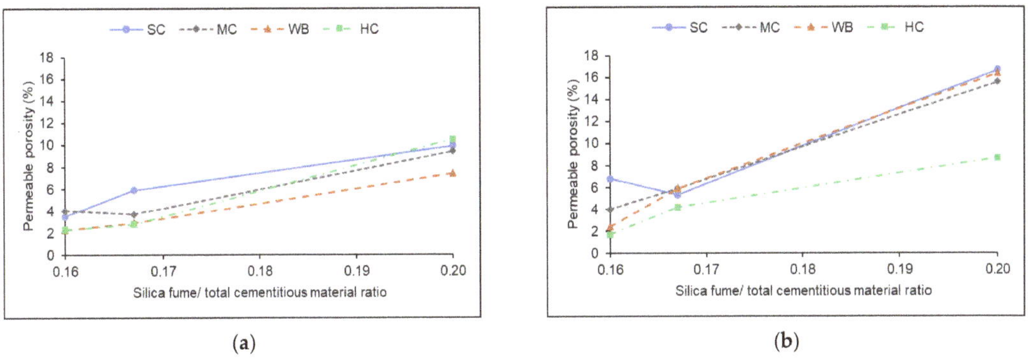

Figure 7. Effect of silica fume content on permeable porosity in (**a**) plain UHPC and (**b**) in fiber reinforced UHPC.

For category A mixtures, w/cm ratio and sf/cm were 0.30 and 0.20, respectively. Sf/cm was reduced to 0.167 in the case of 0.20 w/cm ratio mixtures (category B) and was further reduced to 0.16 for 0.17 w/cm ratio mixtures (category C) to maximize the flowability of UHPC. With the reduction in w/cm, the permeable porosity of the mixtures decreased, consequently; a smaller amount of silica fume seems to be effectively filled up the capillary pores at a very low w/cm ratio.

However, 0.20 sf/cm ratio may not be sufficient to fill up the capillary pores present in the UHPC mixture with a 0.30 w/cm ratio. UHPC mixtures produced using greater w/cm ratios, lower permeable porosity can be achieved by increasing the silica fume content.

3.3. Compressive Strength of UHPC

The Average compressive strengths of 50 mm and 100 mm cube specimens produced from all six mixtures that were cured under different curing regimens are depicted in Figures 8 and 9, respectively. These figures, show that 28 days of curing was required to achieve maximum compressive strength. WB curing regimen helped mixtures AF, BP, and BF to achieve compressive strengths of 138.0 MPa, 138.6 MPa, and 138.0 MPa, respectively. On the other hand, the greatest compressive strengths exhibited by mixtures AP, CP, and CF incurred under HC curing regimen were 127.0 MPa, 134.0 MPa, and 147.2 MPa, respectively. In similar studies, the compressive strength of non-proprietary UHPC mixtures was reported to be within the range of 120.0–190.0 MPa [44–47]. High volume of cementitious material content, use of silica fume to enhance secondary hydration reaction, improved packing density, and application of post-setting heat treatment can be attributed as the reasons for achieving ultra-high compressive strength [1].

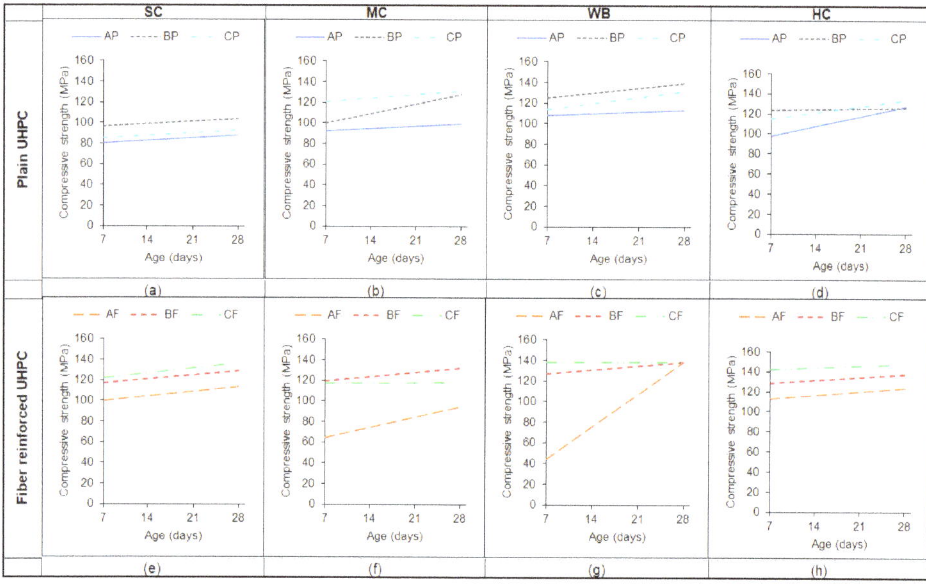

Figure 8. Compressive strengths of 50 mm cube specimens of plain UHPC mixtures when cured under (**a**) SC, (**b**) MC, (**c**) WB, and (**d**) HC regimens and compressive strengths of 50 mm cube specimens of fiber reinforced UHPC mixtures when cured under (**e**) SC, (**f**) MC, (**g**) WB, and (**h**) HC regimens.

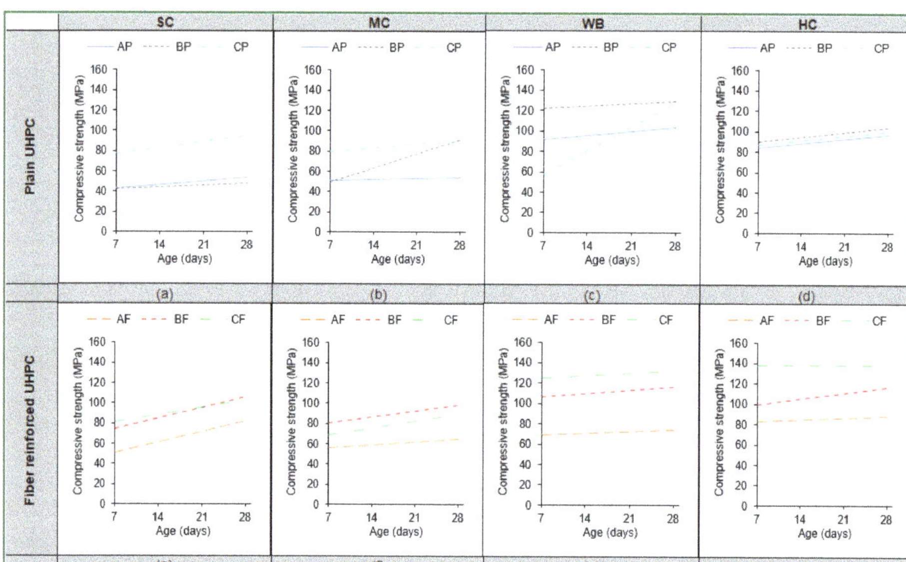

Figure 9. Compressive strengths of 100 mm cube specimens of plain UHPC mixtures when cured under (**a**) SC, (**b**) MC, (**c**) WB, and (**d**) HC regimens and compressive strengths of 100 mm cube specimens of fiber reinforced UHPC mixtures when cured under (**e**) SC, (**f**) MC, (**g**) WB, and (**h**) HC regimens.

3.3.1. Effect of Curing Regimen on Compressive Strength

Development of compressive strength was influenced by the curing regimen in both plain and fiber-reinforced UHPC mixtures (Figures 10 and 11).

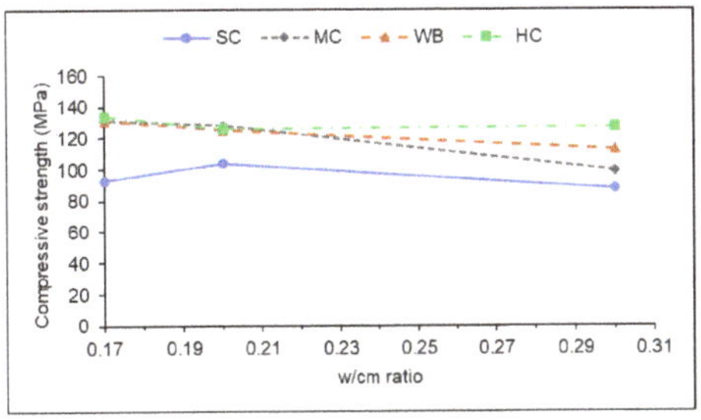

(a)

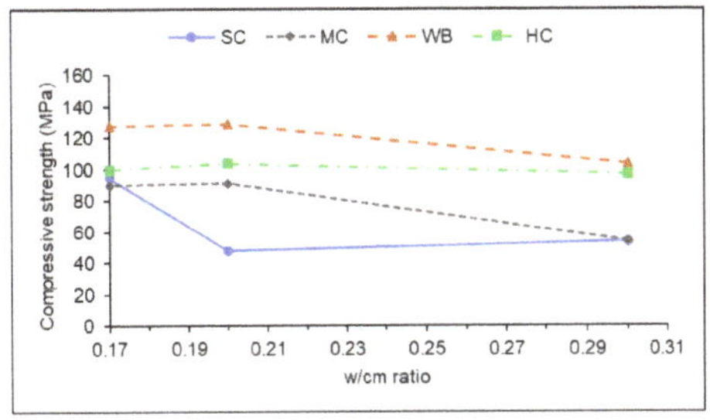

(b)

Figure 10. Effect of different curing regimens on compressive strength of plain UHPC mixtures (**a**) 50 mm cube specimen and (**b**) 100 mm cube specimens.

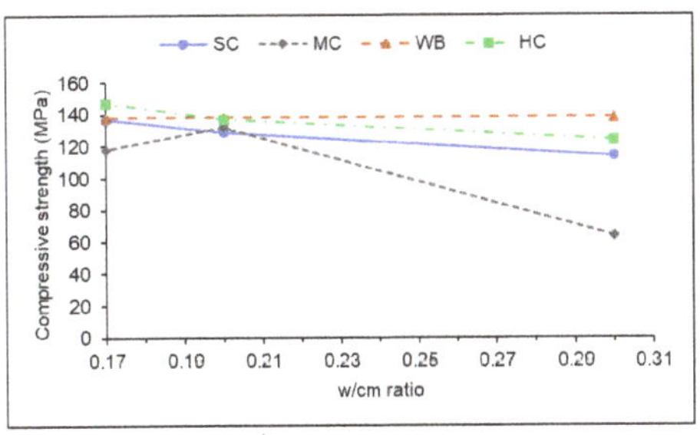

(a)

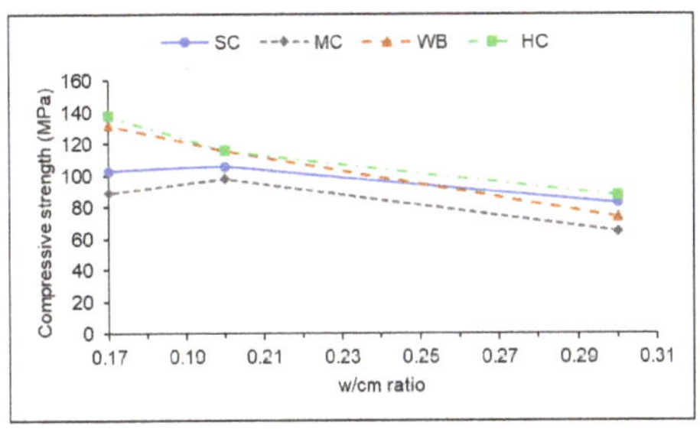

(b)

Figure 11. Effect of different curing regimen on compressive strength of fiber-reinforced UHPC mixtures; (**a**) 50 mm cube specimens and (**b**) 100 mm cube specimens.

Plain UHPC: As can be seen from Figure 10a, the SC curing regimen was not effective in producing compressive strength greater than 120 MPa in plain mixtures. Compared to the SC curing regimen, the MC curing regimen helped increase the 50 mm cube compressive strengths by 24% and 42% for BP and CP mixtures, respectively. 100 mm cube compressive strength, specimens of AP and CP mixtures, when cured under SC and MC conditions, exhibited similar strengths (Figure 10b). Between the two accelerated curing regimens, WB cured specimens exhibited greater strengths than those of HC cured specimens. When compared to the SC curing regimen, the HC regimen helped increase compressive strengths by 45%, 22%, and 44% in CP, BP, and AP mixtures, respectively, while these increases were 41%, 34%, and 28%, respectively for specimens cured under WB curing regimen.

Fiber-reinforced UHPC: In the case of fiber reinforced UHPC mixtures, the target compressive strength of 120 MPa was achieved by 50 mm cube specimens cured under any of the curing regimens, including the SC curing regimen (Figure 11a).

MC curing regimen produced similar strengths as SC curing regimen for BF mixture. However, the compressive strengths of AF and CF mixtures under the MC curing regimen were lower than that of the SC curing regimen by 14% and 17%, respectively. This trend persisted for 100 mm cube specimens (Figure 11b). Again, accelerated curing regimens HC and WB increased 50 mm cube compressive strengths by 7% and 10%, respectively, compared to the SC regimen. However, for 100 mm cube specimens, a 7% increase in compressive strength with the HC curing regimen was observed compared to WB curing regimen.

3.3.2. Effect of Steel Fibers on Compressive Strength

Figure 12 shows the evidence that steel fibers influenced the compressive strength of UHPC cured under the SC curing regimen. This increase is significant in the case of SC cured specimens where fiber reinforced mixtures AF, BF, and CF exhibited 48%, 24%, and 29% greater compressive strengths compared to plain mixtures AP, BP, and CP, respectively. Similarly, 9% and 5% greater compressive strengths were observed for WB and HC curing regimens, respectively. The higher influence factor of steel fibers on compressive strength of SC cured specimens can be attributed to the lower strength gain in that curing regimen. Due to low strength, cracks formed and propagated at lower stress levels due to which the plain mixtures failed. However, the presence of steel fibers might have resisted crack propagation and delayed the failure for which the fiber reinforced mixtures exhibited higher compressive strengths when cured under SC curing regimen.

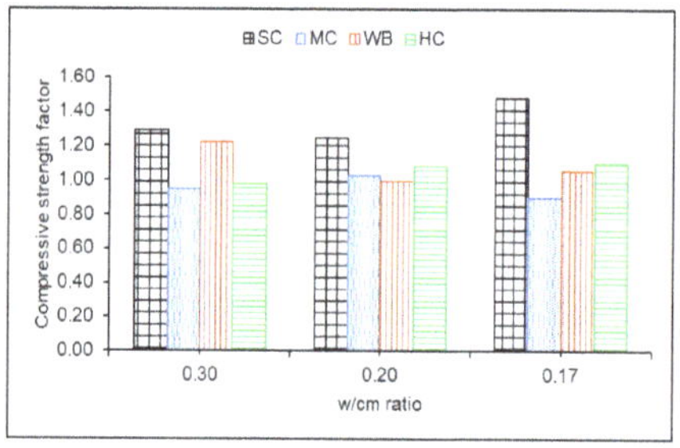

Figure 12. Ratio of compressive strength of fiber-reinforced UHPC mixtures compared to plain UHPC mixtures.

However, a different trend was observed for the fiber reinforced UHPC mixtures under the MC curing regimen. On average, fiber reinforced mixtures exhibited 4% less compressive strength than plain mixtures. The reason for the lower compressive strengths of fiber reinforced mixtures can be attributed to the greater permeable porosity observed under the MC curing regimen (Figure 4b,f). The MC cured fiber reinforced mixtures had 42% more permeable porosity than plain mixtures.

3.3.3. Specimen Size Effect on Compressive Strength

Compressive strength increases as the size of the specimen decrease because the larger volume is more likely to contain an element of low strength. Much research data is available on the effects of specimen size on the compressive strength of normal strength concrete. However, the data on this subject for UHPC mixtures is limited. Therefore, an attempt has been made in the present research to investigate the effect of specimen size on the compressive strength of UHPC. 50 mm and 100 mm cube compressive strengths at 7 and 28 days were compared to investigate the influence of specimen size on compressive strength of UHPC. Conversion factors for the compressive strength of 50 mm cubes produced from each curing regimen were calculated to relate them to the 100 mm cube specimens. These factors are presented in Table 5. The conversion factors obtained here were consistent with earlier research [48–50].

3.3.4. Relationship between Permeable Porosity and Compressive Strength

The compressive strengths of UHPC mixtures increased with the decrease in permeable porosity (Figure 13). Permeable porosity and total porosity have an indirect relationship. Since UHPC contains only fine sand, the porosity within the aggregates may be neglected. Therefore, porosity in UHPC primarily depends on capillary and gel pores. Again, a very low w/cm ratio of UHPC mixtures can result in the reduced size and volume of capillary pores. At the same time, silica fume fills up the micropores in the interfacial transition zone (ITZ) [45]. Silica fume also reacts with $Ca(OH)_2$ at ITZ to produce secondary

C-S-H gel that strengthens the bond strength of ITZ [45]. As a result, the reduction of porosity resulted in denser microstructure, and stronger ITZ increased the stress required to initiate bond cracking. Therefore, UHPC with lower permeable porosity exhibited greater compressive strength.

Table 5. Conversion factors for compressive strength from 100 mm cube strength to 50 mm cube strength at different curing regimen.

Fiber Content (by Volume)	w/cm Ratio	100 mm Cube	50 mm Cube at 7 Day				50 mm Cube at 28 Day			
			SC	MC	WB	HC	SC	MC	WB	HC
0%	0.30	1.00	1.86	1.79	1.18	1.17	1.63	1.83	1.09	1.31
	0.20	1.00	2.32	2.04	1.14	1.39	2.17	1.41	0.97	1.22
	0.17	1.00	1.08	1.52	1.97	1.32	0.98	1.47	1.02	1.33
1.5%	0.30	1.00	1.97	1.69	0.63	1.36	1.37	0.98	1.87	1.41
	0.20	1.00	1.55	1.47	1.20	1.28	1.22	1.34	1.19	1.18
	0.17	1.00	1.49	1.70	1.10	1.03	1.33	1.34	1.05	1.07

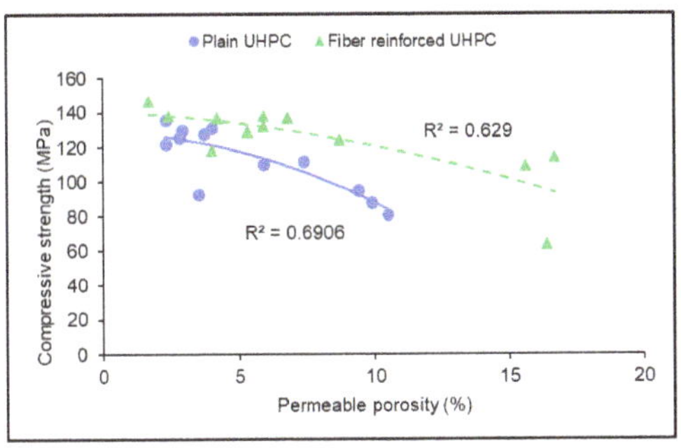

Figure 13. Compressive strength vs. permeable porosity for formulated UHPC mixtures after 28 days of curing.

Figure 13 also shows that fiber reinforced UHPC exhibited greater compressive strength than plain UHPC mixtures, even at higher permeable porosity. The steel fibers might have prevented the crack propagation after their formation under compression, consequently greater strengths were observed. This observation is in agreement with the greater compressive strengths observed for fiber reinforced mixtures under the SC curing regimen as described in Section 3.3.2.

3.4. Splitting Tensile Strength

All the six UHPC mixtures cured for 7 and 28 days under different curing regimens were tested to evaluate the splitting tensile strength and are presented in Figure 14. Plain UHPC mixtures exhibited splitting tensile strength of 3.85–9.30 MPa while fiber reinforced UHPC mixtures showed 8.60–14.70 MPa after 28 days of curing. The splitting tensile strength of fiber reinforced UHPC mixtures was found between the range of 9.0–23.0 MPa [51–54]. Such high tensile strength can be achieved due to improved homogeneity, dense microstructure, as well as secondary hydration resulted from the application of silica fume [1]. Figure 14 shows that splitting tensile strengths were influenced by the w/cm ratio, curing regimen, and steel fibers.

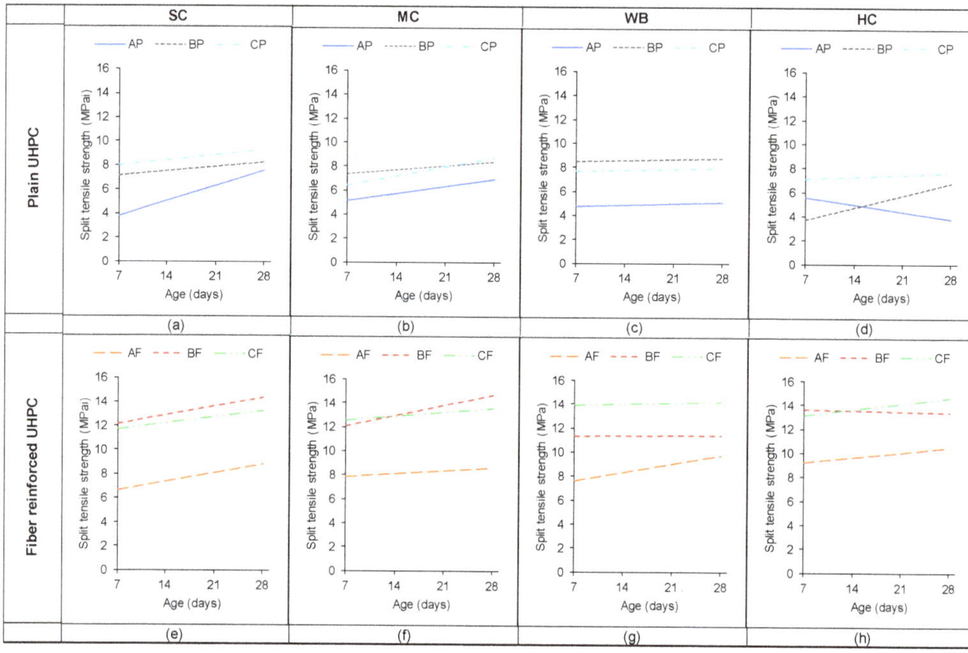

Figure 14. Split tensile strength of plain UHPC mixtures when cured under (**a**) SC, (**b**) MC, (**c**) WB, and (**d**) HC regimens and split tensile strength of fiber reinforced UHPC mixtures when cured under (**e**) SC, (**f**) MC, (**g**) WB, and (**h**) HC regimens.

3.4.1. Effect of w/cm Ratio and Permeable Porosity

Splitting tensile strength was increased with the decrease in the w/cm ratio of UHPC mixtures (Figure 15). For plain mixtures, splitting tensile strength increased by 36.9% as the w/cm ratio decreased from 0.30 to 0.20. Further reduction in w/cm ratio from 0.20 to 0.17 increased the splitting tensile strength by 4.9%. In fiber-reinforced mixtures, these increases were 42.8%, and 3.3%, respectively. This increase in splitting tensile strengths is associated with the reduction in permeable porosity of UHPC as the w/cm ratio decreases (Figure 6).

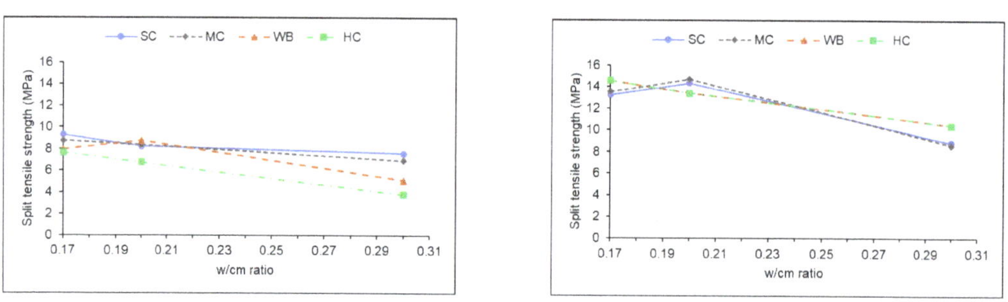

(**a**) (**b**)

Figure 15. Effect of different curing regimens on split tensile strength of (**a**) plain mixtures and (**b**) fiber-reinforced mixtures.

Splitting tensile strength was decreased with permeable porosity increase (Figure 16). For plain UHPC, as the permeable porosity increased by 8.2%, the splitting tensile strength decreased by approximately 35%. For fiber reinforced UHPC, as the permeable porosity increased by 15.0%, the split tensile strength was decreased by approximately 41%. As porosity decreases, the density of UHPC increases. In addition, the pozzolanic reaction of silica fume might have strengthened the ITZ. These two combined actions might have increased the splitting tensile strength at lower permeable porosity.

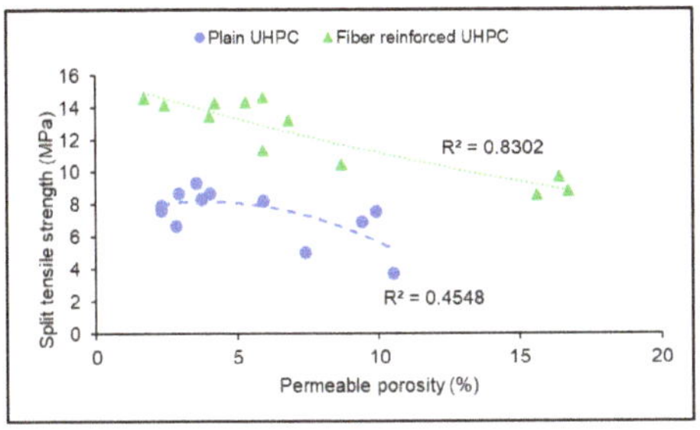

Figure 16. Effect of permeable porosity on split tensile strength.

3.4.2. Effect of Curing Regimens

Plain UHPC: For plain mixtures, SC and MC cured specimens performed better than those cured under WB and HC curing regimens (Figures 15a and 17a). Compared to SC cured specimens, the splitting tensile strength of the MC cured specimens showed an increase by 1.3% for mixture BP while it was decreased by 9.0% and 5.9% for AP and CP mixtures, respectively. Curing at elevated temperatures reduced the split tensile strength of plain UHPC mixtures. Splitting tensile strengths of AP and CP mixtures that were cured under the WB curing regimen were decreased by 33.0% and 14.4%, respectively, compared to those cured in the SC curing regimen. The splitting tensile strength of BP mixture was marginally increased (6.3%) when cured under the WB regimen compared to the SC regimen. However, lower splitting tensile strength was observed under the HC curing regimen in all three mixtures AP, BP, and CP. These decreases were 50.2%, 18%, and 17.8% for AP, BP, and CP, respectively.

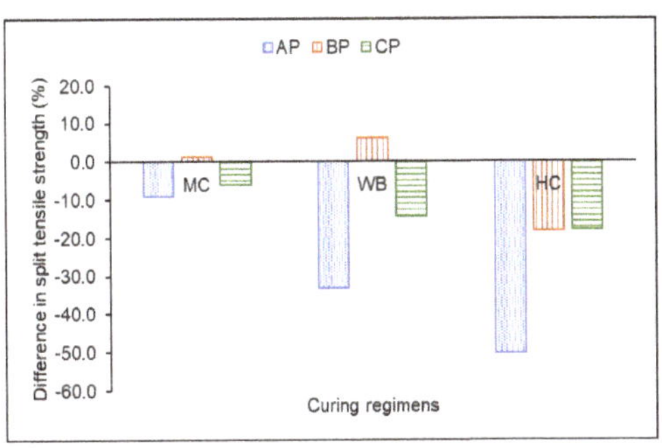

(**a**)

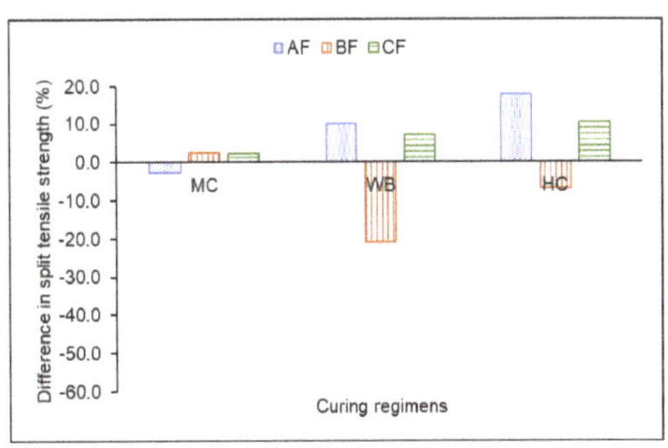

(**b**)

Figure 17. Difference in split tensile strength compared to SC curing regimen for (**a**) plain mixtures and (**b**) fiber−reinforced mixtures.

Curing the specimens at elevated temperatures might have caused the drying of the samples and decreased the bond strength of the ITZ. Heat treatment can also cause thermal cracks within the specimens which might contribute to the lowered splitting tensile strength.

Fiber reinforced UHPC: The effect of curing regimens on splitting tensile strength is depicted in Figures 15b and 17b. Compared to the SC curing regimen, specimens produced from the mixture AP and cured under the MC regimen exhibited a decrease in split tensile strength by 2.7%. However, splitting tensile strengths increased by 2.6% and 2.3% for BP and CP mixtures cured under the MC regimen. However, the WB curing regimen resulted in reduced splitting tensile strength of the BP mixture by 20.7%, while for AP and CP mixtures, it was increased by 10.1% and 7.3%, respectively compared to the SC curing regimen. Similarly, splitting tensile strength under the HC curing regimen was reduced by 6.7% for the BP mixture and increased by 17.9% and 10.4% for AP and CP mixtures, respectively. So, fiber–reinforced mixtures under WB and HC curing regimens performed better than SC and MC curing regimens.

3.4.3. Effect of Steel Fibers

The effect of steel fibers on splitting tensile strength is presented in Figure 18. The addition of 1.5% of steel fibers by volume significantly increased the splitting tensile strength. Specifically, the effect of steel fibers was pronounced with higher permeable porosity is beneficial. Figure 18 depicts the effect of steel fibers on splitting tensile strengths of mixtures produced with different w/cm ratios and by curing under different curing regimens. Splitting tensile strength of the SC cured specimens was increased by 16%, 74%, and 42% for category A, B, and C mixtures, respectively. With the MC curing regimen, these increases were 24%, 76%, and 55% for category A, B, and C mixtures, respectively. At elevated temperature curing, the effect was even more significant. WB curing regimen helped to increase splitting tensile strength by 91%, 30%, and 78% for category A, B, and C mixtures, respectively. For the HC curing regimen, these increases were 176%, 98%, and 91%, respectively. In fiber reinforced mixtures, interfacial shear and interlock between deformed surface of the fibers, and the matrix allowed stress to transfer from matrix to fiber which improved the tensile strength. Ming-zhe et al. [45] explained that the steel fibers impede the cracks from propagating and support the internal forces resulting from the fracture. Additionally, with the increase in stress, the crack-bridging effect of steel fibers, followed by partial debonding and pull out, increases the energy absorption capacity [44].

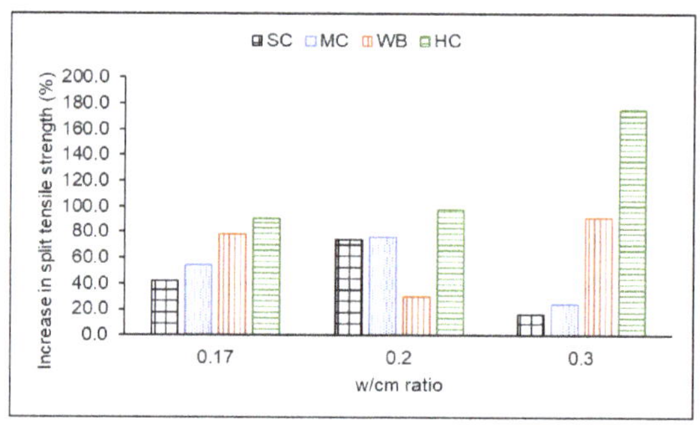

Figure 18. Effect of steel fibers on splitting tensile strength of UHPC mixtures.

4. Conclusions

Six UHPC mixtures were developed in this research using locally available materials. The effect of silica fume content, HRWRA dosage, w/cm ratio, and different curing regimens on UHPC properties were studied. The following conclusions were drawn from this study.

- UHPC with a flow diameter greater than 250 mm can be developed at different w/cm ratios by adjusting silica fume content and HRWRA dosage. At 0.30 w/cm ratio, such a high workability was achieved by using HRWRA/cm ratio of 0.047 at sf/cm ratio of 0.20. At 0.20 w/cm ratio, HRWRA/cm ratio was increased to 0.048 while sf/cm ratio was decreased to 0.17 to achieve similar workability. Both HRWRA/cm ratio and sf/cm ratio were further adjusted to 0.059 and 0.16, respectively to achieve a flow diameter of 250 mm at w/cm ratio of 0.17. Therefore, at a very low w/cm ratio, silica fume content in the UHPC mixtures needs to be reduced with the increase in HRWRA dosage to achieve highly workable mixtures.

- For category A mixtures with a w/cm ratio of 0.30, curing at 90 °C temperature helped to achieve compressive strength as high as 138.0 MPa after 28 days of curing. In similar curing conditions, category B and C mixtures with w/cm ratios of 0.20 and 0.17, achieved compressive strength of 138.6 MPa and 147.2 MPa, respectively. Splitting tensile strengths obtained from category A, B and, C mixtures were 10.45 MPa, 13.40 MPa, and 14.60 MPa, respectively. Therefore, ultra-high strength can be achieved at a w/cm ratio as high as 0.30. This can inspire the local ready-mix concrete companies to produce UHPC as they might not require upgrading their system as warranted for the very low w/cm ratio mixtures.
- Results from the MC cured specimens showed that category B and C mixtures achieved 120 MPa compressive strength after seven days of curing, while the splitting tensile strength was 12.1 MPa for both mixtures. Therefore, UHPC class strength can be achieved in field conditions without using heat treatment which can also reduce the cost associated with the curing.
- Permeable porosity of category A mixtures was as low as 7.4%, which was observed after 28 days of 90 °C curing. However, category B and C mixtures achieved less than 6% permeable porosity after only seven, similar to category B and C mixtures.

5. Recommendation for Future Research

- Durability performance of the presented mixtures in terms of chloride ion ingress, resistance to rapid freezing-thawing cycles, and resistance to carbonation should be evaluated
- Drying shrinkage and autogenous shrinkage of developed mixtures should be investigated.
- Statistical inference using analysis of variance or other statistical tools should be used to validate further the conclusions drawn in this research.
- The cost of the materials and the cost associated with the curing should be investigated.

Author Contributions: The authors confirm the following contribution to this paper: funding acquisition: S.A.; study conception and design: T.M.H., L.G. and S.A.; data collection: T.M.H. and L.G.; analysis and interpretation of results: T.M.H., L.G. and S.A.; draft manuscript preparation: T.M.H., S.A., J.O.-D. and A.T. All authors have read and agreed to the published version of the manuscript.

Funding: This research received no external funding.

Data Availability Statement: Data is contained within the article.

Conflicts of Interest: The authors have no conflict of interest.

References

1. Richard, P.; Cheyrezy, M. Composition of reactive powder concretes. *Cem. Concr. Res.* **1995**, *25*, 1501–1511. [CrossRef]
2. Russell, H.G.; Graybeal, B.A.; Russell, H.G. *Ultra-High Performance Concrete: A State-Of-The-Art Report for the Bridge Community*; (No. FHWA-HRT-13-060); United States, Federal Highway Administration, Office of Infrastructure Research and Development: Mc Lean, VA, USA, 2013.
3. Graybeal, B. *Design and Construction of Field-Cast UHPC Connections*; (No. FHWA-HRT-14-084; HRDI-40/10-14 (750) E); United States, Federal Highway Administration: Mc Lean, VA, USA, 2014.
4. Alsalman, A.; Dang, C.N.; Martí-Vargas, J.R.; Hale, W.M. Mixture-Proportioning of Economical UHPC Mixtures. *J. Build. Eng.* **2020**, *27*, 100970. [CrossRef]
5. Schmidt, M.; Fehling, E. Ultra-High-Performance Concrete: Research, Development and Application in Europe. *ACI Spec. Publ.* **2005**, *228*, 51–78.
6. Yoo, D.Y.; Yoon, Y.S. A Review on Structural Behavior, Design, and Application of Ultra-High-Performance Fiber-Reinforced Concrete. *Int. J. Concr. Struct. Mater.* **2016**, *10*, 125–142. [CrossRef]
7. Boulekbache, B.; Hamrat, M.; Chemrouk, M.; Amziane, S. Flowability of fibre-reinforced concrete and its effect on the mechanical properties of the material. *Constr. Build. Mater.* **2010**, *24*, 1664–1671. [CrossRef]

8. Wang, R.; Gao, X. Relationship Between Flowability, Entrapped Air Content and Strength of UHPC Mixtures Containing Different Dosage of Steel Fiber. *Appl. Sci.* **2016**, *6*, 216. [CrossRef]
9. Khayat, K.H.; Meng, W.; Vallurupalli, K.; Teng, L. Rheological Properties of Ultra-High-Performance Concrete—An Overview. *Cem. Concr. Res.* **2019**, *124*, 105828. [CrossRef]
10. Yu, R.; Spiesz, P.; Brouwers, H. Development of Ultra-High Performance Fibre Reinforced Concrete (UHPFRC): Towards an efficient utilization of binders and fibres. *Constr. Build. Mater.* **2015**, *79*, 273–282. [CrossRef]
11. Yu, R.; Spiesz, P.; Brouwers, H.J.H. Static Properties and Impact Resistance of a Green Ultra-High Performance Hybrid Fiber Reinforced Concrete (UHPHFRC): Experiments and Modeling. *Constr. Build. Mater.* **2014**, *68*, 158–171. [CrossRef]
12. Hung, C.-C.; Chen, Y.-T.; Yen, C.-H. Workability, fiber distribution, and mechanical properties of UHPC with hooked end steel macro-fibers. *Constr. Build. Mater.* **2020**, *260*, 119944. [CrossRef]
13. Liu, J.; Wang, K.; Zhang, Q.; Han, F.; Sha, J.; Liu, J. Influence of superplasticizer dosage on the viscosity of cement paste with low water-binder ratio. *Constr. Build. Mater.* **2017**, *149*, 359–366. [CrossRef]
14. Li, C.-Z.; Feng, N.-Q.; Li, Y.-D.; Chen, R.-J. Effects of polyethlene oxide chains on the performance of polycarboxylate-type water-reducers. *Cem. Concr. Res.* **2005**, *35*, 867–873. [CrossRef]
15. Li, Y.; Yang, C.; Zhang, Y.; Zheng, J.; Guo, H.; Lu, M. Study on dispersion, adsorption and flow retaining behaviors of cement mortars with TPEG-type polyether kind polycarboxylate superplasticizers. *Constr. Build. Mater.* **2014**, *64*, 324–332. [CrossRef]
16. Weyers, M.; Kearsley, E.P. Effect of High Plasticiser Dosage on Ultra High-Performance Fibre Reinforced Concrete (UHPFRC). In *International Conference on Application of Superabsorbent Polymers & Other New Admixtures Towards Smart Concrete*; Springer: Berlin/Heidelberg, Germany, 2019; pp. 248–255.
17. Abdulkareem, O.M.; Ben Fraj, A.; Bouasker, M.; Khelidj, A. Mixture design and early age investigations of more sustainable UHPC. *Constr. Build. Mater.* **2018**, *163*, 235–246. [CrossRef]
18. Muhd Norhasri, M.S.; Hamidah, M.S.; Fadzil, A.M.; Megawati, O. Inclusion of nano metakaolin as additive in ultra high performance concrete (UHPC). *Constr. Build. Mater.* **2016**, *127*, 167–175. [CrossRef]
19. Norhasri, M.M.; Hamidah, M.; Fadzil, A.M. Inclusion of nano metaclayed as additive in ultra high performance concrete (UHPC). *Constr. Build. Mater.* **2019**, *201*, 590–598. [CrossRef]
20. Van Tuan, N.; Ye, G.; van Breugel, K.; Fraaij, A.L.; Dai Bui, D. The study of using rice husk ash to produce ultra high performance concrete. *Constr. Build. Mater.* **2011**, *25*, 2030–2035. [CrossRef]
21. Shi, C.; Wang, D.; Wu, L.; Wu, Z. The hydration and microstructure of ultra high-strength concrete with cement–silica fume–slag binder. *Cem. Concr. Compos.* **2015**, *61*, 44–52. [CrossRef]
22. Yazıcı, H.; Yiğiter, H.; Karabulut, A.; Baradan, B. Utilization of fly ash and ground granulated blast furnace slag as an alternative silica source in reactive powder concrete. *Fuel* **2008**, *87*, 2401–2407. [CrossRef]
23. Wang, D.; Shi, C.; Wu, Z.; Xiao, J.; Huang, Z.; Fang, Z. A review on ultra high performance concrete: Part II. Hydration, microstructure and properties. *Constr. Build. Mater.* **2015**, *96*, 368–377. [CrossRef]
24. Sbia, L.A.; Peyvandi, A.; Lu, J.; Abideen, S.; Weerasiri, R.R.; Balachandra, A.M.; Soroushian, P. Production methods for reliable construction of ultra-high-performance concrete (UHPC) structures. *Mater. Struct.* **2016**, *50*, 7. [CrossRef]
25. Holland, T.C. *Silica Fume User's Manual*; Federal Highway Administration: Washinton, DC, USA, 2005.
26. Duval, R.; Kadri, E. Influence of Silica Fume on the Workability and the Compressive Strength of High-Performance Concretes. *Cem. Concr. Res.* **1998**, *28*, 533–547. [CrossRef]
27. Aldahdooh, M.; Bunnori, N.M.; Johari, M.A.M. Evaluation of ultra-high-performance-fiber reinforced concrete binder content using the response surface method. *Mater. Des.* **2013**, *52*, 957–965. [CrossRef]
28. Khatri, R.; Sirivivatnanon, V.; Gross, W. Effect of different supplementary cementitious materials on mechanical properties of high performance concrete. *Cem. Concr. Res.* **1995**, *25*, 209–220. [CrossRef]
29. *ASTM C1856/C1856M-17*; Standard Practice for Fabricating and Testing Specimens of Ultra-High Performance Concrete. ASTM International: West Conshohocken, PA, USA, 2017. Available online: www.astm.org (accessed on 5 October 2022).
30. *ASTM C33/C33M-18*; Standard Specification for Concrete Aggregates. ASTM International: West Conshohocken, PA, USA, 2018. Available online: www.astm.org (accessed on 5 October 2022).
31. *ASTM C128-15*; Standard Test Method for Relative Density (Specific Gravity) and Absorption of Fine Aggregate. ASTM International: West Conshohocken, PA, USA, 2015. Available online: www.astm.org (accessed on 5 October 2022).
32. *ASTM C136/C136M-19*; Standard Test Method for Sieve Analysis of Fine and Coarse Aggregates. ASTM International: West Conshohocken, PA, USA, 2019. Available online: www.astm.org (accessed on 5 October 2022).
33. Allena, S.; Newtson, C.M. Ultra-High Strength Concrete Mixtures Using Local Materials. *J. Civ. Eng. Arch.* **2011**, *5*, 322–330.
34. *ASTM C1437-15*; Standard Test Method for Flow of Hydraulic Cement Mortar. ASTM International: West Conshohocken, PA, USA, 2015. Available online: www.astm.org (accessed on 5 October 2022).
35. Yu, R.; Spiesz, P.; Brouwers, H.J.H. Mix design and properties assessment of Ultra-High Performance Fibre Reinforced Concrete (UHPFRC). *Cem. Concr. Res.* **2014**, *56*, 29–39. [CrossRef]
36. *ASTM C109/C109M-16a*; Standard Test Method for Compressive Strength of Hydraulic Cement Mortars (Using 2-in. or [50-mm] Cube Specimens). ASTM International: West Conshohocken, PA, USA, 2016. Available online: www.astm.org (accessed on 5 October 2022).

37. *ASTM C496/C496M-17*; Standard Test Method for Splitting Tensile Strength of Cylindrical Concrete Specimens. ASTM International: West Conshohocken, PA, USA, 2017. Available online: www.astm.org (accessed on 5 October 2022).
38. Dyer, T. *Concrete Durability*; CRC Press: Boca Raton, FL, USA, 2014.
39. Pyo, S.; Kim, H.-K. Fresh and hardened properties of ultra-high performance concrete incorporating coal bottom ash and slag powder. *Constr. Build. Mater.* **2017**, *131*, 459–466. [CrossRef]
40. Funk, J.E.; Dinger, D.R. *Predictive Process Control of Crowded Particulate Suspensions: Applied to Ceramic Manufacturing*; Springer Science & Business Media: New York, NY, USA, 2013.
41. Jeong, D.G.; Lee, H.S. The effect of Pozzolanic reaction under different curing temperatures in strength development of RPC. In Proceedings of the International Conference on Sustainable Building Asia, Seoul, Korea, 24–26 February 2010.
42. Grünewald, S. Performance-Based Design of Self-Compacting Fibre Reinforced Concrete. Doctoral Dissertation, Delft University, Delft, The Netherlands, 2004. Available online: http://resolver.tudelft.nl/uuid:07a817aa-cba1-4c93-bbed-40a5645cf0f1 (accessed on 5 October 2022).
43. Mindess, S.; Young, J.F.; Darwin, D. *Concrete*, 2nd ed.; Pearson Education Inc.: Upper Saddle River, NJ, USA, 2002.
44. Yang, I.H.; Joh, C.; Kim, B.-S. Structural behavior of ultra high performance concrete beams subjected to bending. *Eng. Struct.* **2010**, *32*, 3478–3487. [CrossRef]
45. Kang, S.-T.; Kim, J.-K. The relation between fiber orientation and tensile behavior in an Ultra High Performance Fiber Reinforced Cementitious Composites (UHPFRCC). *Cem. Concr. Res.* **2011**, *41*, 1001–1014. [CrossRef]
46. Visage, E.T.; Weldon, B.D.; Jauregui, D.V.; Newtson, C.M. Flexural Performance of Ultrahigh-Performance Concrete Developed Using Local Materials. *J. Mater. Civ. Eng.* **2019**, *31*, 04019050. [CrossRef]
47. Nasrin, S.; Ibrahim, A. Flexural response of Ultra-High-Performance Concrete (UHPC) hybrid bridge deck connections made with local materials. *Constr. Build. Mater.* **2021**, *270*, 121451. [CrossRef]
48. An, M.; Zhang, L.J.; Yi, Q.X. Size Effect on Compressive Strength of Reactive Powder Concrete. *J. China Univ. Min. Technol.* **2008**, *18*, 279–282. [CrossRef]
49. Graybeal, B.; Davis, M. Cylinder or Cube: Strength Testing of 80 to 120 MPa (11.6 to 29 ksi) Ultra-High-Performance Fiber-Reinforced Concrete. *ACI Mater. J.* **2008**, *110*, 603–609.
50. Kazemi, S.; Lubell, A.S. Influence of Specimen Size and Fiber Content on Mechanical Properties of Ultra-High-Performance Fiber-Reinforced Concrete. *ACI Mater. J.* **2012**, *109*, 675–684.
51. Graybeal, B. *Material Property Characterization of Ultra-High Performance Concrete*; Federal Highway Administration: McLean, VA, USA, 2006.
52. Graybeal, B.; Baby, F. Development of a Direct Tension Test Method for UHPFRC. *ACI Mater. J.* **2013**, *110*, 177–186.
53. Shafieifar, M.; Farzad, M.; Azizinamini, A. Experimental and numerical study on mechanical properties of Ultra High Performance Concrete (UHPC). *Constr. Build. Mater.* **2017**, *156*, 402–411. [CrossRef]
54. Gurusideswar, S.; Shukla, A.; Jonnalagadda, K.N.; Nanthagopalan, P. Tensile strength and failure of ultra-high performance concrete (UHPC) composition over a wide range of strain rates. *Constr. Build. Mater.* **2020**, *258*, 119642. [CrossRef]

buildings

MDPI

Article

Utilization of Different Forms of Demolished Clay Brick and Granite Wastes for Better Performance in Cement Composites

Jeonghyun Kim [1], Donwoo Lee [2], Alena Sičáková [3] and Namho Kim [2,*]

[1] Faculty of Civil Engineering, Wrocław University of Science and Technology, Wybrzeże Wyspiańskiego 27, 50-370 Wrocław, Poland
[2] School of Industrial Design & Architectural Engineering, Korea University of Technology & Education, 1600 Chungjeol-ro, Byeongcheon-myeon, Cheonan 31253, Republic of Korea
[3] Civil Engineering Faculty, Institute for Sustainable and Circular Construction, Technical University of Košice, Vysokoskolska 4, 042 00 Kosice, Slovakia
* Correspondence: nhkim@koreatech.ac.kr

Abstract: Clay brick and granite waste are part of the waste generated by construction and demolition activities. The amount of these wastes generated is enormous, but on the one hand, they can be used as a raw material for cement mixtures; thus, it is important to find ways to utilize them efficiently. In this study, clay brick and granite waste were crushed and screened into two size fractions (0.15–2.36 mm for sand replacement and smaller than 0.15 mm for cement replacement), and a total of four different forms of recycled materials were obtained (recycled brick aggregate, recycled brick powder, recycled granite aggregate and recycled granite powder) and used in cement mortar. Various properties (workability, mechanical strength and drying shrinkage) of the mortars were assessed according to standardized test methods. The results showed that the various material forms had different effects on the various properties of cement mortar. At replacement ratios of 10% and 20%, recycled granite showed better workability when used as powder, whereas recycled brick used as aggregate had higher workability. In common, using recycled brick and recycled granite in the form of aggregate was advantageous for the strength development of mortar, while using them in the form of powder helped to mitigate drying shrinkage.

Keywords: recycled brick; recycled granite; cement; recycled aggregate; recycled powder

Citation: Kim, J.; Lee, D.; Sičáková, A.; Kim, N. Utilization of Different Forms of Demolished Clay Brick and Granite Wastes for Better Performance in Cement Composites. *Buildings* **2023**, *13*, 165. https://doi.org/10.3390/buildings13010165

Academic Editor: Xiaoyong Wang

Received: 15 December 2022
Revised: 2 January 2023
Accepted: 6 January 2023
Published: 9 January 2023

1. Introduction

Concrete has made a major contribution to industrial development and economic growth, such as the construction of road networks and urban skyscrapers. However, the enormous consumption of concrete, which is mainly made of cement, sand and aggregate, places a considerable burden on the environment and ecosystem. Cement is produced at a high temperature of about 1400 °C, and CO_2 is generated from burning fossil fuels to heat a kiln to the temperature required for this calcination process. Moreover, during the calcination reaction, limestone ($CaCO_3$), the main raw material of cement, is decomposed to produce lime (CaO) and additional CO_2. The cement industry is known to account for about 7% of human-generated CO_2 and is the second-largest emission source after power generation [1]. Furthermore, the excessive use of natural sand threatens the ecosystem. Pascal [2] mentioned that the amount of sand consumed is greater than the amount of sand that is naturally deposited, and the UN Environment Programme (UNEP) [3] noted that sand, which plays a role in ecosystem services, biodiversity maintenance, and economic development support, should be recognized as a strategic resource.

In order to reduce dependence on sand and cement, there have been studies using construction and demolition (C&D) waste as raw materials for concrete. It has often been reported that the main component of C&D waste is concrete, and accordingly, extensive research has been conducted on the recycling of concrete waste. Followed by concrete waste, several studies have shown that bricks are one of the largest types of C&D waste in various regions [4–7]. Ding and Xiao [8] reported that in 2012, bricks and blocks accounted for 38% of construction waste and 63% of demolition waste in Shanghai, China. Granite waste belongs to the category of waste due to its non-biodegradable nature, and it can also be generated by cutting and grinding stones, as well as by damage, demolition and aging of stone products. In particular, Singh et al. [9] stated that the amount of granite waste generated during the production stage is nearly 65% of the total granite production. Therefore, some studies have been conducted using brick waste and granite waste as replacements for cement and sand in cement composites.

Bektas et al. [10] used brick waste as fine aggregate and reported that there were no significant differences in the compressive strength of mortar at 20% replacement. In a study by He et al. [11], they investigated the applicability of brick waste as partial cement replacement, and the optimal replacement level of waste brick was suggested to be 5–15%, considering the performance degradation caused by the increase in waste brick powder. For granite waste, it has been reported that the angular and rough texture of granite particles reduced the workability of cement mixtures but increased the flexural and compressive strengths at a 20% sand replacement ratio [12]. Asadi et al. [13] and Taji et al. [14] reported that 10% replacement of Portland cement by waste granite powder significantly improved microstructure, achieving a mechanical strength comparable to or greater than that of concrete with 100% Portland cement.

With respect to effective recycling methods for waste, Kim et al. [15] reported that the influence differs depending on how recycled materials are used, even if they are obtained from a single source of waste. In the study, concrete waste was crushed to different sizes and used as recycled coarse aggregate (RCA), recycled fine aggregate (RFA) and recycled powder (RP) in concrete, respectively. The results showed that the compressive strength of concrete with 10% and 20% replacement of cement with RP was only 1% and 7% lower than that of concrete without recycled material, but 30% replacement reduced the compressive strength by 29%. This was 17% lower than concrete with 100% RFA and 12% lower than concrete with 100% RCA and RFA simultaneously. These results suggest the need for an in-depth investigation of what form will be most advantageous to achieve the enhanced properties of cementitious materials, which must also be matched with an appropriate method of waste recycling for their preparation.

To the best of our knowledge, the majority of published articles, as mentioned above, have used single waste as a supplementary cementitious material or as a replacement for aggregate. Recycling waste in a single form (e.g., using brick waste as either brick aggregate or brick powder) can provide information about the potential of a certain type of waste, but recycling in various forms can further provide information about the effective recycling of waste. However, few studies have compared and analyzed the effects of using brick and granite wastes as fine aggregate and cement replacements, respectively. Therefore, in this study, waste brick and waste granite were crushed, collected by size, and recycled as fine aggregate for sand and powder for cement replacement in cement mortar, respectively. For the prepared mortar, fluidity, compressive and flexural strength, ultrasonic pulse velocity (UPV) and drying shrinkage were investigated.

The novelty of this study is to use a single type of waste (in this study, clay brick and granite wastes) for both forms of fine aggregate and cement replacements in cement mortar. The significance of this study is to provide knowledge about which form of waste recycling (aggregate or powder) is more beneficial to the performance of cement mortar by comparing and analyzing the influences of different forms of demolished waste clay brick and waste granite on the properties of cementitious materials. This study can contribute to the conservation of the environment and ecosystems by reducing the consumption of sand and cement.

2. Materials and Methods

2.1. Materials

The main cementitious binder used to prepare mortar mixtures was ordinary Portland cement (OPC), which has a specific gravity of 3.15. The physical properties of OPC are shown in Table 1.

Table 1. Physical properties of ordinary Portland cement.

Specific Gravity	28 d Compressive Strength, MPa	Setting Time, min		Loss of Ignition, %
		Initial	final	
3.15	51.8	220	305	2.2

For fine aggregate, silica river sand with a specific gravity of 2.59 and water absorption of 1.48% was used.

Two types of waste, clay brick and granite block, were obtained from a building demolition in Korea. These two wastes were washed with tap water to remove impurities, then dried and crushed in a ball mill [16]. Afterward, recycled brick (RB) and recycled granite (RG) with two different particle sizes, ranging from 150 μm to 2.36 mm and smaller than 150 μm, were collected, respectively. RG and RB with particle sizes from 150 μm to 2.36 mm were named recycled granite aggregate (GA) and recycled brick aggregate (BA) and were used as partial replacements for fine aggregates in mortar mixtures. RG and RB with particle sizes smaller than 150 μm were denominated as recycled granite powder (GP) and recycled brick powder (BP) and used as partial replacements for cement (Figure 1). The particle size distribution curves of the four prepared recycled materials are provided in Figure 2.

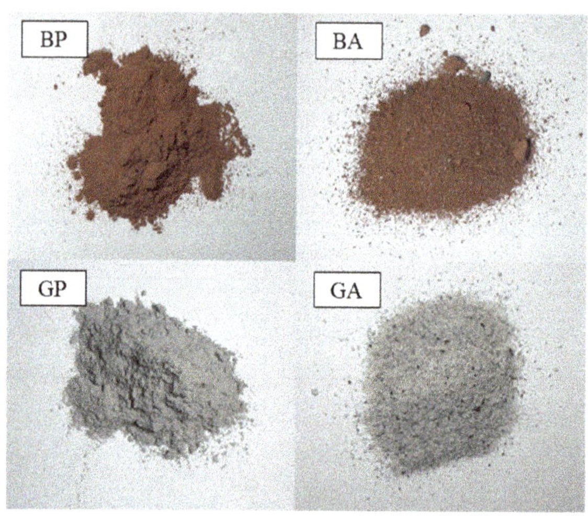

Figure 1. Different forms of recycled materials used in the study.

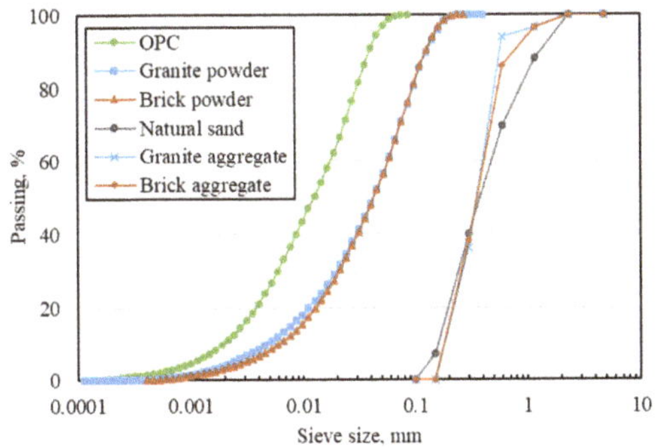

Figure 2. Particle size distribution of natural and recycled materials.

The physical characteristics of RB and RG are given in Table 2. The chemical and mineralogical composition was investigated using XRF and XRD techniques, and the results are shown in Table 3 and Figure 3. For RB, the percentage of $SiO_2 + Al_2O_3 + Fe_2O_3$ is greater than 70%, complying with the chemical composition requirements of pozzolanic materials according to ASTM C618 [17], whereas GB does not. The main minerals constituting RB were quartz (SiO_2), mullite ($3Al_2O_3{\cdot}2SiO_2$), and hematite (Fe_2O_3). For RG, the main minerals were quartz, microcline ($KAlSi_3O_8$) and albite ($NaAlSi_3O_8$).

Table 2. Physical characteristics of recycled brick and recycled granite aggregates.

	BA	GA
Water absorption, %	7.70	0.58
Specific gravity	2.26	2.62

Table 3. Chemical composition of recycled brick and recycled granite powders by percentage.

	SiO_2	Al_2O_3	Fe_2O_3	K_2O	CaO	TiO_2	MgO	ZrO_2	SO_3
OPC	13.11	2.84	4.62	1.48	73.39	0.28	1.05	-	2.82
BP	64.22	18.98	7.57	5.10	1.75	1.21	0.77	0.12	0.11
GP	52.48	10.62	3.03	5.54	26.17	0.31	0.37	0.07	1.12

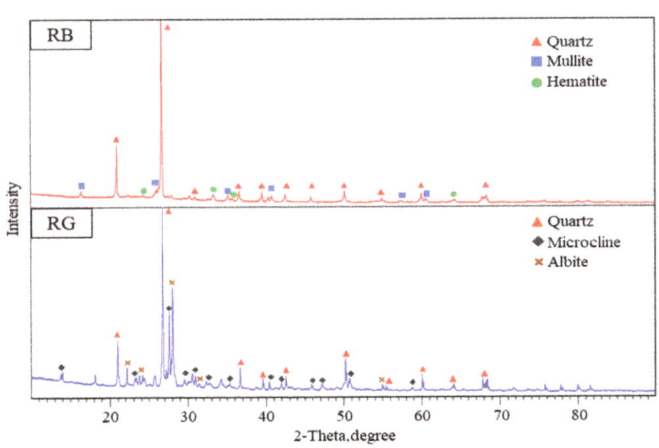

Figure 3. X-ray diffraction of recycled brick and recycled granite powders.

2.2. Mix Proportion and Specimen Preparation

The mix proportion was determined according to ASTM C109 [18]. The material ratios of water, cementitious binders (cement, BP and GP) and fine aggregates (river sand, BA and GA) for all mortars produced were 0.485:1:2.75 by weight. Each recycled material replaced cement and sand, respectively, at 10%, 20% and 30% by weight. Detailed proportions are presented in Table 4. The materials were mixed using an automatic mixer programmed with the mixing procedure provided in ASTM C305 [19].

Table 4. Mix proportion of recycled mortar.

No.	ID	Water	Cement	BP	GP	Sand	BA	GA
1	Reference	0.485	1	0	0	2.75	0	0
2	BP10	0.485	0.9	0.1	0	2.75	0	0
3	BP20	0.485	0.8	0.2	0	2.75	0	0
4	BP30	0.485	0.7	0.3	0	2.75	0	0
5	BA10	0.485	1	0	0	2.475	0.275	0
6	BA20	0.485	1	0	0	2.2	0.55	0
7	BA30	0.485	1	0	0	1.925	0.825	0
8	GP10	0.485	0.9	0	0.1	2.75	0	0
9	GP20	0.485	0.8	0	0.2	2.75	0	0
10	GP30	0.485	0.7	0	0.3	2.75	0	0
11	GA10	0.485	1	0	0	2.475	0	0.275
12	GA20	0.485	1	0	0	2.2	0	0.55
13	GA30	0.485	1	0	0	1.925	0	0.825

2.3. Test Methods

Fluidity, compressive and flexural strength, UPV and drying shrinkage tests were conducted on the prepared mortar.

A flow table test was carried out to determine the fluidity of each mortar mixture in accordance with ASTM C1437 [20].

For the mechanical strength, a hydraulic universal testing machine (WJ-100S, Woojin, Republic of Korea) was utilized. The compressive strength test was carried out as per ASTM C109 [18]. A load of 100 kN/min was applied to three 50-mm cube specimens per mortar mixture.

Flexural strength was performed according to ASTM C348 [21]. At 28 days of age, a load of 2.5 kN/min was applied to three prism specimens with a size of 40 mm × 40 mm × 160 mm.

Drying shrinkage was performed based on ASTM C596 [22], and the length change was measured for four 25 mm × 25 mm × 285 mm bar specimens for each mortar. After 24 h of moisture curing, the specimens were cured in lime-saturated water for 48 h. The initial length (L_0) of the specimen and the reading at a length comparator (L_i) were taken, and the specimens were stored in a chamber maintained at 23 ± 2 °C and 60 ± 4% relative humidity. The length change of the specimens was recorded to the nearest 0.001 mm at 4, 11, 18 and 25 days of storage in the chamber (L_d). The length change was calculated by the following Formula (1):

$$Length\ change,\ \% = \frac{L_d - L_i}{L_0} \tag{1}$$

The UPV can provide predictable information about concrete quality. Using a portable device (Pundit-Plus, CNS Farnell Ltd., Hertfordshire, UK), consisting of a pair of receivers and transmitters, the UPV was measured on the longitudinal ends of three prism specimens cured for 28 days. The results were calculated as the path length divided by the time it takes for an ultrasonic pulse to travel from the transmitter to the receiver.

3. Results and Discussion

3.1. Flow

The flow table test results of mortars in which cement and sand were partially replaced by RG and RB are shown in Figure 4a. It is worth mentioning that the water-to-binder ratio was kept constant to evaluate only the influence of RG and RB in the mortar. As shown in Figure 4a, the increase in the replacement ratios of RG and RB (both powders and aggregates) decreased the fluidity of the mortar. The flow of the reference mortar, which does not contain any recycled materials, was 159 mm, and the GP10, GP20 and BA10 achieved a similar flow to that of the reference mortar. The flow of mortars, except for these three series, was in the range of 68–95% of that of the reference mortar. It has been reported that this decrease is associated with the irregular surface texture and angular shape of RG and RB and the high water absorption of RB [23–25]. Nevertheless, there was no significant difference between the flow values of GA-based mortars and BA-based mortars. Although mortars with BA would be expected to have a poorer consistency due to its greater absorption, the rough and irregular surface of GA due to the crushing process is likely to have a comparable effect on consistency.

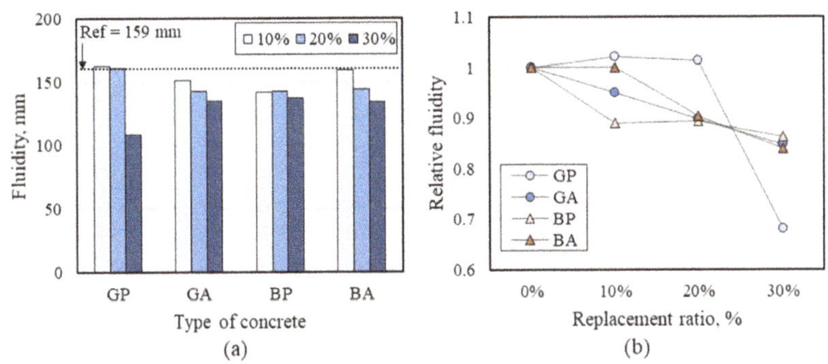

Figure 4. Fluidity (**a**) and relative fluidity (**b**) of mortars made with recycled brick and recycled granite.

Figure 4b shows the relative flow of the prepared mortars, in which the flow of the reference mortar is represented as a benchmark (i.e., the flow of the reference mortar is indicated by 1). For the mortars that replaced fine aggregates with GA and BA, gradual decreases in fluidity were observed with increasing replacement ratios. When the replacement ratios were increased to 10%, 20% and 30%, the flow of GA-based mortar decreased by 5%, 10% and 15% compared to the reference mortar, and the flow of BA-based mortar decreased by 0%, 10% and 16% at the given replacement ratios. Mortars containing BP changed their flow value minimally when the replacement ratio was increased from 10% to 30%; the flow change was from 1% to 3%. However, the consistency of mortars with GP was noticeably reduced by 33% at 30% replacement.

It is worth noting that the material forms of RG and RB are different for better fluidity. At the 10% and 20% replacement ratios, it is favorable to use RG as powder rather than fine aggregate, while RB is recommended as a substitute for fine aggregate. At a 30% replacement ratio, mixtures may require the aid of plasticizers to achieve adequate workability, particularly for GP.

3.2. Compressive Strength

The 28-day compressive strength test results for various mortars are shown in Figure 5. Figure 5a clearly shows that the material forms of RG and RB had different effects on the compressive strength, and also that the proper use of RG and RB can achieve compressive strength exceeding that of the reference mortar.

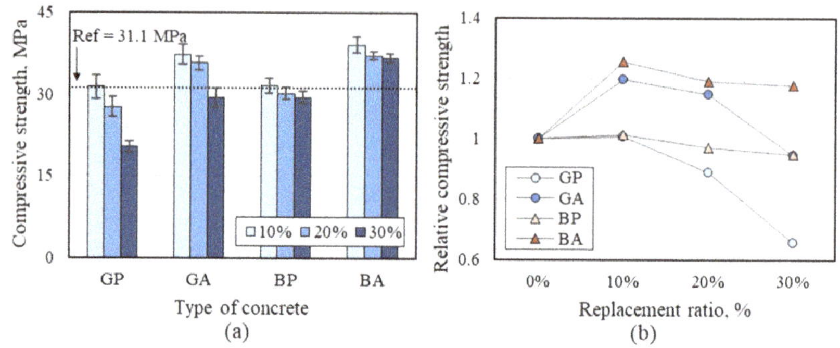

Figure 5. Compressive strength (**a**) and relative compressive strength (**b**) of mortars made with recycled brick and recycled granite.

The compressive strengths of mortars in which sand was replaced by GA and BA ranged from 95–120% and 118–126% of that of the reference mortar, respectively, and all the GA- and BA-based mortars exceeded the compressive strength of the reference mortar, except for GA30, which was 5.4% (1.6 MPa) lower. This increase in strength can be explained by several reasons. For the GA-based mortars, it has been reported that the rough, angular grain shape of GA can increase strength by improving the bonds between the aggregates and binder [26]. As it seems, the benefit is more prevalent at 10% and 20% replacement of sand; however, at the 30% replacement ratio, a greater reduction in strength has been noted. The difference in intrinsic strength between the natural silica sand and the GA can be considered a factor influencing the strength of mortar, and the GA may contain microcracks and other damages during its lifetime, including the crushing for recycling [27]. For the BA-based mortars, the effective water-to-cement (w/c) ratio and internal curing caused by the high porosity of BA need to be considered. The porous BA absorbs a large amount of water during the mixing process. Since the same w/b ratio (=0.485) was applied to all mixtures in this study, the effective w/c ratio for the actual hydration reaction is lower than 0.485, specifically, the effective w/c ratios of 0.464 for BA10, 0.443 for BA20 and 0.422 for BA30. Consequently, the lowered w/c ratio may have contributed to the improved strength of the BA-based mortars. Moreover, the water absorbed into BA during the mixing process acts as a moisture source, causing an internal curing effect. As presented in previous studies [28,29], clay bricks are known to absorb mixing water during the mixing process of cement-based materials in proportion to their absorption capacity. This absorbed water acts as an internal curing agent that internally supplies the water needed for hydration in the later stages of curing, increasing the density and reducing cracking in mortar.

At the given replacement ratios, the compressive strength of the GP-based mortar was 65–101% of that of the reference mortar, and that of the BP-based mortar was 95–102%. According to ASTM C618 [17], the compressive strength of mortars containing pozzolanic materials should achieve a minimum of 75% of that of mortar made with 100% OPC. Namely, the GP- and BP-based mortars, except for GP30, meet the strength requirements. However, as the replacement ratios increase, the decrease in strength of the GP-based mortar is more pronounced than that of the BP-based mortar. This may be because the BP, which met the chemical composition requirements as a pozzolanic material, contributed to the development of compressive strength, whereas the GP, which did not satisfy the chemical requirements, did not have a noticeable effect on the strength development [30]. The pozzolanic reactions are generally greater when the particle size is smaller. Therefore, when RB is used in powder form, the pozzolanic reaction is more active, forming CSH gels that contribute to the enhancement of strength [31]. In fact, the compressive strength of the BP-based mortars ranged from 95% to 102% of that of the reference mortar, showing that the use of BP as a cement substitute did not cause a noticeable loss of compressive strength within a certain replacement range.

From an environmental point of view, 10% replacement of GP and replacement within 30% of BP show no detrimental effect on the compressive strength of mortar. Therefore, GP and BP can be considered eco-friendly supplementary cementitious materials that can reduce the large amount of CO_2-generated in the cement industry. However, from a performance-based perspective, under the same replacement ratios, the compressive strengths of the GP- and BP-based mortars decreased by 19% and 29% compared to the GA- and BA-based mortars (Figure 5b). This is related to a decrease in the hydration product of the cement mortar due to the 'dilution effect' occurring when a large amount of cement is replaced with GP and BP [32]. Thus, to improve the compressive strength, it is desirable to use RG and RB as substitutes for fine aggregates.

Table 5 shows some examples of the practical utilization of each mortar prepared in this study based on the compressive strength required by various standards. In practice, the specific strength class required for a given structure is prescribed by the designer based on the type of construction and the applied loads. The mortars tested meet the strength criteria and the relevant classification as shown in Table 5. For example, according to EN998-1 [33], which specifies the requirements for mortars for plastering and rendering of building elements such as walls, ceilings, columns and internal partitions, all mortars in this study can be used for these purposes, meeting a strength level of at least 6 MPa. According to EN13813 [34], regarding mortar for floor screeds, the required compressive strength varies depending on the degree of load applied to the floor; thus, the forms and replacement ratios of recycled materials need to be selected in consideration of the design load. For example, for members where high loads are expected, RG and RB should be used in the form of aggregate, and the recommended replacement ratios can be 20% and 30%, respectively. For elements with low applied loads, RG and RB can be used as powder forms, meeting the requirements.

3.3. Flexural Strength

The flexural strength test results for mortars containing RG and RB at various replacement ratios are presented in Figure 6a,b. The 28-day flexural strength of all the prepared mortar mixtures was lower than that of the reference mortar, 7.88 MPa. The material forms had a different effect on the flexural strength properties. Whether RG was recycled as aggregate or powder, the flexural strength was the highest at a replacement rate of 20%, whereas RB decreased in flexural strength as the replacement ratio increased.

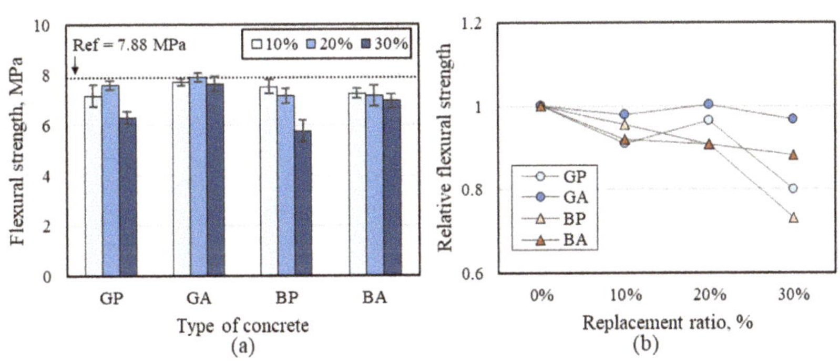

Figure 6. Flexural strength (**a**) and relative flexural strength (**b**) of mortars made with recycled brick and recycled granite.

Table 5. Classification of mortar mixtures for selected practical applications according to compressive strength.

Application	Grade Group	Mix	Remarks
EN 998-1 [33]	CS IV (≥6 MPa) upper level of the classification ranking, meets more demanding strength criteria	All mortars in this study	Mortars for rendering and plastering of construction elements, such as walls, ceilings, columns and internal partition walls
EN 998-2 [35]	M20 (≥20 MPa) upper level of the classification ranking, meets more demanding strength criteria	All mortars in this study	Masonry mortars for brick work (base, bonding and grouting), for bricklaying of walls, columns and internal partition walls, load-bearing and non-load-bearing structures
EN 13813 [34]	C20 (≥20 MPa)	GP30	Levelling of the substrate in the interior, a base layer under the final surface treatment (PVC, carpets, floating floors and wooden floors) for: floors with light load
	C25 (≥25 MPa)	GP20 GA30 BP30	floors with medium load
	C30 (≥30 MPa)	Reference GP10 BP10 BP20	
	C35 (≥35 MPa)	GA10 GA20 BA10 BA20 BA30	floors with high load

Note: the above classification and application remarks refer to the actual parameter only. For a comprehensive assessment of the suitability of the mixture for the purpose stated, all the criteria required by the standard must be fulfilled.

Within a 30% replacement ratio, the flexural strength loss of the GA-based mortar was only 3% of that of the reference mortar, indicating that the effect of GA on the flexural strength was considered negligible. Donza et al. [36] concluded that the effect of crushed sand type on strength was insignificant when the volume of cement paste was the same and the flexural strength did not differ significantly at various replacement ratios. Therefore, GA can be used as an attractive partial substitute for natural sand in terms of flexural strength performance and eco-friendliness. For the BA-based mortars, the flexural strength was progressively decreased with increasing replacement ratios, with the strength losses ranging from 8.0% to 11.7%. This is due to the low intrinsic strength and porosity of BA [31].

Compared to the reference mortar, the flexural strength of the GP-based mortar at 10%, 20% and 30% replacement ratios was 9%, 3.6% and 20.1% lower, respectively. Similarly, the strength loss of the BP-based mortars was 4.5%, 9.1% and 26.8%. As expected, the lowest flexural strengths were observed in GP30 and BP30. In particular, the 30% replacement of cement by both GP and BP dramatically dropped the flexural strength, which is due to the dilution effect caused by the OPC reduction, and this result is consistent with the trend observed for the compressive strength.

As shown in Figure 6b, when RG was used as a substitute for sand, higher flexural strength was achieved than when RG was used as a cement replacement. On the contrary, recycling RB as powder rather than aggregate at 10% and 20% replacement ratios was more favorable for the development of flexural strength.

In practice, values of flexural strength are specified (e.g., for floor mortars), and Table 6 provides a classification of the tested mixtures.

Table 6. Classification of mortar mixtures for selected practical applications according to flexural strength.

Application	Grade Group	Mix	Remarks
EN 13813 [34]	F5 (≥5 MPa)	BP30	Levelling of the substrate in the interior, a base layer under the final surface treatment (PVC, carpets, floating floors and wooden floors) for light levels of loading
	F6 (≥6 MPa)	GP30 BA30	
	F7 (≥7 MPa)	Reference GP10 GP20 GA10 GA20 GA30 BP10 BP20 BA10 BA20	

3.4. Ultrasonic Pulse Velocity

The UPV values for various mortars are presented in Figure 7a,b. Regardless of the material types, the replacement of natural sand and cement by recycled aggregates and recycled powders showed that the UPV values decreased with increasing replacement ratios; however, the UPV of six series out of a total of 12 mortars (i.e., GA10, BP10, BP20, BA10, BA20 and BA30) showed higher values than that of the reference mortar of 4018 m/s. The UPV of cementitious materials is proportional to the volume fraction of the solid phase and inversely proportional to the porosity as ultrasonic pulses are transmitted through the solid phase [37]. Therefore, the UPV results in this study indicate that the use of appropriate amounts of RG and RB can make the pore structure of cement composite denser compared to that made with natural sand and cement. According to the quality classification of mortar by the UPV value presented by Estévez et al. [38], the quality can be considered "excellent" when the UPV exceeds 3800 m/s and "good" when the UPV is between 3500 m/s and 3800 m/s. The lowest UPV of the prepared mortars was 3774 m/s for the GP30, indicating "good" quality, while the mortars other than the GP30 showed "excellent" quality, ranging from 3947 m/s to 4200 m/s.

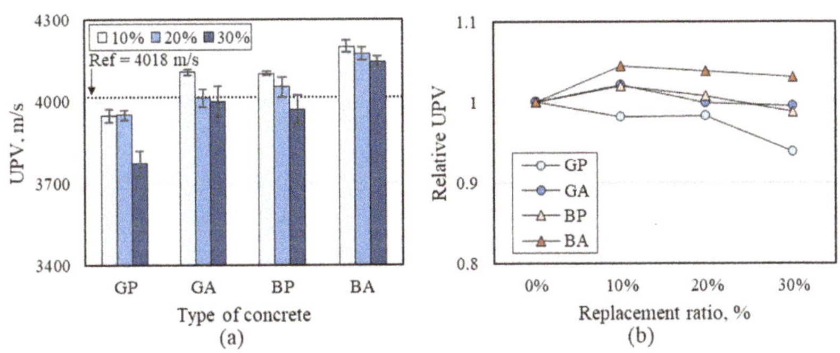

Figure 7. Ultrasonic pulse velocity (**a**) and relative ultrasonic pulse velocity (**b**) of mortars made with recycled brick and recycled granite.

Within the replacement ratios of 30%, the UPV values of the GA-based mortars were in the range of 99.5–102.2% of those of the reference mortar, indicating no significant difference. Unexpectedly, the BA-based mortars, which are generally considered to have porous structures, showed the highest UPV values (103.2–104.5%). According to a previous study [39], the water absorbed into BA during the mixing process due to its high water absorption acts as a self-curing agent in the cement-based materials in later stages, and the absorbed internal water not used for hydration fills the pores to make the BA denser. In fact, Rao [40] reported higher UPV values when brick and stone replaced river sand at 30%.

At the given replacement ratios, the UPV of the GP- and BP-based mortar ranged from 93.9–98.3% and 98.8-102.1% of the reference concrete, respectively. As shown in Figure 7b, the UPV values tend to be higher in the mortars containing recycled aggregate (i.e., GA and BA series) than in the mortars containing recycled powder (i.e., GP and BP series). The UPV results are in good agreement with the trend observed in the compressive strength tests of the corresponding mortars.

Figure 8 shows the correlation between compressive strength and UPV for the mortars prepared in this study. A linear relationship between the two parameters was observed, and the R^2 coefficient of determination was 0.87, indicating that there is very high reliability in predicting the dependent variable. Thus, the UPV technique can be useful for estimating the compressive strength of cement composites containing RG and RB as sand and cement replacements.

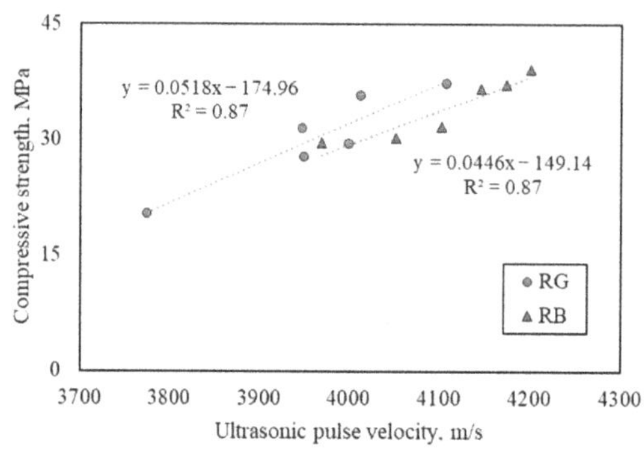

Figure 8. A correlation between UPV and compressive strength.

3.5. Drying Shrinkage

The length changes due to drying shrinkage of the mortar containing RG and RB are plotted in Figure 9a,b, respectively. The drying shrinkage of all mortars increased over time, but the change in length was not as large as that of the reference mortar.

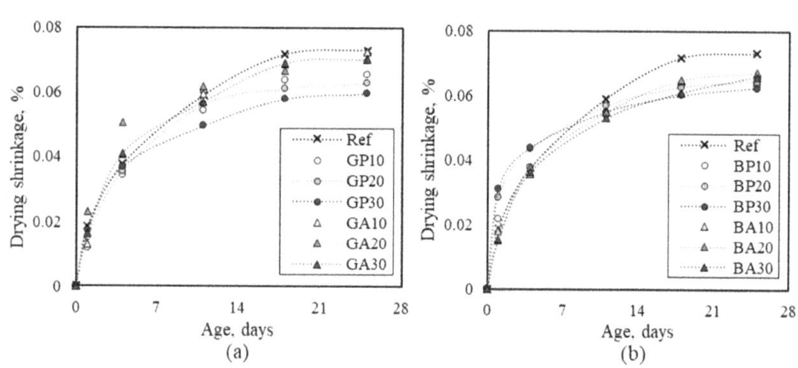

Figure 9. Drying shrinkage of mortars made with recycled granite (**a**) and recycled brick (**b**).

Since GA and BA used in this study are finer than sand, drying shrinkage may have been reduced due to improved pore structure and reduced evaporation of water through capillary pores. These results have been frequently reported in studies using RB and RG as partial substitutes for fine aggregates [41–43].

In addition, in both the GP and BP series, the replacement of cement was more effective in mitigating drying shrinkage than the replacement of sand. The 28-day drying shrinkage of the GP-based mortars ranged from 0.06% to 0.063%, and that of the GA-based mortars was from 0.07% to 0.073%. For RB, no marked difference in shrinkage was observed. For example, the BP-based mortars showed 28-day drying shrinkage of 0.062% to 0.064%, and the BA-based mortars showed shrinkage of 0.066% and 0.067%. Dang et al. [44] found that there was no remarkable difference in drying shrinkage at 50% replacement of sand by BA. The authors concluded that this was due to the combined beneficial effects of the pozzolanic reaction and internal curing. The lowest drying shrinkage was for GP30 and BP30, in which 30% of the cement was replaced by GP and BP, which had about 15–18% less shrinkage than the reference mortar. This is related to the fact that one parameter that causes drying shrinkage is the cement content [45]. One of the causes of drying shrinkage is due to the evaporation of the moisture from the cement matrix, and there is a positive correlation between cement content and moisture loss. Therefore, the drying shrinkage deformation of mortar with low cement content is relatively small, and when RG and RB are recycled as a substitute for cement rather than as a substitute for sand, drying shrinkage can be suppressed by 9–14% and 4–5%, respectively. This trend is in line with previous studies [46,47].

4. Conclusions

In this study, the effects of different forms of recycled materials, waste clay brick and waste granite, both in aggregate and powdery forms, were investigated. Based on the experimental results, the following conclusions can be drawn:

- The effect of the application of RG and RB on cement mortar depends on the material forms in which the wastes are processed. With optimal forms and mix proportions, even better properties of cementitious mixtures can be achieved with recycled materials than with standard ingredients.
- At replacement ratios of 10% and 20%, it is favorable for RG to be recycled as a cement replacement, while RB needs to be recycled as a replacement for sand to obtain a more workable mixture.
- In terms of strength development, RG and RB are advantageous for recycling as alternatives to sand. At the same replacement ratio, the compressive strength of GA was 19% to 45% higher than that of GP, and BA showed approximately 23% higher compressive strength than BP.
- The flexural strength of mortars containing RB decreased with the increase in replacement ratios, whereas that of mortars containing RG was highest at 20% replacement (i.e., GA20 and GP20). Moreover, the replacement of natural sand by GA within 30% resulted in only a 3% loss of flexural strength compared to the reference mortar.
- The UPV values showed an appropriate correlation with the compressive strength ($R^2 = 0.87$), indicating that the UPV technique could be utilized to estimate the compressive strength of mortar containing RB and RG.
- Both forms of RG and RB contributed to a reduction in shrinkage and an improvement in the quality of mortars, with more favorable results in the powder form. Recycled powder contributed to drying shrinkage mitigation by reducing the cement content in cement composites. The lowest shrinkage was observed in GP30 and BP30, where 30% of the cement was replaced by RG and RB.

During the production of recycled materials from waste, residues, mainly of fine size, are inevitably generated. Moreover, the advanced technologies of the present time make it possible to turn waste into recycled materials with the desired sizes. Therefore, further research needs to be conducted on various recycling approaches that can achieve better performance with less energy consumption.

Author Contributions: Conceptualization, J.K. and D.L.; methodology, J.K.; formal analysis, J.K. and D.L.; investigation, J.K. and N.K.; resources, D.L. and N.K.; data curation, J.K. and A.S.; writing—original draft preparation, J.K.; writing—review and editing, A.S.; supervision, A.S. and N.K.; funding acquisition, A.S. and N.K. All authors have read and agreed to the published version of the manuscript.

Funding: This research has been carried out within the project of Slovak Scientific Grant Agency VEGA (Grant No. 1/0336/22) "Research on the effects of Lean Production/Lean Construction for improving the efficiency of on-site and off-site construction technologies".

Data Availability Statement: The data presented in this study are available on request from the corresponding author.

Acknowledgments: This work was supported by NAWA STER Programme Internationalisation of Wroclaw University of Science and Technology Doctoral School. The authors would like to thank the Cooperative Equipment Center at KOREATECH for assistance with particle size, XRF and XRD analysis. The financial support of the National Scholarship Programme of the Slovak Republic (ID 37273) is also acknowledged.

Conflicts of Interest: The authors declare no conflict of interest.

References

1. UNEP-GEAS. *Greening Cement Production Has a Big Role to Play in Reducing Greenhouse Gas Emissions;* UNEP-GEAS: Nairobi, Kenya, 2010.
2. Pascal, P. Sand, Rarer than One Thinks. *Environ. Dev.* **2014**, *11*, 208–218. [CrossRef]
3. UNEP. *Sand and Sustainability: 10 Strategic Recommendations to Avert a Crisis;* UNEP: Geneva, Switzerland, 2022.
4. Kim, J. Construction and Demolition Waste Management in Korea: Recycled Aggregate and Its Application. *Clean Technol. Environ. Policy* **2021**, *23*, 2223–2234. [CrossRef]
5. Islam, R.; Nazifa, T.H.; Yuniarto, A.; Shanawaz Uddin, A.S.M.; Salmiati, S.; Shahid, S. An Empirical Study of Construction and Demolition Waste Generation and Implication of Recycling. *Waste Manag.* **2019**, *95*, 10–21. [CrossRef]
6. Wu, H.; Duan, H.; Zheng, L.; Wang, J.; Niu, Y.; Zhang, G. Demolition Waste Generation and Recycling Potentials in a Rapidly Developing Flagship Megacity of South China: Prospective Scenarios and Implications. *Constr. Build. Mater.* **2016**, *113*, 1007–1016. [CrossRef]
7. Zhao, W.; Leeftink, R.B.; Rotter, V.S. Evaluation of the Economic Feasibility for the Recycling of Construction and Demolition Waste in China—The Case of Chongqing. *Resour. Conserv. Recycl.* **2010**, *54*, 377–389. [CrossRef]
8. Ding, T.; Xiao, J. Estimation of Building-Related Construction and Demolition Waste in Shanghai. *Waste Manag.* **2014**, *34*, 2327–2334. [CrossRef]
9. Singh, S.; Khan, S.; Khandelwal, R.; Chugh, A.; Nagar, R. Performance of Sustainable Concrete Containing Granite Cutting Waste. *J. Clean. Prod.* **2016**, *119*, 86–98. [CrossRef]
10. Bektas, F.; Wang, K.; Ceylan, H. Effects of Crushed Clay Brick Aggregate on Mortar Durability. *Constr. Build. Mater.* **2009**, *23*, 1909–1914. [CrossRef]
11. He, Z.; Shen, A.; Wu, H.; Wang, W.; Wang, L.; Yao, C.; Wu, J. Research Progress on Recycled Clay Brick Waste as an Alternative to Cement for Sustainable Construction Materials. *Constr. Build. Mater.* **2021**, *274*, 122113. [CrossRef]
12. Jain, A.; Gupta, R.; Chaudhary, S. Performance of Self-Compacting Concrete Comprising Granite Cutting Waste as Fine Aggregate. *Constr. Build. Mater.* **2019**, *221*, 539–552. [CrossRef]
13. Asadi Shamsabadi, E.; Ghalehnovi, M.; de Brito, J.; Khodabakhshian, A. Performance of Concrete with Waste Granite Powder: The Effect of Superplasticizers. *Appl. Sci.* **2018**, *8*, 1808. [CrossRef]
14. Taji, I.; Ghorbani, S.; de Brito, J.; Tam, V.W.Y.; Sharifi, S.; Davoodi, A.; Tavakkolizadeh, M. Application of Statistical Analysis to Evaluate the Corrosion Resistance of Steel Rebars Embedded in Concrete with Marble and Granite Waste Dust. *J. Clean. Prod.* **2019**, *210*, 837–846. [CrossRef]
15. Kim, J.; Grabiec, A.M.; Ubysz, A. An Experimental Study on Structural Concrete Containing Recycled Aggregates and Powder from Construction and Demolition Waste. *Materials* **2022**, *15*, 2458. [CrossRef]
16. Kim, J.; Kim, N. Recycling Waste Paver Blocks in the Manufacture of New Concrete Paver Blocks and Building Bricks. *Appl. Sci.* **2022**, *12*, 10970. [CrossRef]
17. *ASTM C618;* Standard Specification for Coal Fly Ash and Raw or Calcined Natural Pozzolan for Use in Concrete. ASTM International: West Conshohocken, PA, USA, 2022.
18. *ASTM C109;* Standard Test Method for Compressive Strength of Hydraulic Cement Mortars. ASTM International: West Conshohocken, PA, USA, 2021.
19. *ASTM C305;* Standard Practice for Mechanical Mixing of Hydraulic Cement Pastes and Mortars of Plastic Consistency. ASTM International: West Conshohocken, PA, USA, 2020.
20. *ASTM C1437;* Standard Test Method for Flow of Hydraulic Cement Mortar. ASTM International: West Conshohocken, PA, USA, 2020.

21. *ASTM C348*; Standard Test Method for Flexural Strength of Hydraulic-Cement Mortars. ASTM International: West Conshohocken, PA, USA, 2021.
22. *ASTM C596*; Standard Test Method for Drying Shrinkage of Mortar Containing Hydraulic Cement. ASTM International: West Conshohocken, PA, USA, 2018.
23. de Azevedo, A.R.G.; Marvila, M.T.; da Silva Barroso, L.; Zanelato, E.B.; Alexandre, J.; de Castro Xavier, G.; Monteiro, S.N. Effect of Granite Residue Incorporation on the Behavior of Mortars. *Materials* **2019**, *12*, 1449. [CrossRef]
24. Schackow, A.; Stringari, D.; Senff, L.; Correia, S.L.; Segadães, A.M. Influence of Fired Clay Brick Waste Additions on the Durability of Mortars. *Cem. Concr. Compos.* **2015**, *62*, 82–89. [CrossRef]
25. Singh, S.; Nagar, R.; Agrawal, V. A Review on Properties of Sustainable Concrete Using Granite Dust as Replacement for River Sand. *J. Clean. Prod.* **2016**, *126*, 74–87. [CrossRef]
26. Gupta, L.K.; Vyas, A.K. Impact on Mechanical Properties of Cement Sand Mortar Containing Waste Granite Powder. *Constr. Build. Mater.* **2018**, *191*, 155–164. [CrossRef]
27. Dong, L.; Zhang, Y.; Ma, J. Micro-Crack Mechanism in the Fracture Evolution of Saturated Granite and Enlightenment to the Precursors of Instability. *Sensors* **2020**, *20*, 4595. [CrossRef]
28. Chen, F.; Wu, K.; Ren, L.; Xu, J.; Zheng, H. Internal Curing Effect and Compressive Strength Calculation of Recycled Clay Brick Aggregate Concrete. *Materials* **2019**, *12*, 1815. [CrossRef]
29. Sicakova, A.; Figmigova, E. Two Variants of Internal Curing of RBA-Based Concrete and Their Influence on the Properties of Concrete. *Rev. Rom. Mater.* **2021**, *51*, 221–227.
30. Bacarji, E.; Toledo Filho, R.D.; Koenders, E.A.B.; Figueiredo, E.P.; Lopes, J.L.M.P. Sustainability Perspective of Marble and Granite Residues as Concrete Fillers. *Constr. Build. Mater.* **2013**, *45*, 1–10. [CrossRef]
31. Wong, C.L.; Mo, K.H.; Yap, S.P.; Alengaram, U.J.; Ling, T.C. Potential Use of Brick Waste as Alternate Concrete-Making Materials: A Review. *J. Clean. Prod.* **2018**, *195*, 226–239. [CrossRef]
32. Zhao, Y.; Gao, J.; Xu, Z.; Li, S.; Luo, X.; Chen, G. Long-Term Hydration and Microstructure Evolution of Blended Cement Containing Ground Granulated Blast Furnace Slag and Waste Clay Brick. *Cem. Concr. Compos.* **2021**, *118*, 103982. [CrossRef]
33. *SK EN 998-1:2019*; Specification for Mortar for Masonry—Part 1: Rendering and Plastering Mortar. The Office for Standardisation, Metrology and Testing of the Slovak Republic: Bratislava, Slovakia, 2019.
34. *SK EN 13813:2009*; Screed Material and Floor Screeds—Screed Material—Properties and Requirements. The Office for Standardisation, Metrology and Testing of the Slovak Republic: Bratislava, Slovakia, 2009.
35. *SK EN 998-2:2019*; Specification for Mortar for Masonry—Part 2: Masonry Mortar. The Office for Standardisation, Metrology and Testing of the Slovak Republic: Bratislava, Slovakia, 2019.
36. Donza, H.; Cabrera, O.; Irassar, E. High-Strength Concrete with Different Fine Aggregate. *Cem. Concr. Res.* **2002**, *32*, 1755–1761. [CrossRef]
37. Hong, G.; Oh, S.; Choi, S.; Chin, W.-J.; Kim, Y.-J.; Song, C. Correlation between the Compressive Strength and Ultrasonic Pulse Velocity of Cement Mortars Blended with Silica Fume: An Analysis of Microstructure and Hydration Kinetics. *Materials* **2021**, *14*, 2476. [CrossRef]
38. Estévez, E.; Martín, D.A.; Argiz, C.; Sanjuán, M.Á. Ultrasonic Pulse Velocity—Compressive Strength Relationship for Portland Cement Mortars Cured at Different Conditions. *Crystals* **2020**, *10*, 133. [CrossRef]
39. Cachim, P.B. Mechanical Properties of Brick Aggregate Concrete. *Constr. Build. Mater.* **2009**, *23*, 1292–1297. [CrossRef]
40. Chakradhara Rao, M. Influence of Brick Dust, Stone Dust, and Recycled Fine Aggregate on Properties of Natural and Recycled Aggregate Concrete. *Struct. Concr.* **2021**, *22*, E105–E120. [CrossRef]
41. Vardhan, K.; Siddique, R.; Goyal, S. Influence of Marble Waste as Partial Replacement of Fine Aggregates on Strength and Drying Shrinkage of Concrete. *Constr. Build. Mater.* **2019**, *228*, 116730. [CrossRef]
42. Khyaliya, R.K.; Kabeer, K.I.S.A.; Vyas, A.K. Evaluation of Strength and Durability of Lean Mortar Mixes Containing Marble Waste. *Constr. Build. Mater.* **2017**, *147*, 598–607. [CrossRef]
43. Hu, L.; Hao, J.; Wang, L. Laboratory Evaluation of Cement Treated Aggregate Containing Crushed Clay Brick. *J. Traffic Transp. Eng.* **2014**, *1*, 371–382. [CrossRef]
44. Dang, J.; Zhao, J.; Pang, S.D.; Zhao, S. Durability and Microstructural Properties of Concrete with Recycled Brick as Fine Aggregates. *Constr. Build. Mater.* **2020**, *262*, 120032. [CrossRef]
45. Abdalhmid, J.M.; Ashour, A.F.; Sheehan, T. Long-Term Drying Shrinkage of Self-Compacting Concrete: Experimental and Analytical Investigations. *Constr. Build. Mater.* **2019**, *202*, 825–837. [CrossRef]
46. Wu, H.; Xiao, J.; Liang, C.; Ma, Z. Properties of Cementitious Materials with Recycled Aggregate and Powder Both from Clay Brick Waste. *Buildings* **2021**, *11*, 119. [CrossRef]
47. Ge, Z.; Feng, Y.; Yuan, H.; Zhang, H.; Sun, R.; Wang, Z. Durability and Shrinkage Performance of Self-Compacting Concrete Containing Recycled Fine Clay Brick Aggregate. *Constr. Build. Mater.* **2021**, *308*, 125041. [CrossRef]

buildings

MDPI

Article

Method for Determining the Fracture Parameters of Fully Graded Dam Concrete

Xiaofeng Gao [1,2,*], Jiong Wu [1], Mengxia Zhou [3,4], Tao Xu [1], Chunfeng Liu [3,4], Yaosheng Tan [3,4], Ning Yang [3,4] and Yu Qiao [3,4]

1 College of Civil Engineering, Zhejiang University of Technology, Hangzhou 310023, China
2 State Key Laboratory of Hydroscience and Engineering, Tsinghua University, Beijing 100084, China
3 China Three Gorges Construction Engineering Corporation, Chengdu 610041, China
4 China Three Gorges Group Corporation, Beijing 100038, China
* Correspondence: gaoxf@zjut.edu.cn; Tel.: +86-571-85290813

Abstract: This paper describes a method for determining the initiation and unstable toughness of fully graded concrete of arbitrary specimen size. The method first predicts the initiation and peak loads of concrete specimens of any size, as well as crack length-to-height ratios based on the fracture test results of concrete specimens with limited sizes or crack length-to-height ratios. Then, combined with the fracture extreme theory, the fracture toughness of concrete with varying size or crack length-to-height ratios is determined. Finally, in order to verify the applicability of the method, it is used to calculate the fracture toughness of small aggregate concrete and fully graded concrete with different sizes or crack length-to-height ratios, and its prediction accuracy is evaluated through indices such as mean absolute percentage error, root mean square error and reliability index a15. The results show that the proposed method can meet the needs of practical engineering applications and can provide theoretical basis for the optimization of the fracture test method of fully graded concrete and the determination of fracture parameters in crack stability or propagation analysis.

Keywords: hydraulic structure; fully graded; fracture parameters; size effect; fracture extreme theory

Citation: Gao, X.; Wu, J.; Zhou, M.; Xu, T.; Liu, C.; Tan, Y.; Yang, N.; Qiao, Y. Method for Determining the Fracture Parameters of Fully Graded Dam Concrete. *Buildings* **2023**, *13*, 24. https://doi.org/10.3390/buildings13010024

Academic Editor: Tomáš Dvorský

Received: 15 November 2022
Revised: 13 December 2022
Accepted: 19 December 2022
Published: 22 December 2022

1. Introduction

In order to ensure the safe operation of dam structures, it is necessary to study the mechanical parameters of dam concrete under actual service conditions. Existing studies have shown that factors such as environment [1–3], concrete raw materials [4,5], and curing conditions [6] affect the toughness, strength, and elastic modulus of dam concrete. Among all of the above mechanical properties, fracture toughness is an important index to evaluate the crack resistance of dam concrete [7]. Generally, it can be determined by the wedge-splitting method [5,8,9] or the three-point bending-beam method [10–12]. Based on the double-*K* fracture theory proposed by Xu et al. [13–15], the National Development and Reform Commission of China [16] and RILEM [17] developed a concrete fracture test standard. The double-*K* fracture theory uses the initiation and unstable toughness to describe the whole process of concrete structure crack from initiation and crack propagation to unstable failure. Existing research shows that the double-*K* fracture toughness of concrete measured by this test may have a size effect [18–21], and the larger the maximum particle size of concrete aggregate, the larger the minimum specimen size required to determine the stable fracture toughness. For fully graded concrete with a maximum aggregate size of 150 mm, when the stable fracture toughness is determined by the wedge-splitting method, the required minimum specimen height should be 1.5 m [7,21,22]. The fracture testing of large-size specimens is associated with problems such as high equipment requirements, difficult testing, and inaccurate testing [22,23], so it is difficult to carry out normal tests at engineering sites. If a method can be found to determine the double-*K* fracture toughness of any size of specimen based on the fracture test results of smaller concrete specimens, it can provide a basis for the optimization of the fracture test of fully graded concrete. According to the double-*K* fracture toughness calculation formula, the initiation toughness is related to the initiation load and the initial crack length, whereas the calculation of the unstable toughness requires the peak load and the effective crack length. Therefore, the key is to solve the problem of how to determine the initiation load, peak load and effective crack length of specimens of any size through the fracture test results of smaller specimens.

In order to predict the fracture failure of quasi-brittle material specimens, scholars have proposed different theoretical models. The most famous ones are the boundary effect model proposed by Hu et al. [24–26] and the size effect model proposed by Bažant et al. [27–29]. Based on the boundary effect model, Guan et al. [30] established a linear formula between the peak load of concrete with different sizes or α_c and tensile strength, where α_c is the crack length-to-height ratio. Furthermore, Yao and Guan et al. [31–33] proposed the concept of cracking strength and established a linear expression between the initiation load and the cracking strength of concrete specimens when the size or α_c changes. Guan et al. [10,34] also established the relationship between fracture toughness and the failure curve based on the improved boundary effect model. The above initiation and peak load prediction methods based on the boundary effect model are currently used for small aggregate concrete. For fully graded concrete, Gao et al. [7] established a linear relationship between the nominal initiation strength, nominal unstable strength and the height of the specimen based on the type 2 size effect model. By introducing the equivalent geometric parameters, the prediction formula of initiation and peak loads of concrete specimens is proposed when the size or α_c changes, so as to realize the prediction of initiation and peak loads of specimens with any size or α_c. Although the basic assumptions of the boundary effect and the size effect model are different, the predicted results obtained by the two are basically consistent. In this study, the size effect model is selected as the method to determine the initiation and peak loads of concrete specimens when the size or α_c changes.

After determining the initiation and peak loads of any size of specimen, the initiation toughness can be directly obtained by the initiation load and the initial crack length, whereas the calculation of the unstable toughness requires the critical crack length (a_c). Because the predicted fracture results based on the size effect model cannot yield a_c, it needs to be obtained by establishing and solving the equations of the critical state of concrete specimens. Based on a fictitious crack model, Wu et al. [35,36] established the equilibrium equation of force and moment in the midspan section of a three-point bending beam and obtained a_c with the flexural tensile strength of concrete as the known quantity. In order to determine the fracture parameters of concrete more easily and reliably, Qing et al. proposed a fracture extreme theory that can determine a_c and other fracture parameters only based on the peak load of fracture test [37–47]. The theory assumes that the partial derivative of the F_H to crack length is continuous when F_H reaches the peak load during the fracture process of concrete, and the derivative of F_H to crack length is zero at the extreme point. Furthermore, based on the relationship between F_H and a under the peak load state, combined with several other equations, the a_c and other fracture parameters of specimens of different sizes can be determined. The accuracy of fracture extreme value theory for predicting fracture parameters of concrete has been widely verified. Accordingly, a_c can be easily and reliably determined by fracture extreme value theory with only the peak load. Therefore, after the peak load is determined, the fracture extreme theory proposed by Qing et al. is an effective method to solve for the critical crack length a_c of any size specimen.

In summary, based on the fracture test results of concrete specimens with limited size and α_c, the size effect model can be used to predict the initiation and peak loads of concrete specimens with any size or α_c, and combined with the fracture extreme theory, the initiation and unstable toughness of concrete specimens can be calculated when the size or α_c changes. Although this method is theoretically feasible, its applicability and accuracy still need to be verified by test results. In this paper, we first introduce the determination method of fracture parameters and then verify the applicability and accuracy of the method using the fracture test results of small aggregate concrete and fully graded large aggregate concrete. The research results presented in this paper can provide a theoretical basis for the optimization of the fracture test at the engineering site, as well as for the determination of fracture parameters in crack stability or propagation analysis.

2. Determination Method of Fracture Parameters

The fracture parameter determination method proposed in this paper is based on the test results of concrete specimens with limited size or α_c. The size effect model is used to predict the initiation and peak loads of concrete specimens with an arbitrary size or α_c. Then, the predicted value of the peak load is taken as the known condition, and the critical crack length is determined according to the fracture extreme theory. Finally, the initiation and unstable toughness of concrete specimens are determined based on the predicted initiation load and initial crack length and the predicted peak load and critical crack length, respectively, when the specimen size or α_c changes. The calculation flow chart is shown in Figure 1. This method can be used to analyze the fracture test results of different specimens, such as those obtained by the wedge-splitting method, the three-point bending-beam method and the compact tension method. Considering that the self-weight of fully graded concrete specimens is generally large, the wedge-splitting method is used to determine the fracture parameters, which can overcome the influence of the self-weight of the specimen on the fracture process. Therefore, the applicability and accuracy of the method are verified by the wedge-splitting fracture test results in the literature. The formula for calculating the initiation and unstable toughness of wedge-splitting specimens given by the "Norm for fracture test of hydraulic concrete" [16] in China is as follows:

$$K_{IC}^{Q} = \frac{F_{HQ} \times 10^{-3}}{th^{1/2}} f(\alpha_0) = Y(\alpha_0)\sigma_N^Q \sqrt{\pi a_0} \quad (1)$$

$$\sigma_N^Q = \frac{4F_{HQ} \times 10^{-3}}{th} \quad (2)$$

$$K_{IC}^{S} = \frac{F_{Hmax} \times 10^{-3}}{th^{1/2}} f(\alpha_c) = Y(\alpha_c)\sigma_N^S \sqrt{\pi a_c} \quad (3)$$

$$\sigma_N^S = \frac{4F_{Hmax} \times 10^{-3}}{th} \quad (4)$$

$$Y(\alpha) = \frac{f(\alpha)}{4\sqrt{\pi\alpha}} = \frac{3.675[1 - 0.12(\alpha - 0.45)]}{4\sqrt{\pi\alpha}(1-\alpha)^{3/2}} \quad (5)$$

where K_{IC}^{Q} and K_{IC}^{S} are the initiation toughness and unstable toughness, respectively (MPa·m$^{1/2}$); F_{HQ} and F_{Hmax} are the initiation load and peak load, respectively (kN); $f(\alpha_0)$ and $Y(\alpha_0)$ are the dimensionless geometric parameters of the specimen; a_c is the critical crack length when the specimen reaches the peak load (m); α_c is the crack length-to-height ratio ($\alpha_c = a_c/h$); and σ_N^Q and σ_N^S are the nominal initiation and unstable strength, respectively, i.e., the maximum tensile stress of the fracture surface caused by the initiation load and peak load without considering the existence of an initial crack (MPa).

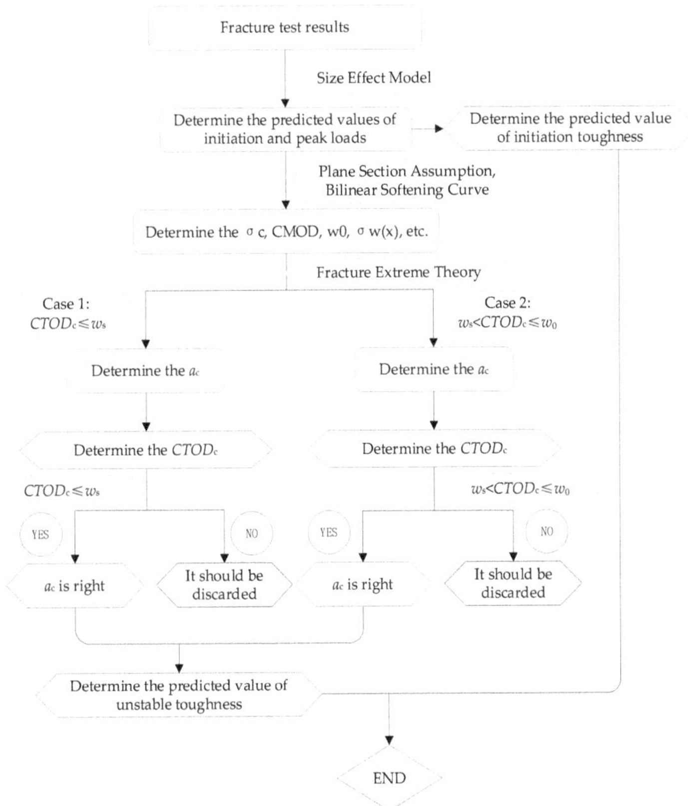

Figure 1. Calculation flow chart.

2.1. *Determination of Initiation Load and Peak Load*

In this paper, the size effect model is selected as the method to determine the initiation and peak loads of concrete specimens when the size or α_c changes. The evolution form of the type 2 size effect model proposed by Gao et al. [48] can be used to analyze the size effect of fracture test results in geometrically similar specimens. On this basis, in order to analyze the test results when the specimen size or α_c changes, the linear equations of nominal strength and equivalent crack length as shown in Equations (6) and (7) can be established [48].

$$\frac{1}{\left(\sigma_{\rm Ne}^{\rm Q}\right)^2} = \frac{1}{\left(K_{\rm IC}^{\rm ini}\right)^2} a_{\rm e} + \frac{c_{\rm f}^{\rm ini}}{\left(K_{\rm IC}^{\rm ini}\right)^2} = A_1 a_{\rm e} + C_1 \tag{6}$$

$$\frac{1}{\left(\sigma_{\rm Ne}^{\rm S}\right)^2} = \frac{1}{\left(K_{\rm IC}^{\rm un}\right)^2} a_{\rm e} + \frac{c_{\rm f}^{\rm un}}{\left(K_{\rm IC}^{\rm un}\right)^2} = A_2 a_{\rm e} + C_2 \tag{7}$$

where $\sigma_{\rm Ne}^{\rm Q}$ and $\sigma_{\rm Ne}^{\rm S}$ are equivalent nominal initiation strength and equivalent nominal unstable strength, respectively ($\sigma_{\rm Ne}^{\rm Q} = \sigma_{\rm N}^{\rm Q} H(\alpha)\pi^{1/2}$, $\sigma_{\rm Ne}^{\rm S} = \sigma_{\rm N}^{\rm S} H(\alpha)\pi^{1/2}$) (MPa); $a_{\rm e}$ is the equivalent crack length ($a_{\rm e} = Y^2(\alpha_0)/H^2(\alpha_0) \times a_0$, m, $H(\alpha_0) = \left[Y^2(\alpha_0) + 2Y(\alpha_0) \times Y'(\alpha_0) \times \alpha_0\right]^{1/2}$); and A_1, A_2, C_1 and C_2 are the parameters of two linear equations, which can be determined by fitting the test results.

Based on the experimental $F_{\rm HQ}$ and $F_{\rm Hmax}$, the linear relationship between $1/\left(\sigma_{\rm Ne}^{\rm Q}\right)^2$, $1/\left(\sigma_{\rm Ne}^{\rm S}\right)^2$ and $a_{\rm e}$ can be established by Equation (6) and Equation (7), respectively. According to the slope and intercept of the two linear equations, the initiation toughness without the size effect (K_{IC}^{ini}), the effective length of the initiation fracture process zone (c_f^{ini}), the unstable toughness without the size effect (K_{IC}^{un}), and the effective length of the unstable fracture process zone (c_f^{un}) can be calculated; $K_{IC}^{ini} = 1/(A_1)^{1/2}$, $c_f^{ini} = C_1/A_1$, $K_{IC}^{un} = 1/(A_2)^{1/2}$, $c_f^{un} = C_2/A_2$.

Furthermore, Gao et al. [7] introduced the equivalent geometric parameters into the size effect model and established the prediction formulas of initiation and peak loads when the size or α_c of a concrete specimen changed, as shown in Equations (8) and (9).

$$F_{HQ}^{Pre} = \frac{ht \times 10^3}{4\sqrt{g'(\alpha_0)c_f^{ini} + g(\alpha_0)h}} K_{IC}^{ini} = A_e^{ini} K_{IC}^{ini} \tag{8}$$

$$F_{Hmax}^{Pre} = \frac{ht \times 10^3}{4\sqrt{g'(\alpha_0)c_f^{un} + g(\alpha_0)h}} K_{IC}^{un} = A_e^{un} K_{IC}^{un} \tag{9}$$

where F_{HQ}^{Pre} and F_{Hmax}^{Pre} are the initiation and peak load prediction, respectively; $g(\alpha_0)$ and $g'(\alpha_0)$ are the dimensionless geometric parameters of the specimen ($g(\alpha_0) = Y^2(\alpha_0)\pi\alpha_0$, $g'(\alpha_0) = H^2(\alpha_0)\pi$); and A_e^{ini} and A_e^{un} are equivalent geometric parameters related to initiation and instability, respectively (mm·m$^{1/2}$). Other parameters are the same as the previous definition.

2.2. Determination of Initiation Toughness and Unstable Toughness

After the F_{HQ}^{Pre} and F_{Hmax}^{Pre} are determined, the K_{IC}^{Q} can be calculated directly according to Equation (1). For K_{IC}^{S}, a_c should be calculated first according to fracture extreme theory and substituted into Equation (3) to determine K_{IC}^{S}. According to the fracture extreme theory, at the extreme point, the derivative of F_H to a is zero [37], namely:

$$\left.\frac{\partial F_H}{\partial a}\right|_{a=a_c} = 0 \tag{10}$$

The functional relationship between F_H and a can be established by fracture propagation criterion equation in the critical state of concrete or the equilibrium equation of a ligament section. By combining the established function relation and Equation (10), a_c can be solved. In this paper, the functional relationship between F_H and a is established by using the equilibrium equation of force and moment of ligament section. Based on a fictitious crack model, Wu [35] and Qing [39] obtained the stress and strain distributions of ligament sections of three-point bending-beam specimens and wedge-splitting specimens, as shown in Figure 2.

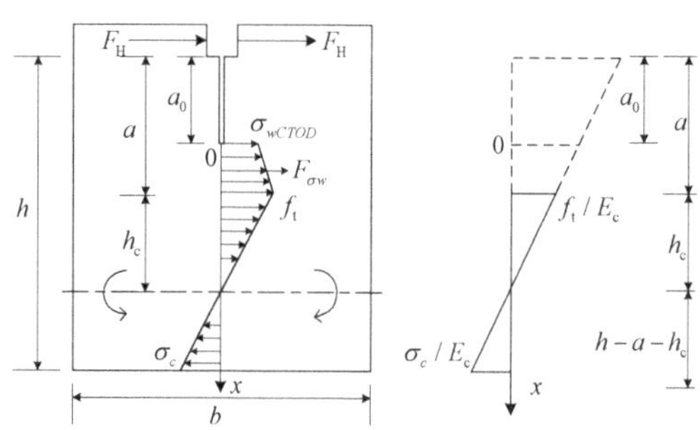

Figure 2. Stress and strain distribution in a ligament cross section of a wedge-splitting specimen.

According to the stress distribution shown in Figure 2, with the initial crack tip as the coordinate origin, the force and moment equilibrium equation of the ligament section shown in Equations (11) and (12), respectively, can be obtained [35,39]:

$$\frac{1}{2}\sigma_c t(h-a-h_c)+F_H\times 10^{-3}=\frac{1}{2}f_t t h_c+\int_0^{a-a_0}\sigma_w t dx \tag{11}$$

$$\frac{1}{3}\sigma_c t(h-a-h_c)^2+\frac{1}{3}f_t t h_c^2+\int_0^{a-a_0}\sigma_w t(h_c+a-a_0-x)dx=F_H(a+h_c)\times 10^{-3} \tag{12}$$

where σ_c is the compressive stress at the bottom of the specimen (MPa); t is the thickness of specimen (m); h is the effective height of the specimen (m); a_0 and a are initial crack length and effective crack length, respectively (m); h_c is the distance from the crack tip to the neutral axis marked by the dotted line in Figure 2 (m); f_t is the equivalent tensile stress at the crack tip, namely the equivalent tensile strength (MPa); and σ_w is the cohesive force in the fracture process zone (MPa).

According to the plane section assumption and the strain distribution shown in Figure 2, the expression of σ_c is obtained as follows [35,39]:

$$\sigma_c=\frac{f_t(h-a-h_c)}{h_c} \tag{13}$$

According to the assumption that the crack-opening surface remains a plane [35,39], the relationship between the virtual crack-opening displacement ($w(x)$) and the crack-tip opening displacement ($CTOD$, $w(x)$), as well as the crack-mouth-opening displacement ($CMOD$) can be established as:

$$w(x)=\left(1-\frac{x}{a-a_0}\right)CTOD \tag{14}$$

$$CTOD=\frac{a-a_0}{a}CMOD \tag{15}$$

where $CMOD$ can be obtained according to the empirical calculation proposed by Xu et al. [15].

$$CMOD=\frac{F_H}{tE_c}\left[13.18\left(1-\frac{a}{h}\right)^{-2}-9.16\right] \tag{16}$$

The relationship between the cohesive force in the fracture process zone and the virtual fracture-opening displacement can be characterized by the tensile softening curve [49], which often adopts a bilinear, exponential, power function, curve or other form. The bilinear softening curve shown in Figure 3 is used in this paper.

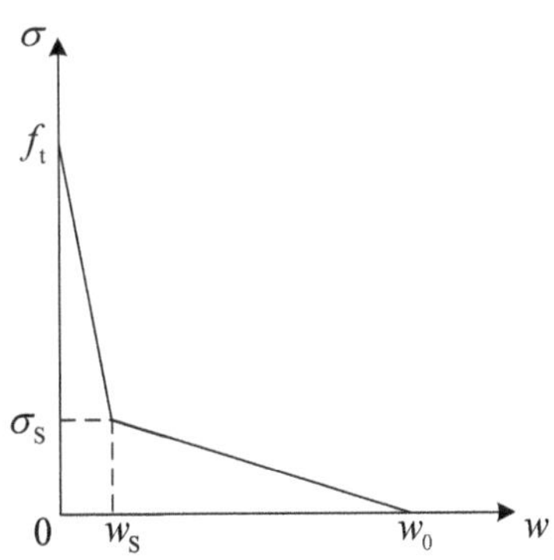

Figure 3. Bilinear softening curve.

The function expression of the bilinear softening curve is expressed as follows:

$$\sigma_w(x) = \begin{cases} f_t - (f_t - \sigma_s)w(x)/w_s & 0 \le w(x) \le w_s \\ \sigma_s(w_0 - w(x))/(w_0 - w_s) & w_s \le w(x) \le w_0 \end{cases} \tag{17}$$

where $w(x)$, w_s and w_0 are the crack-opening displacement at x, the crack-opening displacement at the inflection point of the bilinear softening relationship and the crack-opening displacement at zero cohesion, respectively (μm); and $\sigma_w(x)$ and σ_s are the cohesive force at x and the inflection point, respectively (MPa). The values of σ_s, w_s and w_0 in the formula can be calculated using the assumption proposed by Wittmann et al. [50]:

$$\begin{cases} \sigma_s = f_t/4 \\ w_s = 0.75G_f/f_t \\ w_0 = 5G_f/f \end{cases} \tag{18}$$

In order to reduce the parameters used in the calculation process, in combination with Equation (18), Equation (17) can be rewritten as follows:

$$\sigma_w(x) = \begin{cases} f_t(1 - 5w(x)/w_0) & 0 \le w(x) \le w_s \\ \frac{5}{17}f_t(1 - w(x)/w_0) & w_s \le w(x) \le w_0 \end{cases} \tag{19}$$

In the tensile softening curve, the crack-opening displacement (w_0), which characterizes the zero cohesion, is the material constant, and its value can be generally determined by the direct tensile test [51] or empirical formula [52]. For concrete materials, the variation range of w_0 provided by multiple scholars is large [51–53], mostly between 100 and 1000 μm. Owing to the difficulty of conducting a direct tensile test on fully graded concrete [54] and because the specific empirical formula may not be applicable to different types of concrete materials, in this paper, we determine w_0 using the best fitting method. First, w_0 is selected from 100 to 1000 μm at 50 μm intervals; then, a_c and K_{IC}^S can be calculated according to the fracture extreme theory. When the overall error between the calculated K_{IC}^S and the test results is the smallest, the corresponding w_0 is the best crack-opening displacement at zero cohesion of concrete.

Equation (14) is substituted into (19) to obtain the expression of the cohesion distribution function in the fracture process area, and a_c can be calculated. The $\sigma_w(x)$ selected in this paper is the bilinear softening curve. Therefore, the relationship between $w(x)$ and w_s at the inflection point of the bilinear softening relationship needs to be discussed when determining the stress distribution in the fracture process area. Obviously, under the peak load state, the maximum value of $w(x)$ is $CTOD$, which is located at the initial crack tip, whereas the minimum value of $w(x)$ is 0, which is located at the equivalent crack tip. When $CTOD_c \le w_s$, the cohesion from the initial crack tip to the equivalent crack tip is a linear distribution, whereas when $w_s < CTOD_c \le w_0$, the cohesion is a bilinear distribution. The following two situations are discussed.

Case 1: $CTOD_c \le w_s$. Figure 4 shows the cohesive stress distribution and the corresponding softening relationship in the fracture process zone when $CTOD_c \le w_s$.

Substituting Equation (14) into (19), the cohesive stress distribution function in the fracture process zone of Case 1 can be obtained as:

$$\sigma_w(x) = f_t\left[1 - 5\left(1 - \frac{x}{a-a_0}\right)\frac{CTOD}{w_0}\right] \quad (0 \le x \le a - a_0) \tag{20}$$

Substituting Equations (13) and (20) into (11), the distance from the equivalent crack tip to the neutral axis (h_c) can be obtained as:

$$h_c = \frac{(h-a)^2}{2\left[(a-a_0)\left(1 - \frac{5CTOD}{2w_0}\right) + h - a - \frac{F_H \times 10^{-3}}{f_t t}\right]} \tag{21}$$

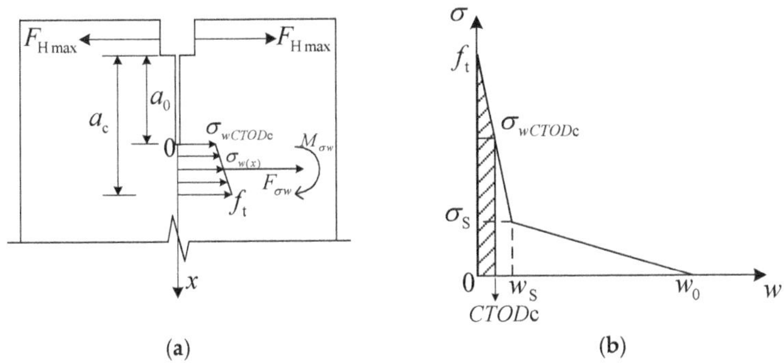

Figure 4. Case 1: (**a**) cohesive stress distribution; (**b**) softening curve.

The functional relationship between F_H and a can be obtained by substituting Equations (13) and (20) into (12):

$$\frac{F_H(a+h_c)\times 10^{-3}}{f_t t} = (a-a_0)\left[(a-a_0)\left(\frac{1}{2}-\frac{5CTOD}{3w_0}\right)+h_c\left(1-\frac{5CTOD}{2w_0}\right)\right]+ \frac{(h-a)^3}{3h_c}-(h-a)^2+(h-a)h_c \quad (22)$$

Furthermore, by substituting Equations (15), (16) and (21) into (22), a function containing only F_H, a and f_t is obtained. According to the fracture extreme theory, the partial derivative of F_H to a is zero at the extreme point. Substituting $a = a_c$ and $F_H = F_{Hmax}$ into Equations (10) and (22), a_c can be obtained; then, $CTOD_c$, K_{IC}^S and other parameters can be calculated. If the calculated $CTOD_c \le w_s$, a_c is the real solution, and on the contrary, it should be discarded.

Case 2, $w_s < CTOD_c \le w_0$. Figure 5 shows the cohesive force distribution and the corresponding softening relationship in the fracture process zone when $w_s < CTOD_c \le w_0$.

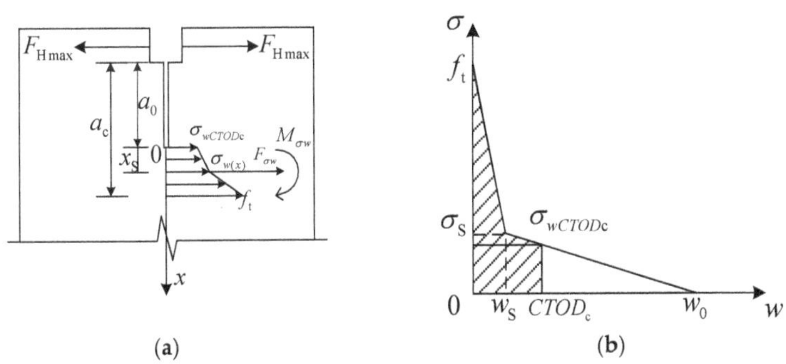

Figure 5. Case 2: (**a**) cohesive stress distribution; (**b**) softening curve.

Figure 5a shows that the cohesive force is a bilinear distribution, and the coordinate (x_s) of the turning point can be calculated according to the following formula:

$$x_s = (a-a_0)\left(1-\frac{3w_0}{20CTOD}\right) \quad (23)$$

Similar to Case 1, the cohesive force distribution function in the fracture process zone of Case 2 can be obtained by substituting Equation (14) into (19):

$$\sigma_w(x) = \begin{cases} \frac{5}{17} f_t\left[1-\left(1-\frac{x}{a-a_0}\right)\frac{CTOD}{w_0}\right] & 0 \le x \le x_s \\ f_t\left[1-5\left(1-\frac{x}{a-a_0}\right)\frac{CTOD}{w_0}\right] & x_s \le x \le a-a_0 \end{cases} \quad (24)$$

Furthermore, the distance from the crack tip to the neutral axis is:

$$h_{c2} = \frac{(h-a)^2}{2\left[-\frac{12}{17}x_s + \frac{80CTODx_s}{17w_0} - \frac{40CTODx_s^2}{17w_0(a-a_0)} - \frac{5(a-a_0)CTOD}{2w_0} + h - a_0 - \frac{F_H \times 10^{-3}}{f_t t}\right]} \quad (25)$$

By substituting Equations (13) and (24) into (12), the relation between the Case 2 F_H and a is obtained:

$$\begin{aligned} \frac{F_H(a+h_c)\times 10^{-3}}{f_t t} &= \frac{(h-a)^3}{3h_c} - (h-a)^2 + (h-a)h_c + \frac{5}{17}x_s\left(1-\frac{CTOD}{w_0}\right)(h_c+a-a_0) - \\ &\frac{5}{34}x_s^2\left[1-\frac{2CTOD}{w_0}-\frac{h_cCTOD}{(a-a_0)w_0}\right] - \frac{5x_s^3CTOD}{51(a-a_0)w_0} + (a-a_0-x_s)(h_c+a-a_0)\left(1-\frac{5CTOD}{w_0}\right) - \\ &\frac{5\left[(a-a_0)^3-x_s^3\right]CTOD}{3(a-a_0)w_0} - \frac{1}{2}\left[(a-a_0)^2-x_s^2\right]\left[1-\frac{10CTOD}{w_0}-\frac{5h_cCTOD}{(a-a_0)w_0}\right] \end{aligned} \quad (26)$$

By substituting Equations (15), (16), (23) and (25) into (26), a functional equation containing only F_H, a and f_t can be obtained. By substituting $a = a_c$ and $F_H = F_{Hmax}$ into Equations (10) and (26), respectively, parameters such as a_c and K_{IC}^S can be obtained. If the calculated $CTOD_c$ satisfies condition 2, a_c is the real solution, and on the contrary, it should be discarded.

According to the above calculation process, a_c and K_{IC} corresponding to a specific w_0 can be obtained. By repeating the above process, the best w_0 value with the smallest overall error between K_{IC}^S and test results can be determined. The error or prediction accuracy can be evaluated using the indicators commonly used in engineering.

2.3. Evaluation Index of Model Prediction Accuracy

The root mean square error (*RMSE*), mean absolute percentage error (*MAPE*) and reliability index a_{15} were used to evaluate the prediction accuracy of the model. The calculation formula of each indicator is as follows:

$$RMSE = \sqrt{\frac{1}{n}\sum_{i=1}^{n}(\hat{y}_i - y_i)^2} \quad (27)$$

$$MAPE = \frac{1}{n}\sum_{i=1}^{n}\left|\frac{\hat{y}_i - y_i}{y_i}\right| \times 100\% \quad (28)$$

$$a15 = \frac{m15}{M} \quad (29)$$

$$Cov(y_i, \hat{y}_i) = E[y_i\hat{y}_i] - E[y_i]E[\hat{y}_i] \quad (30)$$

where n is the amount of data points; i is the number of the data point; y_i and $\hat{y}_i$ are the measured and predicted values of the i-th data point, respectively; a_{15} is the proportion of samples with deviation within ±15% compared with the test results; $m15$ and M are the number of samples with prediction error less than ±15% and the number of samples in the dataset, respectively; $Cov(y_i, \hat{y}_i)$ is the covariance of the measured and predicted values; and $E[y_i]$ and $E[\hat{y}_i]$ are the expected values of the measured and predicted values, respectively.

3. Method Validation and Discussion

In this section, four groups of fracture test results of different sizes and the α_c of small aggregate concrete and Wudongde fully graded concrete are selected to verify the applicability and accuracy of the proposed method.

3.1. Small Aggregate Concrete

3.1.1. Same Crack Length-to-Height Ratio and Different Specimen Sizes

(1) $\alpha_0 = 0.4$, $h = 0.2$~1 m

In order to study the influence of specimen size on the fracture performance of concrete, a fracture test was carried out in [55] on wedge-splitting tensile specimens with a maximum aggregate particle size of 25 mm, an effective specimen height ranging from 0.2 m to 1 m and an initial α_c of 0.4. Table 1 shows the specimen size information and test results. The elastic modulus of concrete (E_c = 30.7 GPa) was obtained by the conversion of cube compressive strength. Additional experimental information can be found in reference [55].

Table 1. Fracture test results reported in reference [55].

Specimen Number	h/m	b/m	α_0	F_{HQ}/kN	F_{Hmax}/kN
200-1	0.20	0.20	0.40	7.85	10.02
200-2	0.20	0.20	0.40	7.81	10.69
200-3	0.20	0.20	0.40	7.65	9.62
200-4	0.20	0.20	0.40	9.27	10.60
200-6	0.20	0.20	0.40	8.65	10.93
300-1	0.30	0.20	0.40	9.97	15.33
300-2	0.30	0.20	0.40	10.45	14.78
300-3	0.30	0.20	0.48	10.00	12.56
300-4	0.30	0.20	0.40	11.31	15.12
300-5	0.30	0.20	0.40	8.90	12.06
600-1	0.60	0.21	0.40	20.47	26.68
600-2	0.61	0.22	0.44	19.20	28.13
600-3	0.60	0.22	0.40	15.84	27.02
600-4	0.60	0.22	0.40	22.90	29.12
600-5	0.60	0.21	0.48	28.17	28.66
600-6	0.61	0.21	0.44	26.49	31.50
800-1	0.81	0.24	0.43	31.47	34.69
800-2	0.81	0.25	0.40	30.64	34.99
800-3	0.80	0.24	0.40	20.65	33.72
800-4	0.80	0.24	0.40	22.56	33.96
800-5	0.80	0.23	0.45	33.90	37.83
800-6	0.80	0.23	0.40	23.72	34.25
1000-1	1.01	0.24	0.40	28.75	43.58
1000-2	1.01	0.24	0.40	28.53	46.63
1000-3	1.01	0.27	0.39	38.51	55.91
1000-4	1.01	0.24	0.40	35.15	48.16
1000-5	1.00	0.25	0.40	30.19	45.52
1000-6	1.01	0.27	0.40	37.32	50.47

The linear equations of $1/\left(\sigma_{\mathrm{Ne}}^{\mathrm{Q}}\right)^2$, $1/\left(\sigma_{\mathrm{Ne}}^{\mathrm{S}}\right)^2$ and a_e, as shown in Figure 6, can be obtained by linear analysis of the test results using Equations (6) and (7). According to the slope and intercept, the initiation toughness without a size effect ($K_{\mathrm{IC}}^{\mathrm{ini}}$) is 1.14 MPa·m$^{1/2}$, the effective length of the initiation fracture process zone ($c_{\mathrm{f}}^{\mathrm{ini}}$) is 56.2 mm, the unstable toughness ($K_{\mathrm{IC}}^{\mathrm{un}}$) is 1.81 MPa·m$^{1/2}$ and the effective length of unstable fracture process zone ($c_{\mathrm{f}}^{\mathrm{un}}$) is 108.6 mm.

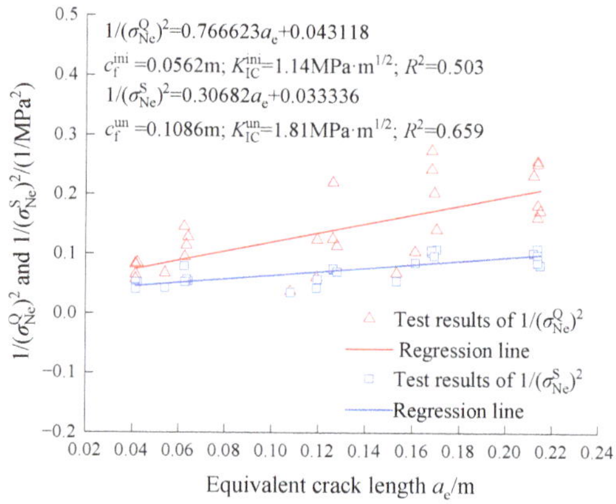

Figure 6. Linear relationships of $1/\left(\sigma_{\mathrm{Ne}}^{\mathrm{Q}}\right)^2$, $1/(\sigma_{\mathrm{Ne}}^{\mathrm{S}})^2$ and a_{e}.

The $K_{\mathrm{IC}}^{\mathrm{ini}}$ and $c_{\mathrm{f}}^{\mathrm{ini}}$ and $K_{\mathrm{IC}}^{\mathrm{un}}$ and $c_{\mathrm{f}}^{\mathrm{un}}$ obtained by linear analysis are substituted into Equations (8) and (9), respectively, to obtain the relationship between the initiation and peak loads with the equivalent geometric parameter (A_{e}), as shown in Figure 7. In the figure, the slope of the initiation load prediction line is $K_{\mathrm{IC}}^{\mathrm{ini}}$ without a size effect, and the slope of the peak load prediction line is $K_{\mathrm{IC}}^{\mathrm{un}}$ without a size effect. Figure 7 shows that the test results of F_{HQ} and F_{Hmax} are nearly linear relative to A_{e}. The *MAPE* of the F_{HQ} prediction is 14.29%, the *RMSE* is 4.54 kN, reliability index *a*15 is 67.86% and the covariance is 83.39 (kN)^2. The *MAPE* of the F_{Hmax} prediction is 7.4%, the *RMSE* is 3.05 kN, *a*15 is 85.71% and the covariance is 187.14 (kN)^2. The above indices show that the size effect model can realize the accurate prediction of initiation load and peak load when the size or α_{c} changes.

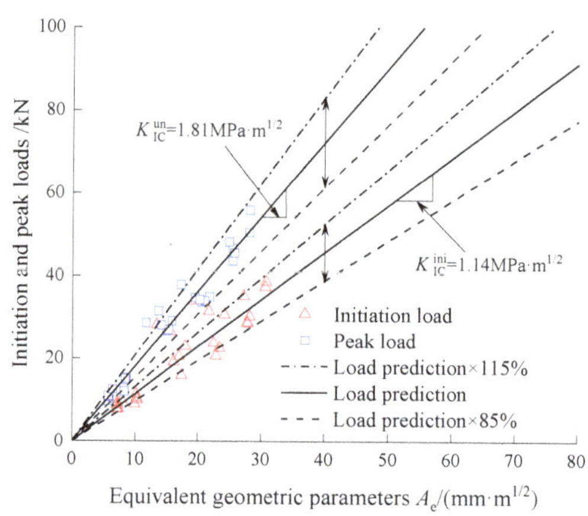

Figure 7. Predicted results of initiation and peak loads.

As mentioned above, in this paper, we use the best fitting method to determine the crack-opening displacement (w_0). In this test, *MAPE* was used to evaluate the prediction accuracy of $K_{\mathrm{IC}}^{\mathrm{S}}$; the calculation results are shown in Figure 8. It can be seen that the minimum *MAPE* is 4.56%, and the corresponding optimal w_0 is 550 μm.

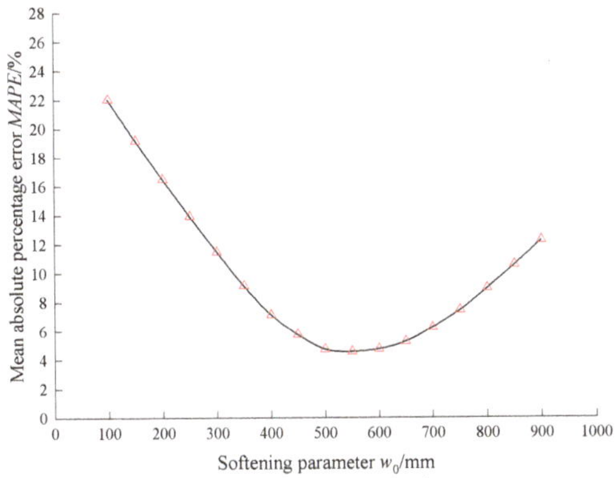

Figure 8. Determination of the softening parameter (w_0).

Based on the optimal w_0 and F^{Pre}_{Hmax} obtained above, the critical crack length of specimens of different sizes with an initial α_c of 0.4 was calculated using Equations (3) and (10). Figure 9 shows that each data point is located near the calculated value, with the following corresponding prediction accuracy evaluation indices: $MAPE = 3.42\%$, $RMSE = 0.02$ m, $a15 = 100\%$ and $Cov = 0.02$ m^2.

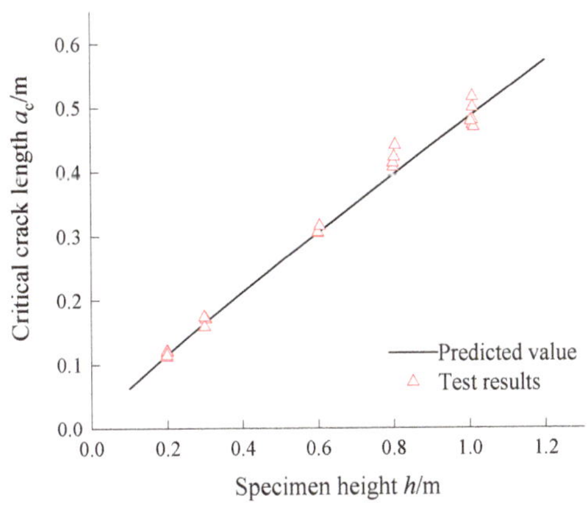

Figure 9. Critical crack length of specimens of different sizes.

Figure 10a shows a comparison between the predicted and test results of unstable toughness (K^S_{IC}), nominal toughness (K^N_{IC}) and initiation toughness (K^Q_{IC}) for specimens of different sizes with an initial α_c of 0.4. The nominal toughness is calculated by substituting the predicted peak load (F^{Pre}_{Hmax}) and the initial crack length (a_0) into Equation (3). It can be seen from Figure 10a that K^S_{IC}, K^N_{IC} and K^Q_{IC} basically increase with the increase in specimen size and gradually stabilize. Among them, K^Q_{IC} tends toward the initiation toughness without a size effect defined in the size effect model. K^S_{IC} and K^N_{IC} tend toward unstable toughness without a size effect, and the difference between them decreases with increased specimen size.

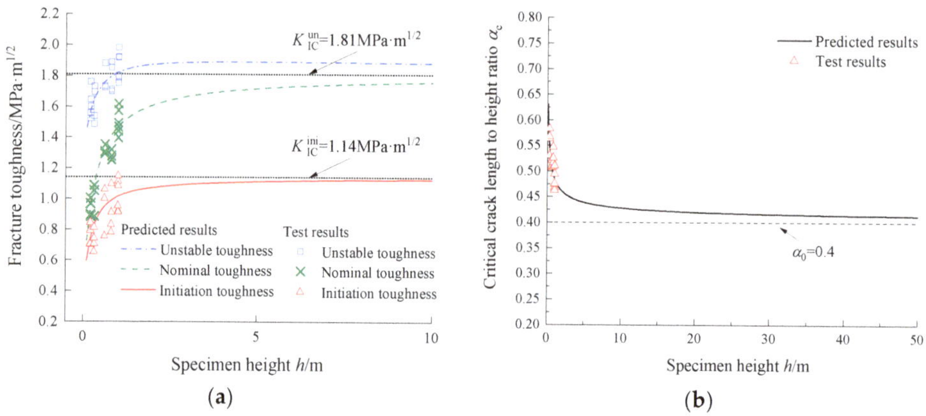

Figure 10. Fracture toughness and crack length-to-height ratio of specimens with different sizes: (**a**) fracture toughness; (**b**) critical crack length-to-height ratios.

This is mainly because the difference between K_{IC}^{S} and K_{IC}^{N} lies in the selection of crack length in toughness calculation. It can be seen from Figure 10b that with the increase in specimen size, α_c gradually approaches α_0. A comparison of the predicted results with the test results shows that the predicted *MAPE* of K_{IC}^{S} is 4.56%, *RMSE* is 0.09 MPa·m$^{1/2}$, *a*15 is 100% and the covariance is 0.01 (MPa)^2 m. Furthermore, $MAPE = 11.46\%$, $RMSE = 0.11$ MPa · m$^{1/2}$, $a15 = 81.82\%$ and $Cov = 0.01$ (MPa)^{2}m of K_{IC}^{Q} indicate that the prediction accuracy of initiation and unstable toughness can meet the requirements of engineering applications.

Figure 11 shows the calculation results of the critical tip-opening displacement ($CTOD_c$) and the cohesive force ($F_{\sigma w}$) in the fracture process zone of specimens with different sizes and an initial α_c of 0.4. It can be seen from Figure 11 that $CTOD_c$ increases rapidly when the specimen height is less than 10 m. When the specimen height reaches 10 m, $CTOD_c$ tends to be stable, and the stable $CTOD_c$ is 47.4 μm. Similarly, $F_{\sigma w}$ first increases rapidly with the increase in specimen size and then tends to be stable at 42.4 kN.

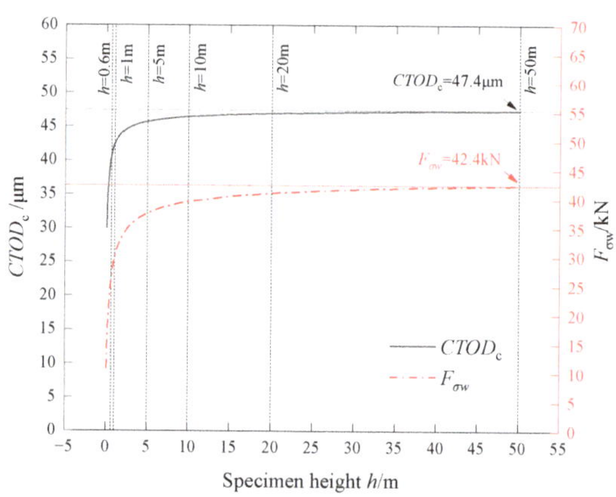

Figure 11. $CTOD_c$ and $F_{\sigma w}$ of specimens with different sizes.

Figure 12 shows the midspan section stress distribution of wedge-splitting tensile specimens with heights of 0.6, 1, 5, 20, 10, 20 and 50 m. In the figure, F_t is the tensile force of the uncracked zone in the midspan section of the specimen (kN), and F_c is the pressure-resultant force on the uncracked area of the midspan section of the specimen (kN).

It can be seen from the figure that with the increase in specimen size, F_t and F_c gradually increase. F_{sw} increases with the increase in size, and its growth rate gradually slows down and tends toward stability increased size.

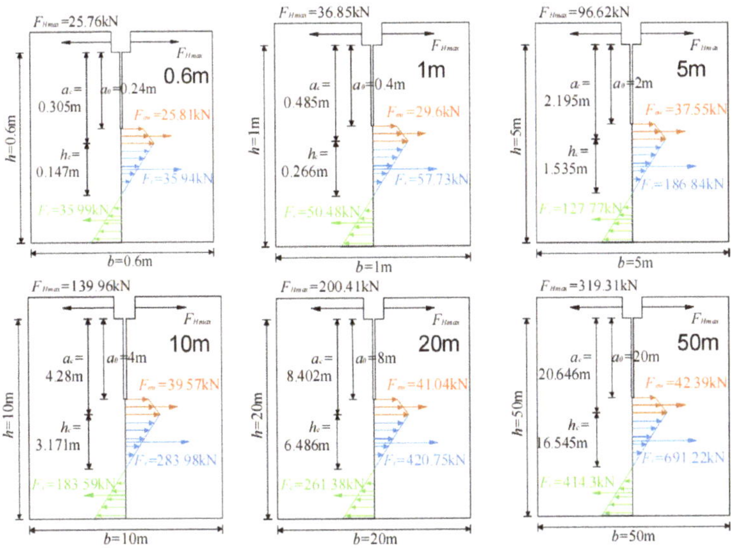

Figure 12. Stress distribution in the midspan section of specimens with different heights.

(2) $\alpha_0 = 0.4$, $h = 0.3$~0.6 m

In reference [56], the fracture properties of wedge-splitting tensile specimens with an α_c of 0.4 and effective heights of 0.3, 0.4, 0.5 and 0.6 m were tested. Table 2 shows the specimen size and test results. Reference [56] reported the test results used to calculate the elastic modulus. The elastic modulus (E_c) required by the method proposed in this chapter was taken as the average value of the calculated elastic modulus, i.e., $E_c = 39.05$ GPa. Detailed test information is provided in reference [56].

Table 2. Fracture test results reported in reference [56].

Specimen Number	h/m	b/m	t/m	α_0	F_{Hmax}/kN	F_{HQ}/kN
WS300	0.3	0.3	0.2	0.4	11.793	7.003
WS400	0.4	0.4	0.2	0.4	15.067	9.342
WS500	0.5	0.5	0.2	0.4	17.766	10.953
WS600	0.6	0.6	0.2	0.4	21.253	13.43

Based on Equations (6) and (7), linear regression analysis of the test results was carried out, and the linear regression equations of $1/\left(\sigma_{Ne}^{Q}\right)^2$, $1/\left(\sigma_{Ne}^{S}\right)^2$ and a_e were obtained, as shown in Figure 13. According to the slope and intercept, K_{IC}^{ini} without a size effect is 1.47 MPa·m$^{1/2}$, c_f^{ini} is 460.6 mm, K_{IC}^{un} is 1.72 MPa·m$^{1/2}$ and c_f^{un} is 193.1 mm.

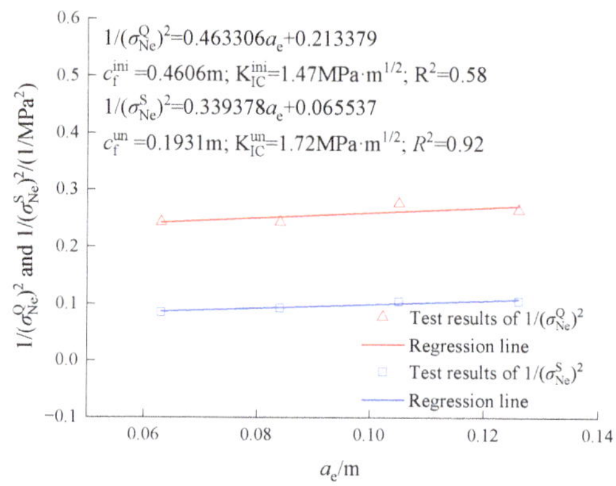

Figure 13. Linear relationships of $1/\left(\sigma_{Ne}^{Q}\right)^2$, $1/\left(\sigma_{Ne}^{S}\right)^2$ and a_e.

The K_{IC}^{ini} and c_f^{ini} and K_{IC}^{un} and c_f^{un} obtained by linear analysis are substituted into Equations (8) and (9), respectively, and the relationship between the initiation and peak loads with the equivalent geometric parameter (A_e) can be obtained as shown in Figure 14. In the figure, the slope of the initiation load prediction line is K_{IC}^{ini} without size effect, and the slope of the peak load prediction line is K_{IC}^{un} without a size effect. It can be seen from Figure 7 that the test results of F_{HQ} and F_{Hmax} are nearly linear relative to A_e. The *MAPE* of F_{HQ} prediction is 0.99%, *RMSE* is 0.22 kN, reliability index *a*15 is 100% and the covariance is 5.43 (kN)^2. The *MAPE* of F_{Hmax} prediction is 1.5%, *RMSE* is 0.19 kN, *a*15 is 100% and the covariance is 12.01 (kN)^2. The above indices show that the size effect model can realize the accurate prediction of initiation and peak loads when the size or α_c changes.

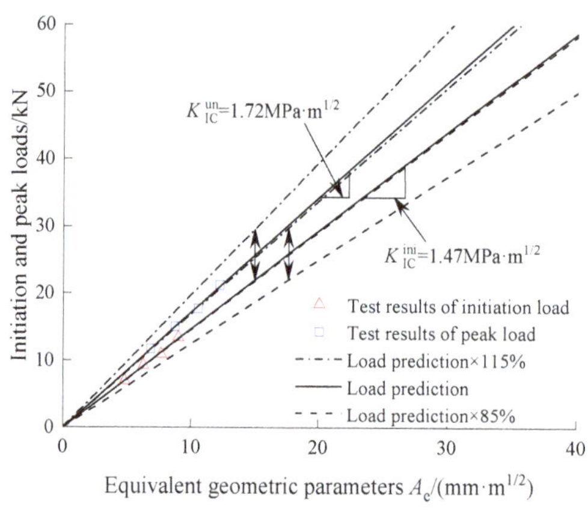

Figure 14. Predicted results of initiation and peak loads.

In order to evaluate the effectiveness of the proposed method, substituting the test results of initiation load into Equation (1), the calculated value is taken as the test value of initiation toughness. The test results of F_{Hmax} and a_c are substituted into Equation (3), and the calculated results are used as the unstable toughness test results. Then, these results are compared with the predicted results of initiation and unstable toughness calculated by the proposed method.

The optimal fitting softening parameter (w_0) was 450 mm, and *MAPE* was 3.46%, as calculated by the proposed method. Figure 15 shows the predicted results of fracture parameters of specimens with an α_c of 0.4 and different sizes. It can be seen from Figure 15a that all data points of critical fracture length fall near the predicted line. It can be seen from Figure 15b that the data points of fracture toughness all fall near the corresponding fracture toughness prediction line; K_{IC}^S, K_{IC}^N and K_{IC}^Q all have a size effect. K_{IC}^Q is close to the initiation toughness without a size effect, i.e., 1.47 MPa·m$^{1/2}$. The predicted value of K_{IC}^S increases with an increase in size and then decreases slightly with a further increase in size. K_{IC}^S and K_{IC}^N approach one another with increased size, with values close to the unstable toughness without a size effect, i.e., 1.72 MPa·m$^{1/2}$. The accuracy evaluation indicators of a_c are $RMSE = 0.007$, $MAPE = 2.22\%$, $a15 = 100\%$ and $Cov = 0.003$ m^2; the accuracy evaluation indicators of K_{IC}^Q are $RMSE = 0.011$, $MAPE = 1.5\%$, $a15 = 100\%$ and $Cov = 0.004$ (MPa)^{2}m; the accuracy evaluation indicators of K_{IC}^S are $RMSE = 0.059$, $MAPE = 3.46\%$, $a15 = 100\%$ and $Cov = 0.004$ (MPa)^{2}m, which shows that the predicted results are more accurate. The proposed method can achieve accurate prediction of specific α_c, critical crack length, crack initiation toughness and unstable toughness of concrete specimens with different sizes and can also determine the initiation and unstable toughness without a size effect.

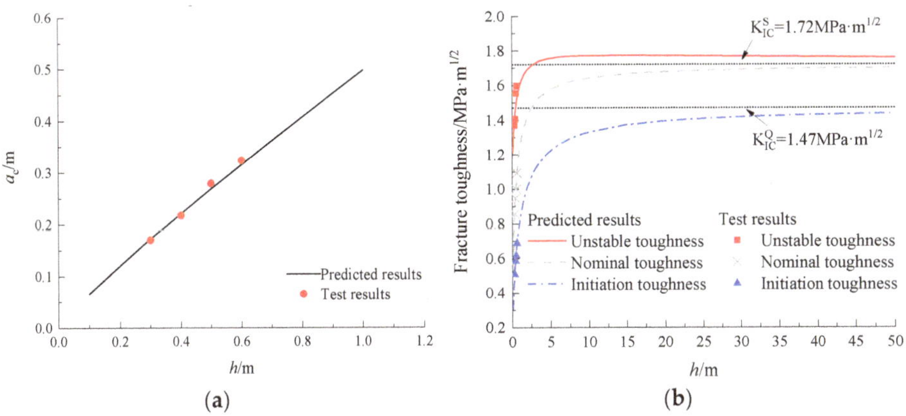

Figure 15. Critical crack length and fracture toughness of specimens with different sizes: (**a**) critical crack length; (**b**) fracture toughness.

3.1.2. Same Specimen Size but Different Crack Length-to-Height Ratio

In order to study the effect of initial α_c on the fracture performance of concrete, a wedge-splitting tensile fracture test was designed and carried out in reference [57] with maximum particle sizes of concrete aggregate of 10 mm and 20 mm, an effective specimen height of specimen of 0.3 m, a thickness of 0.12 m and an initial α_c in the range of 0.2 to 0.7. The specimen size and test results are summarized in Table 3. The compressive strength (f_{cu}) of a concrete cube with a maximum aggregate particle size of 10 mm is 42.2 MPa, and the elastic modulus is $E_c = 27.31$ GPa. The f_{cu} of a concrete cube with a maximum aggregate particle size of 20 mm is 57.04 MPa, with an elastic modulus of $E_c = 31.75$ GPa; the elastic modulus is calculated by f_{cu}. Additional experimental information can be found in reference [57].

Table 3. Fracture test results reported in reference [57].

Specimen Number	d_{max}/mm	h/m	b/m	t/m	α_0	F_{HQ}/kN	F_{Hmax}/kN
R0.2D10-1	10	0.3	0.3	0.12	0.2	10.03	13.32
R0.2D10-2		0.3	0.3	0.12	0.2	9.12	11.23
R0.2D10-3		0.3	0.3	0.12	0.2	11.05	14.27
R0.2D10-4		0.3	0.3	0.12	0.2	9.25	12.34
R0.4D10-1		0.3	0.3	0.12	0.4	6.94	8.72
R0.4D10-2		0.3	0.3	0.12	0.4	5.54	8.39
R0.4D10-3		0.3	0.3	0.12	0.4	5.46	8.38
R0.4D10-4		0.3	0.3	0.12	0.4	6.15	8.87
R0.6D10-1		0.3	0.3	0.12	0.6	2.90	4.30
R0.6D10-2		0.3	0.3	0.12	0.6	2.79	3.45
R0.6D10-3		0.3	0.3	0.12	0.6	3.57	4.43
R0.6D10-4		0.3	0.3	0.12	0.6	2.65	3.86
R0.7D10-1		0.3	0.3	0.12	0.7	1.93	2.54
R0.7D10-3		0.3	0.3	0.12	0.7	2.29	2.64
R0.7D10-4		0.3	0.3	0.12	0.7	2.02	2.74
R0.2D20-1	20	0.3	0.3	0.12	0.2	12.81	16.22
R0.2D20-2		0.3	0.3	0.12	0.2	10.66	15.48
R0.2D20-3		0.3	0.3	0.12	0.2	10.9	14.79
R0.2D20-4		0.3	0.3	0.12	0.2	12.71	16.83
R0.4D20-1		0.3	0.3	0.12	0.4	8.57	10.39
R0.4D20-2		0.3	0.3	0.12	0.4	7.21	9.84
R0.4D20-3		0.3	0.3	0.12	0.4	7.88	11.16
R0.4D20-4		0.3	0.3	0.12	0.4	8.06	11.10
R0.6D20-1		0.3	0.3	0.12	0.6	3.29	5.10
R0.6D20-2		0.3	0.3	0.12	0.6	3.56	4.58
R0.6D20-3		0.3	0.3	0.12	0.6	3.34	4.91
R0.6D20-4		0.3	0.3	0.12	0.6	3.47	4.80
R0.7D20-1		0.3	0.3	0.12	0.7	2.38	3.02
R0.7D20-3		0.3	0.3	0.12	0.7	2.56	3.13
R0.7D20-4		0.3	0.3	0.12	0.7	2.25	3.11

(1) $d_{max} = 10$ mm

Figure 16a,b show the linear relationship between $1/\left(\sigma_{Ne}^{Q}\right)^2$, $1/\left(\sigma_{Ne}^{S}\right)^2$ and a_e, as well as the predicted results of initiation and peak loads when the specimen size or α_c changes. According to the linear relationship shown in Figure 16a, K_{IC}^{ini} and K_{IC}^{un} without a size effect of concrete can be calculated as 0.93 MPa·m$^{1/2}$ and 1.22 MPa·m$^{1/2}$, respectively. Then, using Equations (8) and (9), the predicted lines of initiation and peak loads can be obtained. After calculation, the accuracy evaluation indicators of initiation load are $MAPE = 9.05\%$, $RMSE = 0.56$ kN, $a15 = 86.67\%$ and $Cov = 8.96\ (\text{kN})^2$; the accuracy evaluation indicators of peak load are $MAPE = 6.74\%$, $RMSE = 0.68$ kN, $a15 = 93.33\%$ and $Cov = 15.58\ (\text{kN})^2$. The calculation results show that the initiation and peak loads test results are close to the predicted results.

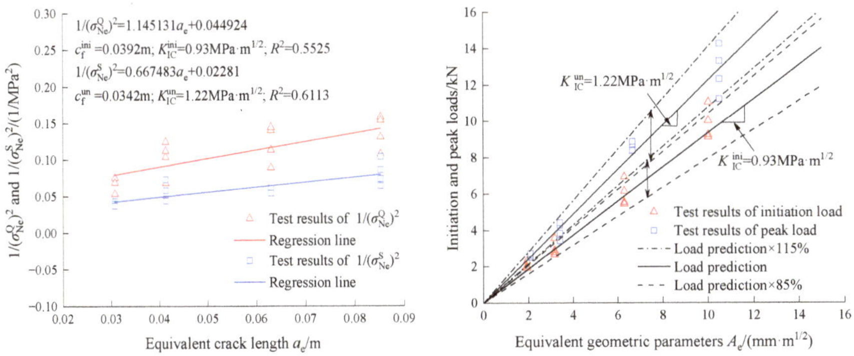

(**a**) (**b**)

Figure 16. (**a**) Linear relationship of the size effect model; (**b**) predictions of initiation and peak loads.

After calculation, the w_0 obtained in this experiment is 150 μm, and the minimum value of *MAPE* of unstable toughness is 8.19%. Figure 17 shows the fracture toughness predicted results of specimens with different initial α_c value when the effective height of the specimen is 0.3 m. It can be seen from Figure 17 that the calculation results of K_{IC}^{Q} and K_{IC}^{S} basically decrease with an increase in the initial α_c, and most of the test results are within the allowable error range of ±15% of the predicted value. $MAPE = 8.19\%$, $RMSE = 0.16\ \text{MPa} \cdot \text{m}^{1/2}$, $a15 = 73.33\%$ and $Cov = 0.0003\ (\text{MPa})^2\text{m}$ of the corresponding K_{IC}^{S} predicted results. The above indices show that the determination method of fracture parameters based on the size effect model and fracture extreme theory can realize the accurate prediction of initiation and unstable toughness when the size or α_c changes.

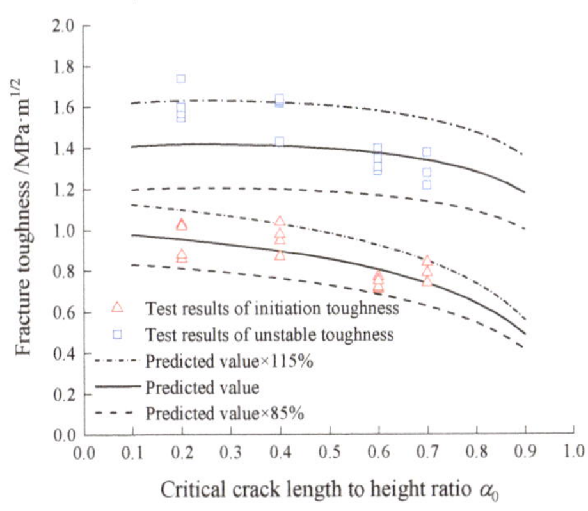

Figure 17. Fracture toughness of specimens with different α_0 values.

(2) d_{max} = 20 mm

Figure 18a shows the linear regression analysis of test results. According to the corresponding slope and intercept, K_{IC}^{ini} and K_{IC}^{un} without a size effect of concrete can be calculated as 1.21 MPa·m$^{1/2}$ and 1.63 MPa·m$^{1/2}$, respectively. Then, Equations (8) and (9) can be used to obtain the predicted line of initiation and peak loads with the slopes of K_{IC}^{ini} and K_{IC}^{un}, as shown in Figure 18b. After calculation, the accuracy evaluation indicators of initiation load are $MAPE = 7.95\%$, $RMSE = 0.66$ kN, $a15 = 93.33\%$ and $Cov = 13.66\ (\text{kN})^2$; the accuracy evaluation indicators of peak load are $MAPE = 4.51\%$, $RMSE = 0.60$ kN, $a15 = 100\%$

and $Cov = 24.75\ (\text{kN})^2$. The calculation results show that the initiation and peak load test results are close to the predicted results.

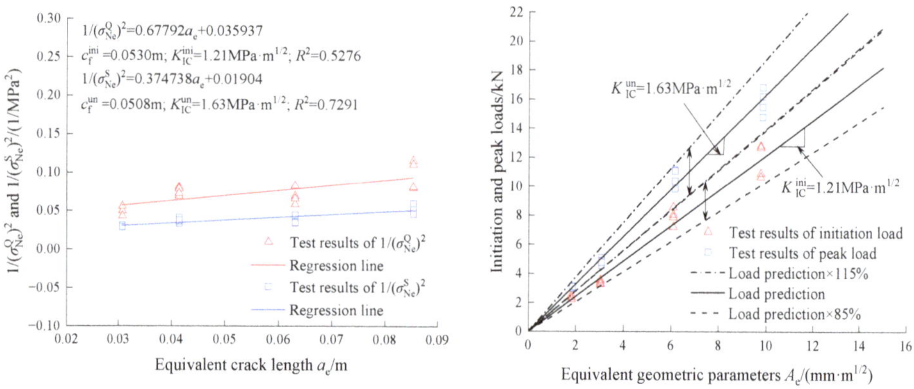

Figure 18. (**a**) Linear relationship of the size effect model; (**b**) predictions of initiation and peak loads.

After calculation, for the concrete with a maximum aggregate size of 20 mm, the optimal w_0 obtained by the test is 150 μm, and the minimum value of *MAPE* of unstable toughness is 7.91%. Figure 19 shows the fracture toughness predicted results of specimens with different initial α_c values when the effective height of the specimen is 0.3 m. It can be seen from Figure 19 that the calculated results of K_{IC}^{Q} and K_{IC}^{S} basically decrease with increased initial α_c, and most of the test results are within the allowable error range of ±15% of the predicted value. The corresponding K_{IC}^{Q} predicted results are $MAPE = 9.06\%$, $RMSE = 0.08\ \text{MPa} \cdot \text{m}^{1/2}$, $a15 = 80.00\%$ and $Cov = 0.007\ (\text{MPa})^2\text{m}$. $MAPE = 7.91\%$, $RMSE = 0.12\ \text{MPa} \cdot \text{m}^{1/2}$, $a15 = 93.33\%$ and $Cov = 0.004\ (\text{MPa})^2\text{m}$ of the predicted results of K_{IC}^{S}. The above indicators show that the fracture parameter determination method based on the size effect model and the fracture extreme theory can achieve more accurate prediction of initiation and unstable toughness when the size or α_c changes.

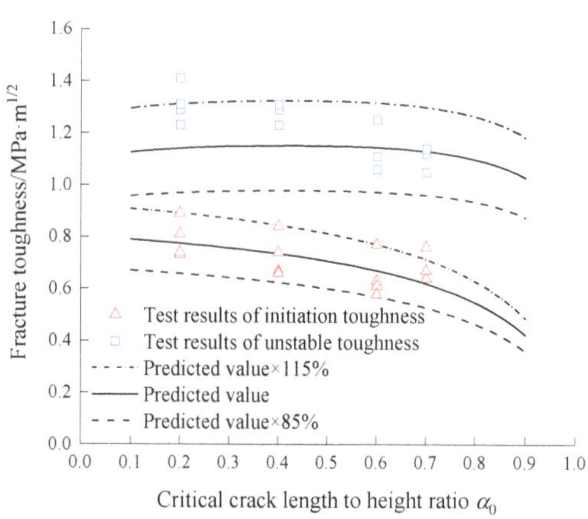

Figure 19. Fracture toughness of specimens with different α_0 values.

3.2. *Fully Graded Concrete*

Same Crack Length-to-Height Ratio and Different Specimen Sizes

In this section, the fracture test results of fully graded concrete of Wudongde Dam are analyzed to verify the applicability of the proposed method for fully graded concrete with a maximum aggregate size of 150 mm. The effective height of the Wudongde fully graded concrete wedge-splitting tensile specimens is 0.75 m, 1.5 m and 2.25 m; the initial α_c is 0.4; and the thickness is 0.45 m. The size and test results of concrete specimens are summarized in Table 4. Other experimental information can be found in reference [7].

Table 4. Test data reported in reference [2].

Specimen Number	h/m	b/m	t/m	α_0	F_{Hmax}/kN	F_{HQ}/kN
FG750-1	0.75	0.75	0.45	0.4	84.78	63.46
FG750-2	0.75	0.75	0.45	0.4	82.78	64.79
FG750-3	0.75	0.75	0.45	0.4	75.62	63.62
FG1500-1	1.5	1.5	0.45	0.4	157.44	117.46
FG1500-2	1.5	1.5	0.45	0.4	158.22	113.25
FG1500-3	1.5	1.5	0.45	0.4	158.43	110.11
FG2250-1	2.25	2.25	0.45	0.4	200.59	/
FG2250-2	2.25	2.25	0.45	0.4	200.67	140.04
FG2250-3	2.25	2.25	0.45	0.4	170.58	147.93

Figure 20a shows the linear relationship between $1/\left(\sigma_{Ne}^{Q}\right)^2$, $1/\left(\sigma_{Ne}^{S}\right)^2$ and a_e. Based on the slope and intercept of the linear equation, K_{IC}^{ini} and K_{IC}^{un} of the fully graded concrete without a size effect can be calculated as 2.12 MPa·m$^{1/2}$ and 2.90 MPa·m$^{1/2}$, respectively. Furthermore, combining Equations (8) and (9) and the obtained fracture toughness without a size effect, the initiation and peak load predicted lines shown in Figure 20b can be obtained. The indices selected in this paper can be used to evaluate the prediction accuracy of initiation and peak loads. After calculation, the evaluation indicators of initiation load are $MAPE = 2.51\%$, $RMSE = 3.62$ kN, $a15 = 100\%$ and $Cov = 1024.63$ (kN)^2; the evaluation indicators of peak load are $MAPE = 6.46\%$, $RMSE = 11.52$ kN, $a15 = 100\%$ and $Cov = 2032.55$ (kN)^2. The calculation results show that the predicted results of initiation and peak loads based on the size effect model are close to the test results.

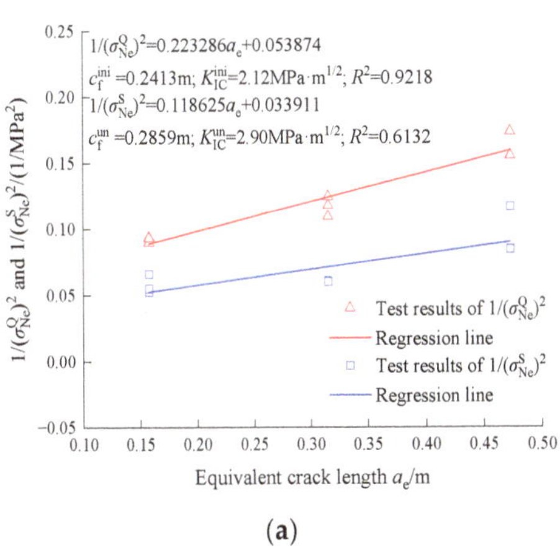

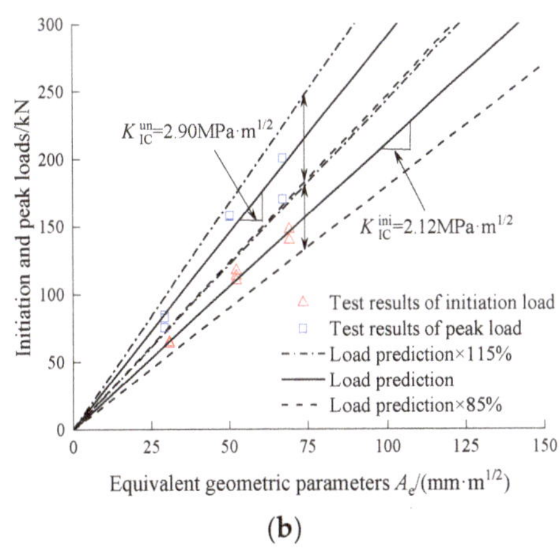

Figure 20. (**a**) Linear relationship of the size effect model; (**b**) predictions of initiation and peak loads.

Figure 21a,b show comparisons between the predicted results and the test results of a_c, K_{IC}^S, K_{IC}^N and K_{IC}^Q of fully graded concrete specimens with different sizes and an initial α_c of 0.4 when the optimum w_0 is 350 μm. It can be seen from Figure 21a that the test results of a_c are close to the predicted results, and the prediction accuracy indices are $MAPE = 2.61\%$, $RMSE = 0.03$ m, $a15 = 100\%$ and $Cov = 0.06$ m^2. From Figure 21b, it can be seen that K_{IC}^Q, K_{IC}^N and K_{IC}^S increase with increased specimen size and tend toward K_{IC}^{ini} and K_{IC}^{un}, respectively without a size effect defined in the size effect model.

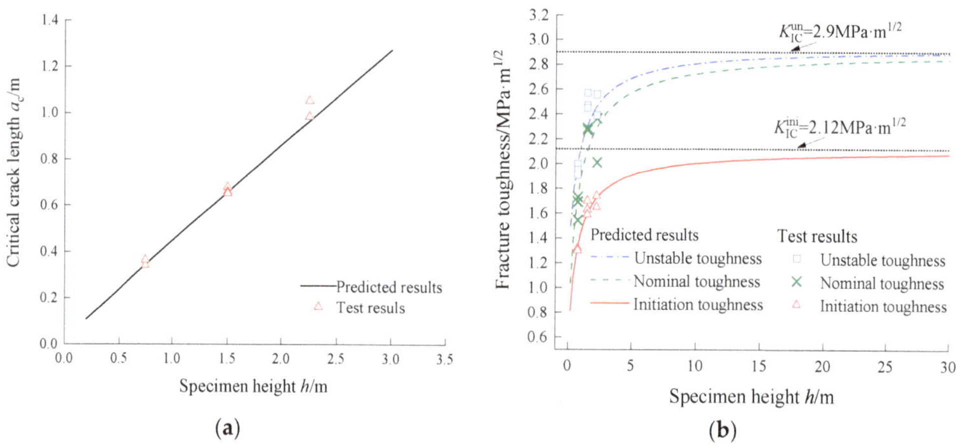

Figure 21. Critical crack length and fracture toughness of specimens with different sizes: (**a**) critical crack length; (**b**) fracture toughness.

Compared with the predicted results and test results of fracture toughness, the prediction accuracy evaluation indices of K_{IC}^Q are $MAPE = 2.44\%$, $RMSE = 0.05\ \text{MPa} \cdot \text{m}^{1/2}$, $a15 = 100\%$ and $Cov = 0.03\ (\text{MPa})^2\text{m}$. The above indicators show that the fracture parameter determination method proposed in this paper is also applicable to fully graded concrete, and the prediction accuracy can meet the needs of dam engineering.

Furthermore, based on the fracture test results of fully graded concrete with limited size, the proposed method can be used to determine the fracture toughness of specimens with a specific size and arbitrary initial α_c. Figure 22 shows a comparison between the predicted fracture toughness and the test results of specimens with different initial α_c values when the effective heights of specimens are 0.75 m, 1.5 m and 2.25 m. From Figure 22, it can be seen that K_{IC}^Q and K_{IC}^S decrease with increased initial α_c, and the test results are close to the predicted results. According to the evaluation index of prediction accuracy, the $MAPE$ of fracture toughness is within 5%, and the maximum deviation of all predicted results is only 7.5%. It can be seen that the fracture parameter determination method proposed in this paper can accurately predict the fracture toughness of specimens with arbitrary size and α_c values only based on the limited size fracture test results.

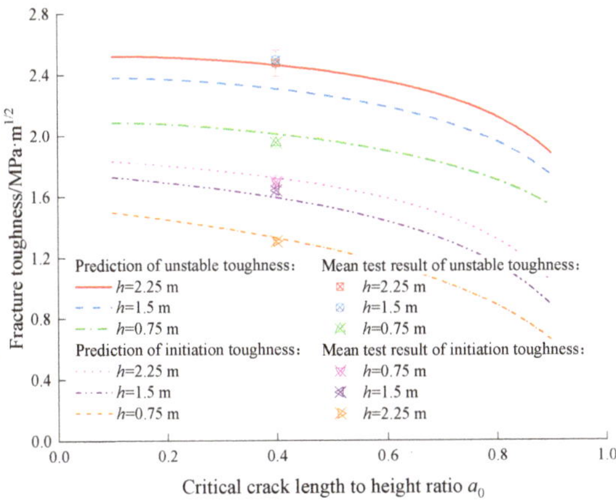

Figure 22. Fracture toughness of specimens with different α_0 values.

4. Conclusions

Combining the size effect model and the fracture extreme theory, in this paper, we propose a method to determine the double-K fracture toughness of concrete specimens with arbitrary size and α_c. The applicability and accuracy of the method were verified by the fracture test results of small aggregate concrete and Wudongde fully graded concrete. The main conclusions are as follows:

(1) Based on the wedge-splitting tensile test results of small aggregate concrete and fully graded concrete with limited size or α_c, the size effect model and fracture extreme theory can be used to accurately predict the initiation and peak loads, as well as the initiation and unstable toughness when the size or α_c changes. Furthermore, the prediction accuracy can meet the needs of engineering applications; the mean absolute percentage error is basically below 10%, most of the reliability indices are above 85% and the maximum can reach 100%. It can also be seen from the covariance that the predicted value is positively correlated with the test value.

(2) The K_{IC}^{Q}, K_{IC}^{N} and K_{IC}^{S} increased with increased specimen size and gradually stabilized. Furthermore, K_{IC}^{Q} tends toward K_{IC}^{ini} as defined in the size effect model. Both K_{IC}^{S} and K_{IC}^{N} tend toward K_{IC}^{un}, and the difference between the two gradually decreases with increased specimen size. The K_{IC}^{ini} and K_{IC}^{un} of the small aggregate concrete in example 1 in this paper are 1.14 MPa·m$^{1/2}$ and 1.81 MPa·m$^{1/2}$, respectively, and the K_{IC}^{ini} and K_{IC}^{un} of fully graded concrete are 2.12 MPa·m$^{1/2}$ and 2.9 MPa·m$^{1/2}$, respectively.

(3) The research results reported in this paper can provide a theoretical basis for the optimization of the fracture test. In dam engineering, we suggest carrying out fracture tests with limited size and α_c directly; the test results can then be analyzed using the method proposed in this paper. Then, the double-K fracture toughness of fully graded concrete specimens with arbitrary size can be determined. This method can be used for the analysis of the cracking risk and crack stability of dam structures in reality.

For dam concrete in actual service, its fracture parameters are not only affected by the size and crack length-to-height ratio of the specimen but also the age, maximum aggregate size, curing humidity and temperature, and other factors may also affect the fracture toughness of concrete. Therefore, in order to accurately evaluate the fracture parameters of dam concrete under actual service conditions, it is necessary to carry out research on the prediction of concrete fracture parameters considering additional factors in the future.

Author Contributions: Conceptualization, X.G.; Formal analysis, X.G., J.W. and T.X.; Funding acquisition, X.G.; Investigation, X.G. and J.W.; Methodology, X.G., M.Z., Y.T., N.Y., Y.Q., C.L., T.X. and J.W.; Supervision, X.G.; Validation, X.G.; Writing—original draft, X.G. and J.W.; Writing—review and editing, X.G., J.W. and T.X. All authors have read and agreed to the published version of the manuscript.

Funding: This research was funded by the National Natural Science Foundation of China (No. 52009123), the Zhejiang Provincial Natural Science Foundation of China (No. LY22E090004), the Open Research Fund Program of the State Key Laboratory of Hydroscience and Engineering (No. sklhse-2020-C-04) and Research Projects of China Three Gorges Corporation (Nos. BHT/0809, BHT/0806).

Data Availability Statement: The data presented in this study are available upon request from the corresponding author.

Conflicts of Interest: The authors declare no conflict of interest.

Nomenclature

K_{IC}^{Q}	Initiation toughness
K_{IC}^{S}	Unstable toughness
F_{HQ}	Initiation load
F_{Hmax}	Peak load
$f(\alpha_0)$	Dimensionless geometric parameter of the specimen
$Y(\alpha_0)$	Dimensionless geometric parameter of the specimen
a_c	Critical crack length
α_c	Crack length-to-height ratio ($\alpha_c = a_c/h$)
σ_N^Q	Nominal initiation strength
σ_N^S	Nominal unstable strength
σ_{Ne}^Q	Equivalent nominal initiation strength
σ_{Ne}^S	Equivalent nominal unstable strength
a_e	Equivalent crack length
K_{IC}^{ini}	Initiation toughness without size effect
c_f^{ini}	Effective length of the initiation fracture process zone
K_{IC}^{un}	Unstable toughness without size effect
c_f^{un}	Effective length of the unstable fracture process zone
F_{HQ}^{Pre}	Initiation load prediction
F_{Hmax}^{Pre}	Peak load prediction
$g'(\alpha_0)$	Dimensionless geometric parameter of the specimen
$g(\alpha_0)$	Dimensionless geometric parameter of the specimen
A_e^{ini}	Equivalent geometric parameter related to initiation
A_e^{un}	Equivalent geometric parameter related to instability
σ_c	Compressive stress at the bottom of the specimen
t	Thickness of specimen
h	Effective height of the specimen
a_0	Initial crack length
a	Effective crack length
h_c	Distance from the crack tip to the neutral axis marked by the dotted line
f_t	Equivalent tensile stress at the crack tip
σ_w	Cohesive force in the fracture process zone
CTOD	Crack-tip-opening displacement
CMOD	Crack-mouth-opening displacement
$w(x)$	Crack-opening displacement at x
w_s	Crack-opening displacement at the inflection point
w_0	Crack-opening displacement at zero cohesion
$\sigma_w(x)$	Cohesive force at x
σ_s	Cohesive force at a point
RMSE	Root mean square error
MAPE	Mean absolute percentage error

References

1. Li, Y.; Zhou, Y.; Wang, R.J.; Li, Y.L.; Wu, X.J.; Si, Z. Experimental investigation on the properties of the interface between RCC layers subjected to early-age frost damage. *Cem. Concr. Compos.* **2022**, *134*, 104745. [CrossRef]
2. Meng, T.; Lian, S.S.; Yu, H.M.; Yang, C.J.; Wang, M.H. Long-term influence of tailings wastewater on mechanical performance and microstructure of dam concrete: A case study in southeastern China. *Case Stud. Constr. Mater.* **2021**, *15*, e00720. [CrossRef]
3. Liu, W.J.; Tan, Y.S.; Li, Q.B.; Yang, N.; Liu, C.F.; Gao, X.F.; Hu, Y.; Niu, X.J.; Zhang, B. Influences of short-term heavy rainfall on interlayer properties and microstructure of concrete dam. *Case Stud. Constr. Mater.* **2022**, *14*, e01544. [CrossRef]
4. Qian, P.; Xu, Q.J. Experimental investigation on properties of interface between concrete layers. *Constr. Build. Mater.* **2018**, *174*, 120–129. [CrossRef]
5. Gao, X.F.; Hu, Y.; Yang, N.; Wu, K.; Li, Q.B. Fracture test and size effect analysis of low-heat cement fully-graded concrete. *Eng. Mech.* **2022**, *39*, 183–193. (In Chinese) [CrossRef]
6. Liu, W.J.; Wang, Y.B.; Li, Q.B.; Gao, X.F.; Tan, Y.S.; Liu, C.F.; Hu, Y.; Niu, X.J. Research on interlayer bonding quality control method of dam concrete based on equivalent age. *Materials* **2021**, *14*, 5192. [CrossRef]
7. Guan, J.F.; Li, Q.B.; Wu, Z.M.; Zhou, S.W. Necessity and Feasible Way of Studying True Fracture Parameters of Super-High Arch Dam Concrete. *J. Hydroelectr. Eng.* **2014**, *33*, 152–158. Available online: http://www.slfdxb.cn/CN/Y2014/V33/I5/152 (accessed on 15 July 2022). (In Chinese)
8. Kim, J.K.; Yun, L.; Yi, S.T. Fracture characteristics of concrete at early ages. *Cem. Concr. Res.* **2004**, *34*, 507–519. [CrossRef]
9. Xu, S.L.; Zhao, Y.H.; Wu, Z.M.; Gao, H.B. The experimental study on the fracture energy of concrete using wedge splitting specimens. *J. Hydroelectr. Eng.* **2003**, 15–22. (In Chinese) [CrossRef]
10. Hoover, C.G.; Bažant, Z.P. Comprehensive concrete fracture tests: Size effects of Types 1 & 2, crack length effect and postpeak. *Eng. Fract. Mech.* **2013**, *110*, 281–289. [CrossRef]
11. Beygi, M.H.A.; Kazemi, M.T.; Nikbin, I.M.; Amiri, J.V. The effect of water to cement ratio on fracture parameters and brittleness of self-compacting concrete. *Mater. Des.* **2013**, *50*, 267–276. [CrossRef]
12. Zhang, G.H.; Li, Z.L.; Nie, K.Y.; Liu, M.H. Experimental study on fracture toughness of concrete with different moisture contents. *J. Hydroelectr. Eng.* **2016**, 109–116. (In Chinese) [CrossRef]
13. Xu, S.L.; Li, Q.H.; Wu, Y.; Dong, L.X.; Lyu, Y.; Reinhardt, H.W.; Christopher, K.Y.L.; Ruiz, G.; Kumar, S.; Hu, S.W. Determination of double-K criterion for crack propagation in quasi-brittle fracture, Part II: Analytical evaluating and practical measuring methods for three-point bending notched beams. *Int. J. Fract.* **1999**, *98*, 151–177. [CrossRef]
14. Xu, S.L.; Reinhardt, H.W. Determination of double-*K* criterion for crack propagation in quasi-brittle fracture Part I: Experimental investigation of crack propagation. *Int. J. Fract.* **1999**, *98*, 111–149. [CrossRef]
15. Xu, S.L.; Reinhardt, H.W. Determination of double-*K* criterion for crack propagation in quasi-brittle fracture, Part III: Compact tension specimens and wedge splitting specimens. *Int. J. Fract.* **1999**, *98*, 179–193. [CrossRef]
16. *Norm for Fracture Test of Hydraulic Concrete*; China Electric Power Press: Beijing, China, 2005. (In Chinese)
17. Xu, S.L.; Li, Q.H.; Wu, Y.; Dong, L.X.; Lyu, Y.; Reinhardt, H.W.; Christopher, K.Y.L.; Ruiz, G.; Kumar, S.; Hu, S.W. RILEM Standard: Testing methods for determination of the double-K criterion for crack propagation in concrete using wedge-splitting tests and three-point bending beam tests, recommendation of RILEM TC265-TDK. *Mater. Struct.* **2021**, *54*, 1–11. [CrossRef]
18. Alam, S.Y.; Zhu, R.; Loukili, A. A new way to analyse the size effect in quasi-brittle materials by scaling the heterogeneity size. *Eng. Fract. Mech.* **2020**, *225*, 106864. [CrossRef]
19. Pirooznia, A.; Moradloo, A. Investigation of size effect and smeared crack models in ordinary and dam concrete fracture tests. *Eng. Fract. Mech.* **2020**, *226*, 106863. [CrossRef]
20. Yin, Y.Y.; Hu, S.W. Double-*K* fracture parameters of concrete in three point bending beams with small span-depth ratios. *Eng. Mech.* **2020**, *37*, 138–146 (In Chinese) [CrossRef]
21. Guan, J.F.; Li, Q.B.; Wu, Z.M.; Zhao, S.B.; Dong, W.; Zhou, S.W. Minimum specimen size for fracture parameters of site-casting dam concrete. *Constr. Build. Mater.* **2015**, *93*, 973–982. [CrossRef]
22. Guan, J.F.; Li, Q.B.; Wu, Z.M.; Dong, W. Determination of minimum size for double *K* fracture parameters of site-casting fully-graded hydraulic concrete. *J. Basic Sci. Eng.* **2016**, *24*, 1219–1231. (In Chinese) [CrossRef]
23. Guan, J.F. Fracture parameters of site-cast dam and sieved concrete. *Mag. Concr. Res.* **2016**, *68*, 43–54. [CrossRef]
24. Hu, X.Z.; Duan, K. Size effect and quasi-brittle fracture: The role of FPZ. *Int. J. Fract.* **2008**, *154*, 3–14. [CrossRef]
25. Hu, X.Z. An asymptotic approach to size effect on fracture toughness and fracture energy of composites. *Eng. Fract. Mech.* **2002**, *69*, 555–564. [CrossRef]
26. Hu, X.Z.; Wittmann, F. Size effect on toughness induced by crack close to free surface. *Eng. Fract. Mech.* **2000**, *65*, 209–221. [CrossRef]
27. Bažant, Z.P. Size effect in blunt fracture: Concrete, rock, metal. *J. Eng. Mech.* **1984**, *110*, 518–535. [CrossRef]
28. Bažant, Z.P. Scaling theory for quasibrittle structural failure. *Proc. Natl. Acad. Sci. USA* **2004**, *101*, 13400–13407. [CrossRef]
29. Hoover, C.G.; Bažant, Z.P. Comparison of the Hu-Duan boundary effect model with the size-shape effect law for quasi-brittle fracture based on new comprehensive fracture tests. *J. Eng. Mech.* **2014**, *140*, 480–486. [CrossRef]
30. Guan, J.F.; Yuan, P.; Hu, X.Z.; Qing, L.B.; Yao, X.H. Statistical analysis of concrete fracture using normal distribution pertinent to maximum aggregate size. *Theor. Appl. Fract. Mech.* **2019**, *101*, 236–253. [CrossRef]

31. Yao, X.H.; Li, L.L.; Guan, J.F.; Zhang, M.; Liu, Z.P.; Han, R.C.; He, S.H. Initial cracking strength and initial fracture toughness from three-point bending and wedge splitting concrete specimens. *Fatigue Fract. Eng. Mater. Struct.* **2021**, *44*, 601–621. [CrossRef]
32. Guan, J.F.; Liu, Z.P.; Yao, X.H.; Li, L.L.; He, S.H.; Zhang, M. Determination of the cracking strength tensile strength and double-K fracture parameters of concrete. *Eng. Mech.* **2020**, *37*, 124–137. (In Chinese) [CrossRef]
33. Guan, J.F.; Yin, Y.N.; Li, Y.; Yao, X.H.; Li, L.L. A design method for determining fracture toughness and tensile strength pertinent to concrete sieving curve. *Eng. Fract. Mech.* **2022**, *271*, 108596. [CrossRef]
34. Guan, J.F.; Li, C.M.; Wang, J.; Qing, L.B.; Song, Z.K.; Liu, Z.P. Determination of fracture parameter and prediction of structural fracture using various concrete specimen types. *Theor. Appl. Fract. Mech.* **2019**, *100*, 114–127. [CrossRef]
35. Wu, Z.M.; Yang, S.T.; Hu, X.Z.; Zheng, J.J. An analytical model to predict the effective fracture toughness of concrete for three-point bending notched beams. *Eng. Fract. Mech.* **2006**, *73*, 2166–2191. [CrossRef]
36. Wu, Z.M.; Yang, S.T.; Zheng, J.J. Analytical method for predicting effective fracture toughness of concrete and its size effect. *J. Hydraul. Eng.* **2006**, 795–800. (In Chinese) [CrossRef]
37. Qing, L.B.; Cheng, Y.H.; Guan, J.F. Simplified extreme method for determining initiation toughness of dam concrete. *J. Hydroelectr. Eng.* **2018**, *37*, 93–100. (In Chinese) [CrossRef]
38. Qing, L.B.; Cheng, Y.H.; Fan, X.Q.; Mu, R.; Ding, S.Q. An arc bending notched specimen for determining the mechanical and fracture parameters of concrete based on the FET. *Eng. Fract. Mech.* **2019**, *220*, 106639. [CrossRef]
39. Li, Y.; Qing, L.B.; Cheng, Y.H.; Ma, G.W. Determining the tensile strength of concrete for wedge splitting specimens based on the fracture extreme theory. *Theor. Appl. Fract. Mech.* **2020**, *108*, 102654. [CrossRef]
40. Qing, L.B.; Dong, M.W.; Guan, J.F. Determining initial fracture toughness of concrete for split-tension specimens based on the extreme theory. *Eng. Fract. Mech.* **2018**, *189*, 427–438. [CrossRef]
41. Qing, L.B.; Cheng, Y.H. The fracture extreme theory for determining the effective fracture toughness and tensile strength of concrete. *Theor. Appl. Fract. Mech.* **2018**, *96*, 461–467. [CrossRef]
42. Nie, Y.T.; Qing, L.B. The relationship between double-K parameters of concrete based on fracture extreme theory. *J. Theor. Appl. Mech.* **2020**, *58*, 59–71. [CrossRef]
43. Li, Y.; Qing, L.B.; Cheng, Y.H.; Dong, M.W.; Ma, G.W. A general framework for determining fracture parameters of concrete based on fracture extreme theory. *Theor. Appl. Fract. Mech.* **2019**, 102259. [CrossRef]
44. Qing, L.B.; Su, Y.M.; Dong, M.W.; Cheng, Y.H.; Li, Y. Size effect on double-K fracture parameters of concrete based on fracture extreme theory. *Arch. Appl. Mech.* **2020**, 427–442. Available online: https://link.springer.com/article/10.1007/s00419-020-01781-5 (accessed on 15 July 2022).
45. Qing, L.B.; Li, Q.B. A theoretical method for determining initiation toughness based on experimental peak load. *Eng. Fract. Mech.* **2013**, 295–305. [CrossRef]
46. Qing, L.B.; Tian, W.L.; Wang, J. Predicting unstable toughness of concrete based on initial toughness criterion. *J. Zhejiang Univ. -Sci. A* **2014**, *15*, 138–148. [CrossRef]
47. Qing, L.B.; Nie, Y.T.; Wang, J.; Hu, Y. A simplified extreme method for determining double-K fracture parameters of concrete using experimental peak load. *Fatigue Fract. Eng. Mater. Struct.* **2017**, *40*, 254–266. [CrossRef]
48. Gao, X.F.; Liu, C.F.; Tan, Y.S.; Yang, N.; Qiao, Y.; Hu, Y.; Li, Q.B.; Koval, G.; Chazallon, C. Determination of Fracture Properties of Concrete Using Size and Boundary Effect Models. *Appl. Sci.* **2019**, *9*, 1337. [CrossRef]
49. Qing, L.B.; Li, Q.B.; Guan, J.F.; Wang, J. Study of concrete fracture process zone based on fictitious crack model. *Eng. Mech.* **2012**, *29*, 112–116,132 (In Chinese) [CrossRef]
50. Wittmann, F.H.; Rokugo, K.; Brühwiler, E.; Mihashi, H.; Simonin, P. Fracture energy and strain softening of concrete as determined by means of compact tension specimens. *Mater. Struct.* **1988**, *21*, 21–32. [CrossRef]
51. Li, Q.B.; Deng, Z.C.; Fu, H. Effect of aggregate type on mechanical behavior of dam concrete. *ACI Mater. J.* **2004**, *101*, 483–492. Available online: https://www.concrete.org/publications/internationalconcreteabstractsportal/m/details/id/13487 (accessed on 15 July 2022).
52. *CEB-FIP Model Code 1990. Programs Usenix Unix Supplementary Documents. UK*; Redwood Books: Coconut Creek, FL, USA, 1993.
53. Reinhardt, H.W.; Cornelissen, H.; Hordijk, D.A. Tensile tests and failure analysis of concrete. *J. Struct. Eng.* **1986**, *112*, 2462–2477. [CrossRef]
54. Ding, X.T.; Ding, X.; Liu, H.X.; Zheng, Y. Comparison study on softening curve of concrete by direct tension test and three-point bending fracture test. *Water Resour. Power* **2014**, *32*, 116–118; 156. (In Chinese)
55. Zhang, X.F.; Xu, S.L. A comparative study on five approaches to evaluate double-*K* fracture toughness parameters of concrete and size effect analysis. *Eng. Fract. Mech.* **2011**, *78*, 2115–2138. [CrossRef]
56. Hu, S.W.; Xie, J.F. Experimental research and size effect analysis of fracture parameter of nonstandard wedge splitting tensile specimen. *Water Resour. Power* **2015**, *33*, 105–108. (In Chinese)
57. Du, M.; Wu, L.; Zhang, J.M. Effect of different initial crack-depth ratios and aggregate size on fracture properties of low-heated concrete. *Concrete* **2020**, 14–17. (In Chinese) [CrossRef]

buildings

MDPI

Article

Effects of CH_2$ and CH on Strength and Hydration of Calcium Sulphoaluminate Cement Prepared from Phosphogypsum

Jixin Zhang [1], Jun Chang [1], Ping Zhang [2,]* and Tong Wang [1]

[1] School of Civil Engineering, Dalian University of Technology, Dalian 116024, China
[2] School of Materials and Architectural Engineering, Guizhou Normal University, Guiyang 550025, China
* Correspondence: pingzhang@gznu.edu.cn

Abstract: Using phosphogypsum (PG) as a raw material to prepare calcium sulphoaluminate cement (CAS) is an effective way of treating phosphogypsum. In order to meet the different application requirements of CSA cement and promote the application of CAS cement, it is necessary to add mineral admixtures to adjust the performance of cement. This paper incorporated two minerals, gypsum dihydrate (CH_2$) and calcium hydroxide (CH), into cement clinker prepared from phosphogypsum. The compressive strength and hydration process of the mixtures with different blending levels were investigated around the C_4A_3-CH_2$ system (SC) and the C_4A_3-CH_2$-CH system (SCC). The optimum dosing level was determined on the basis of the strength and hydration properties. In the SC system, adding CH_2$ promoted the hydration of C_4A_3. The compressive strength of the cement was highest at a CH_2$/$C_4A_3$$ molar ratio of 1.5, with a 7-day compressive strength of 56.5 MPa. AFt was mostly needle-rod and columnar and was tightly cemented to the gel phase, improving the denseness of the matrix. When the molar ratio was 2, the strength of the cement was inverted, and the shape of the AFt changed from needle and rod to columnar, the size of the grains increased, and it could not be filled with the AH_3 phase in an excellent staggered manner. At the same time, CH_2$ was not fully reacted, increasing matrix porosity and inversion of strength. In the SCC system, adding CH reduced the cement's compressive strength, and the compressive strength reduction increased with the increase in admixture. According to the experimental results, CH inhibited the formation of AFt, resulting in the appearance of new hydration products, AFm. As the amount of CH increased, the amount of hydration products, AFm, increased, while the amount of AFt and AH_3 decreased. However, adding CH raised the paste's pH and later facilitated the development of strength. The optimum admixture of CH/$C_4A_3$$ was 0.5 mol.

Keywords: phosphogypsum; calcium sulphoaluminate aluminate cement; gypsum; calcium hydroxide; strength; hydration properties

check for updates

Citation: Zhang, J.; Chang, J.; Zhang, P.; Wang, T. Effects of CH_2$ and CH on Strength and Hydration of Calcium Sulphoaluminate Cement Prepared from Phosphogypsum. *Buildings* **2022**, *12*, 1692. https://doi.org/10.3390/buildings12101692

Academic Editor: Tomáš Dvorský

Received: 11 September 2022
Accepted: 5 October 2022
Published: 14 October 2022

1. Introduction

Phosphogypsum (PG), as an industrial waste product emitted from the production of phosphoric acid, contains about 4–5 tonnes of phosphogypsum per tonne of phosphoric acid produced [1,2]. The main component of phosphogypsum is gypsum dihydrate, with a content of up to 85% or more, making it a gypsum resource that cannot be ignored [3,4]. The phosphate industry continues to develop, and the corresponding phosphogypsum emissions are increasing. According to statistics, about 280 million tonnes of phosphogypsum are emitted worldwide in one year, with only about 10% being used effectively [5,6]. Phosphogypsum contains small amounts of under-reacted sulphuric acid, phosphoric acid, phosphide, fluoride, and other impurities, resulting in its low utilisation rate [7–9]. Large amounts of untreated phosphogypsum are piled on land, polluting the environment and restricting economic development [10–12]. The way in which to deal with phosphogypsum is a current problem to be solved. The application of phosphogypsum to construction materials, with high consumption and low side effects on the environment, has the best prospects for utilisation and is thus the main focus of researchers [13–15].

Phosphogypsum is used to prepare gypsum cementitious materials, and impurities can lead to longer setting times, reduced material density and strength, and poor durability [16]. When used as a silicate cement retarder, phosphogypsum contains phosphorus, fluorine, and other harmful impurities and a lower pH [17], so the application performance is negatively affected; thus, the phosphogypsum must be pretreated. Pretreatment increases not only the cost of the process and the difficulty of use but also causes the treatment to not be good, thus making the phosphogypsum utilisation rate very low [18]. Scholars have attempted to prepare CSA cement from phosphogypsum to improve the utilisation of phosphogypsum [19,20]. Studies have shown that P_2O_5 and F- impurities in phosphogypsum promote the formation of clinker minerals and lower firing temperatures [21,22]. However, pollutants in the clinker lead to increased setting times and reduced pH and strength, which are not conducive to large-scale use of the cement [1,23]. Mineral admixtures such as gypsum and lime are usually added to adjust cement properties to meet the requirements of cement used in different environments [24]. During the hydration of $C_4A_3\$$, both gypsum and pH have an important influence on the hydration reaction [25]. Winnfield et al. found that anhydrous gypsum slowed the initial rate of $C_4A_3\$$ hydration compared to gypsum dihydrate [26]. Hargis et al. studied the hydration system of $C_4A_3\$$-$C\$H_2$-CH. They found that the incorporation of CH retarded the formation of AFt crystals and was associated with ye'elimite surface collocation [27]. However, for CAS cement prepared from phosphogypsum, few studies have been reported on the influence of the mineral admixtures gypsum and lime on their properties and hydration processes. Therefore, it is essential to study the effects of different doping amounts of gypsum and lime on the strength, hydration products, pH, and hydration process of PG-CAS cement for the promotion and application of cement and the resourcefulness of phosphogypsum.

In this paper, CAS cement clinker was prepared using PG as raw material, and the clinker's mineral composition was determined in order to meet the requirements. Then $C\$H_2$ and CH were added in different ratios to construct the $C_4A_3\$$-$C\$H_2$ system and the $C_4A_3\$$-$C\$H_2$-CH system. The compressive strength, pH, and hydration process of the cement of the SC and SCC systems were investigated. XRD and TG-DTG were used to analyse the changes in hydration products. SEM and MIP observed microscopic morphological changes to verify the development of macroscopic properties. It was hoped that the effect of mixing the three minerals $C_4A_3\$$, $C\$H_2$, and CH in different mix ratios on the strength and hydration process of the cement would be obtained. The work contributes to the promotion and application of PG-CAS cement-based materials.

2. Experimental Program

2.1. Materials

The raw materials used in this paper were bauxite, phosphogypsum, and limestone from the Guizhou Urnford Group. All raw materials were ground by a planetary ball mill and passed through an 80 μm sieve with a residue rate of less than 5%. The chemical composition of the raw materials is listed in Table 1. The target phase combination for the calcium sulphoaluminate cement was 70% $C_4A_3\$$, 16% C_2S, 8% C_4AF, and 6% C\$. On the basis of the target phase combination and the chemical composition of the raw materials shown in Table 1, the formulation of the CAS cement raw material was determined, as shown in Table 2. The ground raw materials were accurately weighed and homogenised according to the ratios in Table 2. The raw material mixture was mixed with 10 wt% water and made into Φ 5 × 1.5 cm round cakes under 8 MPa pressure. The clinker was dried in an oven at 40 °C and then placed in a high-temperature furnace for firing at a heating rate of 10 °C/min, calcination temperature of 1250 °C, and a holding time of 1 h [28]. After calcination, the clinker was removed for rapid cooling, and after cooling, the clinker was ground with a planetary ball mill for 5 min and then finely ground with an agate mortar and passed through an 80 μm sieve. As shown in Figure 1, the XRD curves had distinct peaks characteristic of the clinker minerals, mainly $C_4A_3\$$, C_2S, $C_2A\$$, and C$. The Rietveld method was used to quantify the cement clinker minerals and determine the content of each phase in the clinker. The quantification results are shown in Table 3. The XRD profiles had good profile fitting (Rwp < 15%) [29–31], and the quantification results were reliable. The microscopic morphology of the prepared CAS clinker is shown in Figure 2. The clinker minerals were mostly $C_4A_3\$$, in the form of hexagonal plates or round grains, 1–2 μm in size, with a smooth surface and tightly arranged. The prepared calcium sulphoaluminate cement clinker was mixed and ground with $CaSO_4$-$2H_2O$ and $Ca(OH)_2$ in the proportions shown in Table 4. Figure 3 shows the flow chart for this experiment.

Table 1. Chemical compositions of the materials (wt%).

Materials	SO_3	CaO	SiO_2	Al_2O_3	Fe_2O_3	TiO_2	MgO	P_2O_5	F
PG	53.04	41.48	3.32	0.24	0.40	0.07	-	0.81	0.64
Limestone	0.24	89.68	3.75	0.94	0.77	-	4.62	-	-
Bauxite	6.25	0.47	9.82	72.72	5.19	5.33	0.22	-	-

Table 2. Proportioning of the cement raw meal (wt%).

Sample	Limestone	Phosphogypsum	Bauxite
S	39.44	22.67	37.89

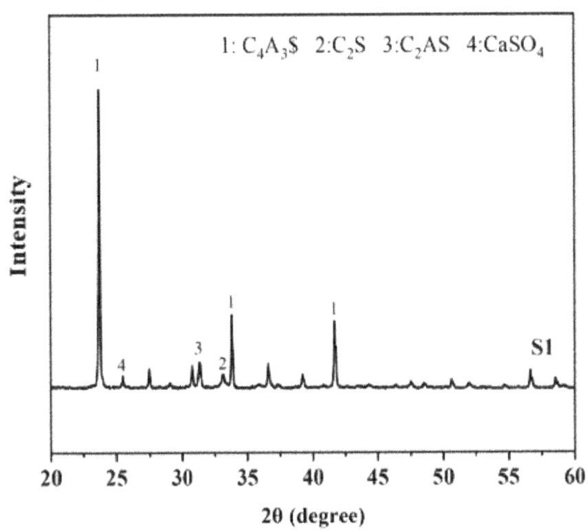

Figure 1. XRD pattern of the clinkers calcined at 1250 °C for 60 min.

Table 3. Contents of cement clinker minerals (wt%).

Sample	Phase Content (wt%)					R_{wp} (%)
	$C_4A_3\$$	C_2S	C_4AF	C$	C_2AS	
S	73.00	13.54	5.42	5.26	2.78	12.76

Figure 2. SEM-SE morphology of cement clinkers.

Table 4. Mineralogical phase in samples SC1-SC3 and SCC1-SCC3.

Sample	The Molar Ratio of $C\$H_2$ to $C_4A_3\$$	The Molar Ratio of CH to $C_4A_3\$$	wt%		
			Clinker/$C_4A_3\$$	$C\$H_2$	CH
SC1	1.0	0	82.10/59.93	17.90	0
SC2	1.5	0	76.90/56.14	23.10	0
SC3	2.0	0	70.90/51.76	29.10	0
SCC1	1.5	0.25	103.97/75.90	22.70	1.40
SCC2	1.5	0.50	102.6/74.90	22.48	2.62
SCC3	1.5	0.75	101.34/73.98	22.19	3.83

Figure 3. Experimental flowchart.

2.2. Test Methods

The samples were prepared using the water-to-cement ratio of 0.5 and a 20 × 20 × 20 mm mould. The moulds were placed in a standard curing room at a temperature of (20 ± 1) °C and (95 ± 1)% humidity for curing and moulding. The moulds were removed after curing age for 1 day, and the samples were placed in the standard curing room. The cube samples were cured in the standard curing room, and the compressive strength of the cement paste was determined at 1, 3, 7, and 28 days. The test blocks were tested for compressive strength using six cube samples for the compression test, and the average value was the final compressive strength.

At the end of the compression test, the hydrated samples were placed in acetone for 24 h to terminate the hydration, after which the blocks were dried in an oven at 40 °C for

24 h. After drying, the samples were ground in a mortar and passed through an 80 μm square-hole sieve. The samples were analysed by XRD (Bruker D8 Advance). The test range was 5–80° (2θ) with a single step scan time of 0.5 s at a step size of 0.02°. The sample phase was detected using EVA software and quantified using TOPAS 4.2 software.

The weight loss of the different samples was tested using an integrated calorimetric analyser (Mettler-Toledo TGA/DSC1, Switzerland) in a dry nitrogen environment at 50 mL/min. The samples were placed in an alumina crucible at a temperature of 50–1000 °C with a heating rate of 10 °C/min. The DTG curve was obtained by solving the TG curve in the first order, which reflects the decomposition peaks corresponding to the decomposition of the different phases. The pore solution was prepared by mixing the sample with deionised water in a 10:1 liquid to solid ratio, then stirred for 2 h with a magnetic mixer and filtered.

The pore size distribution and porosity of the sample in the middle of the 2 g block (3–5 mm) were tested by an automatic pore structure analyser (AutoPore IV 9500, USA). The morphology and particle size of the hydration products were analysed using a scanning electron microscope (FEI Quanta 450, USA). Gold spray coating of samples was at 30 Pa, 15 mA.

3. Results and Discussion

3.1. Compressive Strength

Figure 4 shows the compressive strength of six groups of samples at different curing times. The compressive strengths of SC1 specimens at 1, 3, 7, and 28 days were 34.8, 41.9, 45.4, and 47.8 MPa, respectively. With increasing $C\$H_2$ admixture, the strength of SC2 increased by approximately 14.1, 18.85, 24.23, and 22.6% compared to SC1 specimens. The strength of SC3 specimens at different curing times were 40.9, 52.4, 52.1, and 49.7 MPa, increasing and then decreasing with increasing hydration time. The addition of $C\$H_2$ has a beneficial effect on the compressive strength of the cement. However, the excessive amount of $C\$H_2$ causes a reduction in the strength of the specimens at a later stage, which is related to the micro-expansion of the cement specimens at the later stage of hydration [32,33]. The compressive strengths of specimens in the SCC1 to SCC3 groups were reduced in the early stages of hydration compared to the SC2 specimens. The reduction in compressive strength increased with increasing amounts of CH. The incorporation of CH reduced the early strength of the cement, which may have been related to the acidity and alkalinity in the environment. As the reaction continued, the strength of SCC2 specimens developed the fastest; by 28 days of age, the compressive strength of SCC2 specimens was close to that of SC2, indicating that adding CH reduced the cement's compressive strength. Still, the appropriate amount of CH benefited the development of the later strength of the paste.

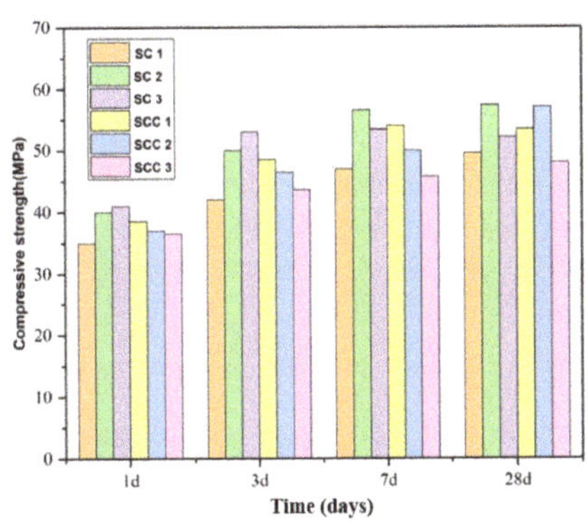

Figure 4. Compressive strength of samples SC1-SC3 and SCC1-SCC3 at different curing times.

3.2. pH Measurement

Figure 5 shows the pH variation at different curing times for samples from the SC2 and SCC1-SCC3 groups. Usually, the pH of calcium sulphoaluminate cement is roughly around 11 to 12; as the raw material, PG has a pH of about 5.8, resulting in a low pH of the prepared cement clinker. The pH values of the four groups of samples were 10.13, 10.32, 10.43, and 10.58. The pH values of the SCC1-SCC3 group increased due to the addition of CH. It can be found that as the hydration reaction proceeded, the pH values of all four groups of samples increased in the early stage of hydration, which may have been due to the large amount of OH^- released by the hydration of clinker minerals and a large amount of water required for AFt generation, increasing solution concentration. As the hydration reaction proceeded, the pH stabilised between the alkali consumption in the solution in equilibrium. Adding CH increased the concentration of Ca^{2+} and OH^- ions in the solution, inhibiting the clinker minerals hydrolysis, resulting in the slow production of AFt in the early stages of hydration, which had a negative impact on the strength development of the specimens. pH analysis results were in good agreement with the compressive strength analysis results.

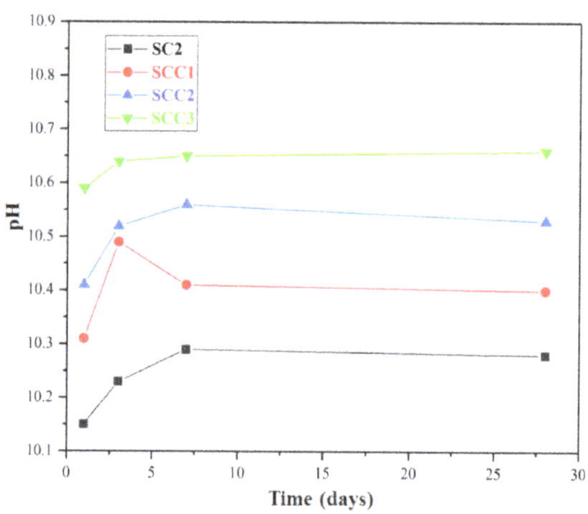

Figure 5. pH value of liquid for cement.

3.3. X-ray Diffraction Analysis

Figure 6 shows the XRD profiles of the six groups of cement samples at 3 days of age. Clear characteristic peaks can be seen in Figure 6, where the main crystalline phases were AFt, AFm, $C\$H_2$, and $C_4A_3\$$, and no characteristic peaks of CH were detected, indicating a complete reaction of CH. In the SC1 to SC3 samples, the diffraction peaks of AFt gradually increased with the increase in $C\$H_2$ content, indicating that the degree of crystallisation of AFt increased due to the rise of $C\$H_2$. Furthermore, in the XRD patterns of the SC1 to SC3 samples, there was no diffraction peak of AFm, which showed that the main hydration reaction of the cement at this time was the typical reaction of $C_4A_3\$$, $C\$H_2$, and water to produce AFt and AH_3; Equation (1) [29] is shown as follows:

$$C_4A_3\$ + 2C\$H_2 + 34H \rightarrow C_6A\$_3H_{32} + 2AH_3 \quad (1)$$

In the samples from SCC1 to SCC3, the content of $C\$H_2$ remained unchanged, and with the increase in CH doping, the AFm diffraction peak was gradually apparent. The diffraction peak of $C_4A_3\$$ gradually decreased. It means that the doping of CH made the generation of AFt limited and played an inhibiting role, while there was AFm generation. The following Equation (2) [27] will occur in the system:

$$C_4A_3\$ + 2C\$H_2 + 6CH + 26H \rightarrow 3C_4A\$H_{12} \quad (2)$$

Figure 7 depicts the content of the material phase at the age of 3 days of hydration for the different samples. In the SC1 to SC3 samples, the AFt content of the hydration products increased as the $C\$H_2$ content gradually increased. However, there was still a tiny amount of $C_4A_3\$$ remaining, indicating that the hydration was not yet complete. In the SCC1 to SCC3 samples, the more CH was incorporated, the lower the AFt content, and the AFm content of the hydration products increased significantly. The $C\$H_2$ was able to promote the hydration of $C_4A_3\$$, causing a large amount of AFt to form in the early stage of hydration. $C\$H_2$ was able to promote the hydration of $C_4A_3\$$, resulting in a large amount of AFt formation in the early stage of hydration. The results of the XRD analysis verified the development of the sample strength.

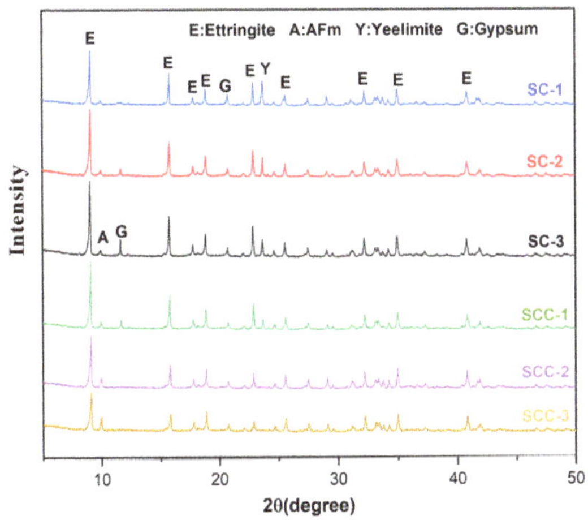

Figure 6. XRD curves of samples SC1-SC3 and SCC1-SCC3 at 3 days of age.

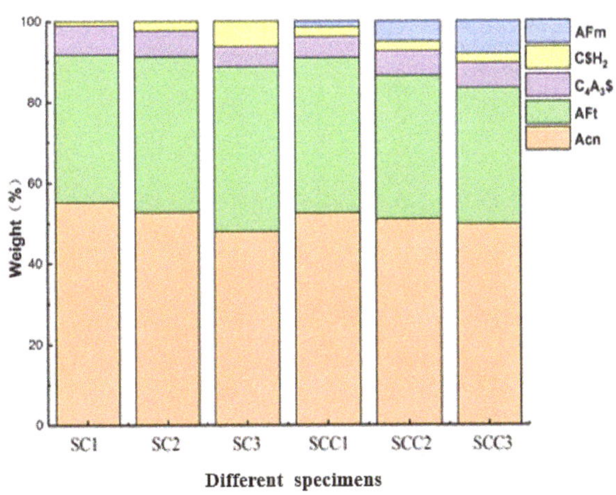

Figure 7. Percentage mass content of samples SC1-SC3 and SCC1-SCC3 at 3 days of age.

3.4. Thermogravimetric Analysis

Figures 8 and 9 show the DTG curves for the hydration products of the SC1 to SC3 and SCC1 to SCC3 samples, respectively. Figure 8 shows three distinct weight loss peaks in the DTG curves. AFt was dehydrated and decomposed in the temperature range of 100–150 °C, $C\$H_2$ was dehydrated and decomposed at around 150 °C, and AH_3 was dehydrated and decomposed in the temperature range of 250–275 °C [34–36]. With the increase in $C\$H_2$ content, the mass loss of AFt tended to increase, indicating that the hydration reaction produced more and more Aft, verifying that $C\$H_2$ can promote the hydration reaction and form Aft.

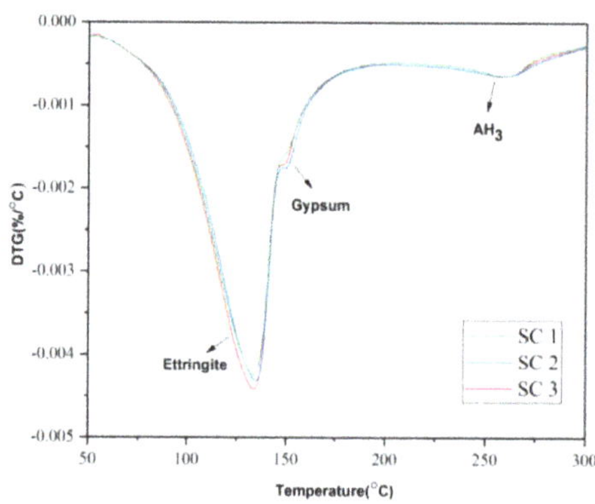

Figure 8. DTG curves of samples SC1-SC3 at 3 days of age.

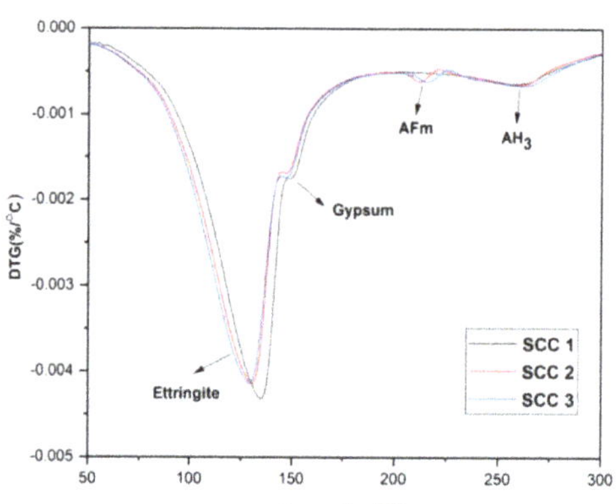

Figure 9. DTG curves of samples SCC1-SCC3 at 3 days of age.

Figure 9 shows the DTG curves of the samples after adding CH. In addition to the weight loss peaks of Aft, $C\$H_2$, and AH_3, a weight loss peak of Afm dehydration decomposition also appeared at around 220 °C. Moreover, the mass loss of Aft decreased, and the mass loss of Afm increased with CH content, indicating that adding CH inhibited the formation of the hydration product Aft and promoted the formation of Afm simultaneously.

3.5. Microstructural Analysis

Figure 10 shows SEM photographs of samples SC1 to SC3 hydrated for 3 days. The AFt and AH_3 can be seen clearly in Figure 10; the AFt crystals were interspersed in a pin-rod or columnar shape, forming a spatial network skeleton. The gel phase appeared to be fibrous, interlacing and filling the pores of the AFt crystals and gluing them to each other, hence its dense nature. This interlaced and densely filled structure can improve the mechanical properties of the cement [37,38]. The AFt in the SC3 sample was relatively large in particle size compared to the other two groups and was not well interleaved and filled with the AH_3 phase. Moreover, the excess $C\$H_2$ reaction left larger pores, which affected the structure of the cement samples and thus reduced the compressive strength of the cementite. Microscopic morphological observations verified the development of macroscopic properties.

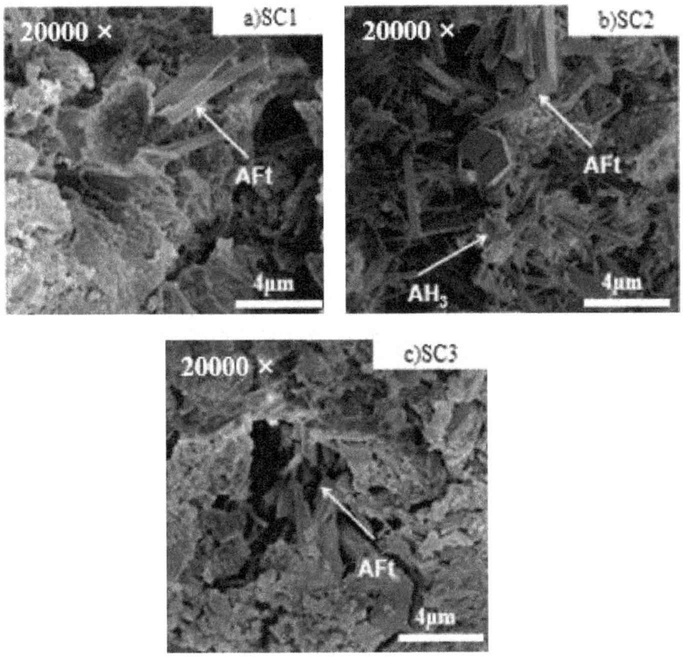

Figure 10. SEM-SE morphology of hydrated sample of SC1-SC3 at 3 days of age.

3.6. *Porosity Analysis*

The influence of porosity and pore size distribution on the strength of cement cannot be ignored and can laterally characterise the degree of hydration of the clinker minerals. The pore structure determines the denseness of cement-hardened slurries, and the pores are divided by the cause of their generation, with the pore types and pore size ranges shown in Table 5 [39–41]. Of these, gel pores and capillary pores are closely associated with the hydration products of the cement, while introduced pores and moulded pores are produced as a result of air-entraining agents or vibrations.

Table 5. Types and the ranges of pores in hardened pastes.

Pore Type	Aperture/nm	Causes
Gel pore	$1–10^2$	Pores of gel-like substances in hydration products
Capillary pore	$10^2–10^4$	Pores between the larger size crystals in the hydration product
Introduced pore	$10^4–10^6$	Pores introduced by air-entraining agents or aluminium powder during hydration
Forming pores	$10^6–10^7$	Pore space due to the poor pounding of the paste

Figure 11 shows the pore size distribution of SC1-SC3 samples hydrated for 3 days. From the test results, the pore types of the samples were mainly gel pores; the median pore sizes of SC1 to SC3 were 16.4 nm, 11.9 nm, and 14.4 nm, respectively. Combined with the compressive strength of the cement test blocks, it can be found that the smaller the median pore size, the larger the compressive strength value of the test blocks; from SC1 to SC2 samples, the porosity decreased, and the compressive strength increased. The internal structure of the specimen gradually tended to become denser, and its contribution to the strength tended to increase. This was due to the gradual increase in AFt and AH_3, which formed a structure that contributes to strength [42]. For the SC3 samples, the compressive strength decreased with increasing porosity. This indicates that the internal structure became loose when the $C\$H_2$ admixture was too high, which may have been related to the accumulation of AFt and the excess of $C\$H_2$. More voids and unreacted gypsum were left in the hardened paste, adversely affecting the pore structure of the cement test blocks. The test results showed a good correspondence between porosity and compressive strength.

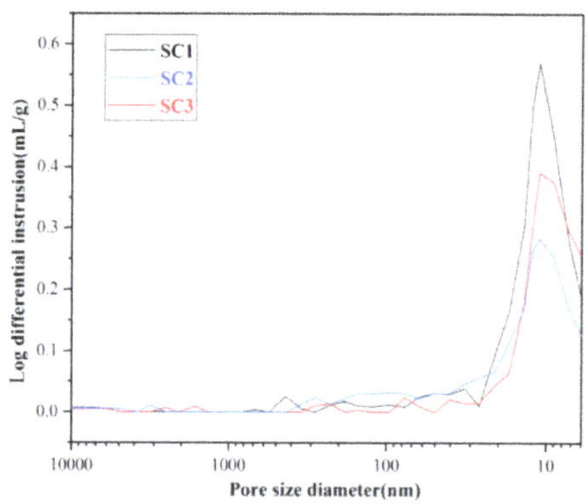

Figure 11. Pore size distribution of samples SC1-SC3 at 3 days of age.

4. Conclusions

This paper focused on calcium sulphoaluminate cement prepared from phosphogypsum, using two mineral admixtures, $C\$H_2$ and CH, to optimise the calcium sulphoaluminate cement-based materials. The development of compressive strength and changes in hydration products were studied, and the effects of clinker mineral composition and $C\$H_2$ and CH adding on the physical behaviour, hydration products, and microstructure of cement hardening pastes were analysed. The following conclusions were drawn from the above research process.

(1) When $C\$H_2$ was added to the cementitious material, the early strength of the cement increased as the amount of $C\$H_2$ increased. The optimum admixture of $C\$H_2/C_4A_3\$$ was 1.5 mol. Adding CH decreased the cement's compressive strength and increased the paste's pH, which facilitated the development of the later strength. The optimum dose of $CH/C_4A_3\$$ was 0.5 mol.

(2) The addition of $C\$H_2$ promoted the early hydration of $C_4A_3\$$ and formed more AFt. The amount of $C\$H_2$ incorporation and the AFt content were positively correlated. The incorporation of CH changed the hydration process. It inhibited the early hydration of $C_4A_3\$$, and the hydration products appeared as AFm. With the increase in CH, the amount of AFt production became less, and the AFm content increased.

(3) The addition of $C\$H_2$ influenced the microstructure of the matrix. With the increase in $C\$H_2$, it was observed that the AFt crystals developed well, and the microscopic morphology changed from needle-rod to columnar or block. Excessive $C\$H_2$ caused AFt crystals to be stacked, not tightly cemented with AH_3. At the same time, the matrix contained unreacted $C\$H_2$, leading to an increase in pore size and porosity.

Author Contributions: Conceptualization, J.C. and P.Z.; Methodology, J.C.; Software, J.Z.; Validation, J.Z., J.C., and P.Z.; Formal analysis, J.C.; Investigation, T.W.; Resources, P.Z.; Data curation, T.W.; Writing—original draft preparation, J.Z.; Writing—review and editing, J.C.; P.Z.; Visualization, J.Z.; Supervision, J.C.; P.Z.; Funding acquisition, J.C.; P.Z. All authors have read and agreed to the published version of the manuscript.

Funding: This work was funded by the National Natural Science Foundation of China (52172015), the Fundamental Research Funds for the Central Universities in China (DUT21ZD201), and the Natural Science Foundation of Guizhou Provincial Department of Education (Guizhou cooperation KY word [2021]301).

Institutional Review Board Statement: Not applicable.

Informed Consent Statement: Not applicable.

Data Availability Statement: Not applicable.

Conflicts of Interest: The authors declare no conflict of interest.

References

1. Rashad, A.M. Phosphogypsum as a construction material. *J. Clean. Prod.* **2017**, *166*, 732–743. [CrossRef]
2. Islam, G.S.; Chowdhury, F.H.; Raihan, M.T.; Amit, S.K.S.; Islam, M.R. Effect of Phosphogypsum on the Properties of Portland Cement. *Procedia Eng.* **2017**, *171*, 744–751. [CrossRef]
3. Cuadri, A.; Navarro, F.; García-Morales, M.; Bolivar, J.P. Valorization of phosphogypsum waste as asphaltic bitumen modifier. *J. Hazard. Mater.* **2014**, *279*, 11–16. [CrossRef] [PubMed]
4. Rashad, A.M. Potential use of phosphogypsum in alkali-activated fly ash under the effects of elevated temperatures and thermal shock cycles. *J. Clean. Prod.* **2015**, *87*, 717–725. [CrossRef]
5. Pérez-López, R.; Macías, F.; Cánovas, C.R.; Sarmiento, A.M.; Pérez-Moreno, S.M. Pollutant flows from a phosphogypsum disposal area to an estuarine environment: An insight from geochemical signatures. *Sci. Total Environ.* **2016**, *553*, 42–51. [CrossRef]
6. Ma, B.; Lu, W.; Su, Y.; Li, Y.; Gao, C.; He, X. Synthesis of α-hemihydrate gypsum from cleaner phosphogypsum. *J. Clean. Prod.* **2018**, *195*, 396–405. [CrossRef]
7. Papaslioti, E.-M.; Pérez-López, R.; Parviainen, A.; Sarmiento, A.M.; Nieto, J.M.; Marchesi, C.; Delgado-Huertas, A.; Garrido, C.J. Effects of seawater mixing on the mobility of trace elements in acid phosphogypsum leachates. *Mar. Pollut. Bull.* **2018**, *127*, 695–703. [CrossRef]
8. Cánovas, C.R.; Macías, F.; López, R.P.; Nieto, J.M. Mobility of rare earth elements, yttrium and scandium from a phosphogypsum stack: Environmental and economic implications. *Sci. Total Environ.* **2018**, *618*, 847–857. [CrossRef]
9. Papastefanou, C.; Stoulos, S.; Ioannidou, A.; Manolopoulou, M. The application of phosphogypsum in agriculture and the radiological impact. *J. Environ. Radioact.* **2006**, *89*, 188–198. [CrossRef]
10. Shukla, V.; Ramachandran, T.; Chinnaesakki, S.; Sartandel, S.; Shanbhag, A. Radiological impact of utilization of phosphogypsum and fly ash in building construction in India. *Int. Congr. Ser.* **2005**, *1276*, 339–340. [CrossRef]
11. Pérez-López, R.; Nieto, J.M.; López-Coto, I.; Aguado, J.L.; Bolívar, J.P.; Santisteban, M. Dynamics of contaminants in phosphogypsum of the fertilizer industry of Huelva (SW Spain): From phosphate rock ore to the environment. *Appl. Geochem.* **2010**, *25*, 705–715. [CrossRef]
12. Kadirova, Z.; Hojamberdiev, M.; Bo, L.; Hojiyev, R.; Okada, K. Ion uptake properties of low-cost inorganic sorption materials in the $CaO–Al_2O_3–SiO_2$ system prepared from phosphogypsum and kaolin. *J. Clean. Prod.* **2014**, *83*, 483–490. [CrossRef]
13. Cui, K.; Chang, J.; Feo, L.; Chow, C.L.; Lau, D. Developments and Applications of Carbon Nanotube Reinforced Cement-Based Composites as Functional Building Materials. *Front. Mater.* **2022**, *9*. [CrossRef]
14. Wu, S.; Yao, Y.; Yao, X.; Ren, C.; Li, J.; Xu, D.; Wang, W. Co-preparation of calcium sulfoaluminate cement and sulfuric acid through mass utilization of industrial by-product gypsum. *J. Clean. Prod.* **2020**, *265*, 121801. [CrossRef]
15. Rosales, J.; Pérez, S.M.; Cabrera, M.; Gázquez, M.J.; Bolivar, J.P.; de Brito, J.; Agrela, F. Treated phosphogypsum as an alternative set regulator and mineral addition in cement production. *J. Clean. Prod.* **2020**, *244*, 118752. [CrossRef]
16. Nizevičienė, D.; Vaičiukynienė, D.; Vaitkevičius, V.; Rudžionisb, Ž. Effects of waste fluid catalytic cracking on the properties of semi-hydrate phosphogypsum. *J. Clean. Prod.* **2016**, *137*, 150–156. [CrossRef]
17. Li, B.; Li, L.; Chen, X.; Ma, Y.; Zhou, M. Modification of phosphogypsum using circulating fluidized bed fly ash and carbide slag for use as cement retarder. *Constr. Build. Mater.* **2022**, *338*, 127630. [CrossRef]
18. Xu, W.; Li, Q.; Haruna, S. The Effect of Calcium Formate, Sodium Sulfate, and Cement Clinker on Engineering Properties of Fly Ash-Based Cemented Tailings Backfill. *Adv. Mater. Sci. Eng.* **2019**, *2019*, 5370360. [CrossRef]
19. Isteri, V.; Ohenoja, K.; Hanein, T.; Kinoshita, H.; Kletti, H.; Rößler, C.; Tanskanen, P.; Illikainen, M.; Fabritius, T. Ferritic calcium sulfoaluminate belite cement from metallurgical industry residues and phosphogypsum: Clinker production, scale-up, and microstructural characterisation. *Cem. Concr. Res.* **2022**, *154*, 106715. [CrossRef]

20. Shen, Y.; Qian, J.; Huang, Y.; Yang, D. Synthesis of belite sulfoaluminate-ternesite cements with phosphogypsum. *Cem. Concr. Compos.* **2015**, *63*, 67–75. [CrossRef]
21. Huang, Y.; Qian, J.; Liu, C.; Liu, N.; Shen, Y.; Ma, Y.; Sun, H.; Fan, Y. Influence of phosphorus impurities on the performances of calcium sulfoaluminate cement. *Constr. Build. Mater.* **2017**, *149*, 37–44. [CrossRef]
22. Huang, Y.; Qian, J.; Kang, X.; Yu, J.; Fan, Y.; Dang, Y.; Zhang, W.; Wang, S. Belite-calcium sulfoaluminate cement prepared with phosphogypsum: Influence of P_2O_5 and F on the clinker formation and cement performances. *Constr. Build. Mater.* **2019**, *203*, 432–442. [CrossRef]
23. Kuryatnyk, T.; da Luz, C.A.; Ambroise, J.; Pera, J. Valorization of phosphogypsum as hydraulic binder. *J. Hazard. Mater.* **2008**, *160*, 681–687. [CrossRef] [PubMed]
24. Madan, C.S.; Panchapakesan, K.; Reddy, P.V.A.; Joanna, P.S.; Rooby, J.; Gurupatham, B.G.A.; Roy, K. Influence on the Flexural Behaviour of High-Volume Fly-Ash-Based Concrete Slab Reinforced with Sustainable Glass-Fibre-Reinforced Polymer Sheets. *J. Compos. Sci.* **2022**, *6*, 169. [CrossRef]
25. Sánchez-Herrero, M.J.; Fernández-Jiménez, A.; Palomo, A. $C_4A_3\check{S}$ hydration in different alkaline media. *Cem. Concr. Res.* **2013**, *46*, 41–49. [CrossRef]
26. Winnefeld, F.; Barlag, S. Calorimetric and thermogravimetric study on the influence of calcium sulfate on the hydration of ye'elimite. *J. Therm. Anal. Calorim.* **2010**, *101*, 949–957. [CrossRef]
27. Hargis, C.W.; Kirchheim, A.P.; Monteiro, P.J.; Gartner, E.M. Early age hydration of calcium sulfoaluminate (synthetic ye'elimite, $C_4A_3\$$) in the presence of gypsum and varying amounts of calcium hydroxide. *Cem. Concr. Res.* **2013**, *48*, 105–115. [CrossRef]
28. Huang, Y.; Qian, J.; Lu, L.; Zhang, W.; Wang, S.; Wang, W.; Cheng, X. Phosphogypsum as a component of calcium sulfoaluminate cement: Hazardous elements immobilization, radioactivity and performances. *J. Clean. Prod.* **2020**, *248*, 119287. [CrossRef]
29. Cui, K.; Lau, D.; Zhang, Y.; Chang, J. Mechanical properties and mechanism of nano-$CaCO_3$ enhanced sulphoaluminate cement-based reactive powder concrete. *Constr. Build. Mater.* **2021**, *309*, 125099. [CrossRef]
30. Cui, K.; Chang, J. Hydration, reinforcing mechanism, and macro performance of multi-layer graphene-modified cement composites. *J. Build. Eng.* **2022**, *57*, 104880. [CrossRef]
31. Le Saoût, G.; Kocaba, V.; Scrivener, K. Application of the Rietveld method to the analysis of anhydrous cement. *Cem. Concr. Res.* **2011**, *41*, 133–148. [CrossRef]
32. Hargis, C.W.; Telesca, A.; Monteiro, P.J.M. Calcium sulfoaluminate (Ye'elimite) hydration in the presence of gypsum, calcite, and vaterite. *Cem. Concr. Res.* **2014**, *65*, 15–20. [CrossRef]
33. Zhang, H.; Li, J.; Kang, F. Real-time monitoring of humidity inside concrete structures utilizing embedded smart aggregates. *Constr. Build. Mater.* **2022**, *331*, 127317. [CrossRef]
34. Yu, C.; Cui, C.; Zhao, J.; Zheng, J. A novel approach to utilizing dredged materials at the laboratory scale. *Constr. Build. Mater.* **2021**, *313*, 125568. [CrossRef]
35. Bullerjahn, F.; Boehm-Courjault, E.; Zajac, M.; Ben Haha, M.; Scrivener, K. Hydration reactions and stages of clinker composed mainly of stoichiometric ye'elimite. *Cem. Concr. Res.* **2019**, *116*, 120–133. [CrossRef]
36. Cui, K.; Liang, K.; Chang, J.; Lau, D. Investigation of the macro performance, mechanism, and durability of multiscale steel fiber reinforced low-carbon ecological UHPC. *Constr. Build. Mater.* **2022**, *327*, 126921. [CrossRef]
37. Xue, W.; Liu, W.; Gao, C.; Fan, H.; Zhang, H. Experimental study of permeability evolution in modified desulfurization gypsum under complete stress–strain process. *J. Clean. Prod.* **2022**, *359*, 132185. [CrossRef]
38. Zhang, L.; Qian, X.; Yu, C.; Fang, M.; Qian, K.; Lai, J. Influence of evaporation rate on pore size distribution, water loss, and early-age drying shrinkage of cement paste after the initial setting. *Constr. Build. Mater.* **2019**, *226*, 299–306. [CrossRef]
39. Zhang, H.; Li, J.; Kang, F.; Zhang, J. Monitoring and evaluation of the repair quality of concrete cracks using piezoelectric smart aggregates. *Constr. Build. Mater.* **2022**, *317*, 125775. [CrossRef]
40. Xue, W.; Fan, H.; Liu, X.; Shen, L. Effects of early-age temperature and salt ion corrosion on the macroproperty deterioration of concrete and corresponding micromechanism. *Arch. Civ. Mech. Eng.* **2022**, *22*, 1–16. [CrossRef]
41. Xue, W.; Liu, X.; Jing, W.; Yao, Z.; Gao, C.; Li, H. Experimental study and mechanism analysis of permeability sensitivity of mechanically damaged concrete to confining pressure. *Cem. Concr. Res.* **2020**, *134*, 106073. [CrossRef]
42. Li, J.; Chang, J.; Wang, T.; Zeng, T.; Li, J.; Zhang, J. Effects of phosphogypsum on hydration properties and strength of calcium aluminate cement. *Constr. Build. Mater.* **2022**, *347*, 128398. [CrossRef]

buildings

MDPI

Article

Mechanical Performance of Date-Palm-Fiber-Reinforced Concrete Containing Silica Fume

Yasser E. Ibrahim [1,*], Musa Adamu [1,*], Mohammad Louay Marouf [1], Omar Shabbir Ahmed [1], Q. A. Drmosh [2] and Mohammad Abdul Malik [1]

1 Engineering Management Department, College of Engineering, Prince Sultan University, Riyadh 11586, Saudi Arabia
2 Interdisciplinary Research Center for Hydrogen and Energy Storage, King Fahd University of Petroleum and Minerals, Dhahran 31216, Saudi Arabia
* Correspondence: ymansour@psu.edu.sa (Y.E.I.); madamu@psu.edu.sa (M.A.)

Abstract: The use of date palm fiber (DPF) as natural fiber in concrete and mortar continues to gain acceptability due to its low-cost and availability. However, the main disadvantage of DPF in cement-based composites is that it reduces compressive strength and increases the porosity of the composite. Hence, for DPF to be efficiently used in concrete, its negative effects must be counteracted. Therefore, in this study, silica fume was employed as supplementary cementitious material to alleviate the negative effects of DPF on the strength and porosity of concrete. The DPF was added in different dosages of 0%, 1%, 2%, and 3% by weight of binder materials. Silica fume was used as a cement replacement material at dosages of 0% to 15% (intervals of 5%) by volume of cement. The unit weights, mechanical strengths, water absorption, and microstructural morphology were all evaluated. The concrete's fresh and hardened densities were reduced with the increment in DPF and silica fume. The compressive strength declined at all ages with the increment in DPF addition, while the flexural and splitting tensile strengths improved with addition of up to 2% DPF. Furthermore, the concrete's water absorption escalated with an increase in DPF content. Silica fume significantly enhanced the mechanical strength of the concrete. The dissipation in compressive strength with the addition of up to 2% DPF was mitigated by replacing up to 10% cement with silica fume, where it densified the microstructure and refined the interfacial transition zone between the fibers and cement matrix, hence significantly decreasing the porosity and enhancing durability.

Keywords: natural fiber; date palm fiber; silica fume; composites; date-palm-fiber-reinforced concrete; mechanical properties

check for updates

Citation: Ibrahim, Y.E.; Adamu, M.; Marouf, M.L.; Ahmed, O.S.; Drmosh, Q.A.; Malik, M.A. Mechanical Performance of Date-Palm-Fiber-Reinforced Concrete Containing Silica Fume. *Buildings* **2022**, *12*, 1642. https://doi.org/10.3390/buildings12101642

Academic Editor: Tomáš Dvorský

Received: 22 September 2022
Accepted: 7 October 2022
Published: 10 October 2022

Publisher's Note: MDPI stays neutral with regard to jurisdictional claims in published maps and institutional affiliations.

1. Introduction

The use of sustainable construction and building materials offers a series of advantages, such as lower production and maintenance cost, environmental friendliness, biodegradability, non-toxicity, and improved mechanical performance [1]. Concrete has been the most utilized building and construction material because of its series of advantages, such as high strength and durability, good quality, lower strength-to-cost ratio, and ability to be easily cast into various shapes compared to other construction materials such as steel and timber [2,3]. However, concrete's high brittleness, poor tensile strength, low strain, and low bending resistance are its main drawbacks. Due to this, concrete is prone to cracking while under tensile loads. To overcome these concrete shortcomings, reinforcement is normally used with concrete to form a composite action i.e., reinforced concrete elements. However, due to the high cost of reinforcement, fibers are incorporated in concrete to enhance its tensile strength and minimize the amount of reinforcement required to achieve tensile strength. Additionally, fibers are used in plain concrete to enhance its tensile strength. Reinforcing fibers of various kinds, classified as either natural or synthetic, have been employed in concrete to enhance its hardened and physical qualities. [4–6]. Glass, steel, organic plastics, and polymers are the best reinforcing materials used in concrete. A few naturally occurring fibers, such as asbestos and cellulose, as well as agricultural fibers such as jute, sisal, date palm mesh, etc., are also utilized in concrete [7,8]. DPF is one of the popular, inexpensive, environmentally beneficial, and sustainable natural fibers. The date palm tree produces vast amounts of waste, much of which is discarded without being properly utilized. Comparing DPF to other natural and synthetic fibers, it has the distinct benefit of having a higher strength-to-cost ratio [9].

Several studies have reported that DPF has been used as natural fiber in cement-based materials such as concrete and mortar. Ozerkan, Ahsan [10] added DPF in dosages of 0%, 0.5%, 1%, and 2% in mortar. The initial and final settings times and drying shrinkage of the mortar increased with the addition of up to 1% DPF and declined with the addition of 2% fiber. Furthermore, DPF escalated the rate of water absorption of the mortar, with 0.5% addition being the optimum. They also reported improvement in mechanical properties and sulfate attack resistance with the addition of up to 1% DPF, which they attributed to better compaction between the fibers and cement matrix resulting in good homogeneity in the mix. Benaniba, Driss [11] developed a DPF–cement mortar composite using different proportions of DPF (6% to 30% at intervals of 6%) by weight. They reported a significant increase in water absorption, and a decrease in density, thermal conductivity, flexural and compressive strengths with the increment in the percentage of DPF. There was an improvement in the flexural strength at a lower dosage of 6% and 12% compared to plain cement mortar. In a related work, Zanichelli, Carpinteri [12] evaluated the fracture behavior of mortar reinforced with DPF at different proportions from 2% to 10% (at intervals of 2%) by volume. They reported a reduction in peak load, elastic modulus, and fracture toughness with an increment in DPF addition. On the contrary, their results showed improved ductile behavior and energy absorption with the increment in DPF content. Based on their findings, they recommended using DPF-reinforced mortar for nonstructural applications where high strength is not important and structures where lightweight and durability are very significant. Vantadori, Carpinteri [1] examined the impact of DPF in cement mortar, where they added DPF in varying proportions from 2% to 10% (interval of 2%) by volume. They observed a decline in density, flexural strength, and fracture toughness with the increment in fiber content. There was a reduction between 9–52% and 7–66% for the modulus of rupture and fracture toughness, respectively, with the addition of 2–10% DPF. The ductility and energy absorption capacity of the mortar enhanced with percentage increase in DPF, where an increase between 27–162% was found with addition of 2–10% DPF compared to the plain mortar. Kareche, Agoudjil [13] investigated DPF's impact on the durability performance of cement mortar, where they incorporated DPF at different proportions of 5%, 10%, and 15% by weight. They reported a surge in porosity and a decline in drying shrinkage in the mortar with DPF addition. The addition of 15% DPF led to an increment in porosity by up to 51% and a decline in drying shrinkage up to 71.4%. Additionally, they reported an improvement in the acid attack resistance (sulfuric attack), and a reduction in the compressive strength and resistance to cyclic wetting and drying with the increment in DPF content. Alatshan, Altlomate [14] prepared several mixes by addition different DPF dosages of 0.5%, 1%, 1.5%, 2% and 2.5% for fiber lengths of 50 mm, 60 mm, and 70 mm. They noticed a reduction in the compressive strength due the high content of DPF dosage and length. However, the concrete's compressive strength reinforced with 0.5%–50 mm DPF was reported to be of a higher strength than plain concrete due to the tensile strength of the added DPF. The flexural strength was moderately enhanced with a lower dosage of DPF of all lengths. The flexural strength of composites with 0.5%–50 mm DPF, 1%–60 mm DPF, and 0.5%–70 mm DPF were higher in comparison to the conventional concrete due to the tensile strength of the fiber. However, for all other dosages and fiber length mix combinations, there was depletion in the flexural strength.

Based on the available studies, concrete and mortar reinforced with DPF have several advantages compared to plain concrete and mortar, which include lower density, lightweight, higher ductility and energy absorption capacity, lower thermal conductivity, higher thermal insulation, and resistivity. These properties improve thermal comfortability and energy savings in buildings through good thermal insulation [15,16]. However, the major challenge and disadvantage of using DPF in concrete and mortar is the reduction in mechanical properties and elastic modulus [17–19]. Additionally, DPF affects the durability performance of cementitious composites through increased water absorption, porosity, drying shrinkage, etc. These negative effects are caused mainly by the increase in pore volume in the composite's microstructure, poor adhesion between fibers and cement paste, and the hydrophilic nature of the DPF, making it absorb more water and swell during mixing [16]. Therefore, for DPF to be effectively added in concrete and mortar to derive its maximum benefits, methods for reducing or mitigating its undesirable effect on the cement composite's performance must be derived. Several studies treated DPF using different methods; some methods were slightly effective, while others further caused reduction in the mechanical properties. SCM such as nano silica and silica fume has been used in fiber-reinforced concrete to overcome the fiber's negative effects on the composite's performance. Gencel, Nodehi [20] used silica fume as an SCM in basalt-fiber-reinforced foam concrete (BFRFC). They added basalt fiber in three volumes, i.e., 0%, 1.5%, and 3%, and partially substituted cement with silica fume at 0% and 15%. They reported a decrease in the workability (flowability) and density of the concrete with silica fume addition, where a decrease by 16% was observed with 15% silica fume addition. They reported an increased in thermal conductivity, elastic modulus, flexural, and compressive strengths for the BFRFC with the inclusion of silica fume. The addition of 15% silica fume to the BFRFC containing 0%, 1.5%, and 3% improved the 28 days compressive strength by 54%, 70.5%, and 66.8%, respectively, and the 28 days flexural strength by 5.8%, 9.4%, and 8.9%, respectively. Furthermore, they reported a decrease in the porosity, water absorption, and splitting tensile strength of the BFRFC with silica fume addition. For 15% silica fume addition to the BFRFC containing 0%, 1.5%, and 3% fiber, the porosity decreased by 51.2%, 23.2%, and 35.2%, respectively, the water absorption decreased by 52.2%, 24.4%, and 35.7%, respectively, and the tensile strength diminished by 16.7%, 12.5%, and 4.6%, respectively. Sadrmomtazi, Tahmouresi [21] examined the effects of silica fume as a SCM on basalt-fiber-reinforced cementitious composite (BFRCC)'s properties. They prepared several mixes by adding basalt fiber at different proportions by volume (0%, 1%, and 1.5%) and substituted cement using silica fume at different percentages by weight (5%, 10%, and 15%). The increment in silica fume reduced the workability, where a 15% silica fume addition decreased the flowability of the BFRCC by 28%, 19%, and 21% for 0%, 1%, and 1.5% fiber content, respectively, while the flexural and compressive strengths of the BFRCC were enhanced with the inclusion of silica fume for any fiber content. For a 5% silica fume addition, there was enhancement in compressive strength by 23%, 37%, and 12%, and flexural strength by 4%, 12%, and 5% for BFRCC containing 0%, 1%, and 1.5% fiber, respectively. Khan, Rehman [22] utilized silica fume as an SCM in coconut-fiber-reinforced concrete (CFRC). They prepared the CFRC by adding 2% by weight coconut fiber and replaced cement with different proportions of silica fume between 5–20%. The modulus of rupture, splitting tensile, and compressive strengths of the CFRC improved substantially with the increment in silica fume content. Replacement of cement with 5%, 10%, 15%, and 20% in the CFRC resulted in an increase in compressive strength by 5.8%, 10.8%, 24.6%, and 3%, respectively, and the modulus of elasticity by 10.8%, 22%, 33.6%, and 1.2%, respectively. The splitting tensile strength increased by 25.9%, 29.6%, and 33.5%, while the flexural strength increased by 26.9%, 50%, and 59.6% for 5%, 10%, and 15% silica fume addition to the CFRC, respectively, when compared to plain concrete.

Other materials have been used to improve the performance of both plain and fiber-reinforced concrete. Zhang, Wang [23] added nano silica to improve the performance of fly ash (30%) and metakaolin (70%)-based polyvinyl-alcohol (PVA)-fiber-reinforced geopolymer/alkali-activated mortar. They added PVA fiber at, 0.2%, 0.4%, 0.6%, 0.8%, 1.0%, and 1.2% by volume of mortar and 1% nano silica by weight of binder. They reported a reduction in sump of 26.4% with the addition of 0 to 1.2% PVA, while 1% nano silica slightly increased the slump. The compressive strength, flexural strength, and elastic modulus of the mortar increased with the addition of nano silica and PVA fiber. The 1% nano silica increased the strength by up to 29.1%, while up to 0.8% PVA increased the strength by 27.6%. They achieved the highest strength with addition of 1% nano silica and 0.8% PVA. Up to a 68% increase in flexural strength was reported with addition of PVA. The elastic modulus increased by 20–130% with addition of 0.2% to 1.2% PVA. Furthermore, the addition of up to 1% PVA enhanced the fracture toughness of the mortar, and nano silica also enhanced the load-carrying capacity of the mortar with up to 0.6% PVA. Golewski [24] studied the effects of different SCMs on the properties of concrete. They used fly ash, silica fume, and nano silica as the SCM. They replaced cement with a constant dosage of silica fume and nano silica at 10% and 5%, respectively, while for fly ash they used different replacement proportions of 0%, 5%, and 15%. When compared to the plain concrete, all the quaternary blend concretes demonstrated a higher strength. For the blended concrete, they reported a reduction in compressive strength with the increment in fly ash content. The quaternary blend with 5% fly ash, 10% silica fume, and 5% nano silica gave the best strength and fracture toughness [25]. In a similar study, Gil and Golewski [26] investigated the effects of the interaction of silica fume and fly ash as SCMs on the properties of concrete. They produced different blends using combinations of different proportions of fly ash between 0% and 20% and silica fume proportions between 0% and 10%. Their findings showed that the concrete blend containing 10% silica fume with 0% fly ash had the best performance in terms compressive and tensile strengths, water absorption, and water penetration, while the fly ash decreased the compressive and tensile strengths and increased the water absorption and depth of penetration of the concrete.

Different types of natural fibers have been used to improve the performance of cementitious composites. However, there are few studies that have utilized DPF as a natural fiber in cementitious composites. The available studies that used DPF mostly used it in mortar, hence there is need for more and also comprehensive studies on the use of DPF in concrete. This is due to the aforementioned advantages of DPF. Based on the available studies, the use to silica fume has proven to significantly enhance the properties of fiber-reinforced concrete. To the best of the authors knowledge, there are limited studies that have investigated the effects of DPF addition and silica fume as a partial replacement for cement to improve the mechanical performance and mitigate the negative effects of the fiber on the strength of the concrete. Hence, in this study, DPF was treated using alkaline solution (NaOH), and silica fume was used as an SCM in the concrete to reduce the porosity and mitigate the undesirable effects of the DPF on the concrete's performance. The DPF was added at different dosages of 1%, 2%, and 3% by weight of binder materials, and silica fume was used to partially replace cement at 5%, 10%, and 15%. The effects of the DPF and silica fume on the fresh, mechanical, and durability performance of the DPF-reinforced concrete was then investigated.

2. Materials and Methods

2.1. Materials

Type I OPC was used as the main binder material. The properties of the cement conformed to the standard guidelines outlined in ASTM C150/150M [27] and are presented in Table 1, and its X-ray diffraction (XRD) is shown in Figure 1a. Densified silica fume was used as an SCM. The silica fume was obtained from Xiamen All Carbon Company, Xiamen China, through Al-Rashad Cement Company, Saudi Arabia. The silica fume had a specific gravity (SG) of 2.25, bulk density of 630 kg/m^3, and a specific surface area of 18,000 m^2/kg. The chemical properties and XRD pattern of the silica fume are presented in Table 1 in Figure 1b, respectively. Natural sand was used as fine aggregate. The properties of the aggregates were in line with ASTM C33 [28] specifications. The fine aggregate was well-graded, with a particle size distribution falling within the limits specified by ASTM C33 [28] as highlighted in Figure 2. Additionally, the fine aggregate had an SG value of

2.63, 1.87% water absorption, 1565 kg/m^3 bulk density, fineness modulus of 2.26, and mud content of 1.1%. The coarse aggregate utilized was crushed gravel of 19 mm size, which was well-graded. As demonstrated in Figure 3, the coarse aggregate's particle size distribution also complied with ASTM C33 [28] guidelines. The coarse aggregate had an SG, bulk density, and water absorption of 2.67, 1455 kg/m^3, and 0.65%, respectively. A polycarboxylate-based water-reducing admixture with a density of 1.060 g/cm^3 was used as a superplasticizer so as reduce the required quantity of mixing water.

Table 1. Properties of binder materials.

Oxide Composition (%)	Cement	Silica Fume
CaO	65.18	0.21
Al_2O_3	5.39	0.26
Fe_2O_3	3.40	0.05
SiO_2	19.71	95.85
MgO	0.91	0.45
Na_2O	0.17	–
K_2O	1.22	–
TiO_2	0.24	–
SO_3	3.51	1.00
P_2O_5	0.09	–
MnO	0.18	–
LOI	2.38	2.80

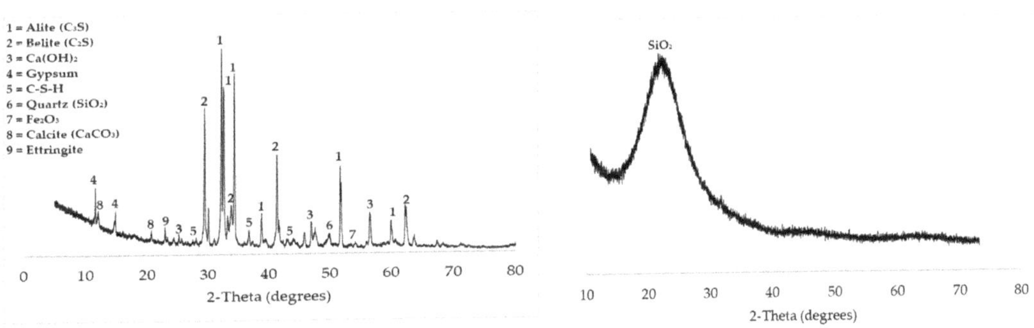

(**a**) Cement (**b**) Silica fume

Figure 1. XRD pattern of binder materials.

The DPF used in this study was obtained from the Aldebass farm in Majmaah, Saudi Arabia. The fiber was plucked from the trunks of date palm trees in the form of a natural rectangular woven mesh in three superposing layers and was about 200–300 mm in width and 300–500 mm in length as shown in Figure 4a. Before treatment, the fiber was immersed in water, separated into single fibers of about 0.2–1.0 mm diameter, and cut to a fiber length of 20–30 mm. The DPF was then treated using alkaline solution, i.e., a 3% NaOH solution for 3 h based on the recommendation of Ali-Boucetta, Ayat [29]. After treatment, the DPF was then thoroughly washed with clean water and air dried for 48 h before use in the concrete. Figure 4b shows the processed DPF ready to be used. Table 2 presents the physical and mechanical properties of the DPF.

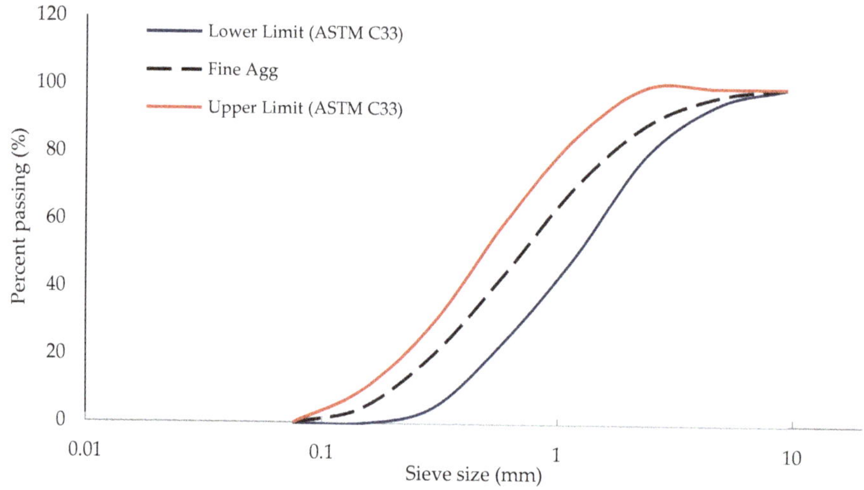

Figure 2. Fine aggregate gradation.

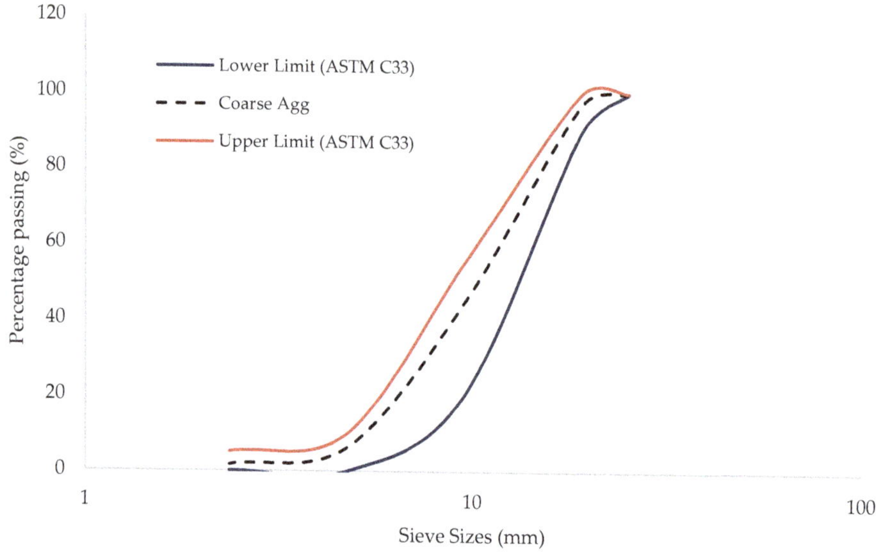

Figure 3. Coarse aggregate gradation.

Table 2. Properties of DPF (1).

Property	Value
Diameter (mm)	0.2–1.0
Length (mm)	20–30
Bulk density (kg/m^3)	877.43 ± 4.8
Natural moisture content (%)	10.2 ± 0.4
Water absorption to saturation (%)	102.65 ± 3.3
Tensile strength (MPa)	203.24 ± 30
Elongation at break (%)	13.5 ± 2
Modulus of elasticity (GPa)	3.35 ± 1.5

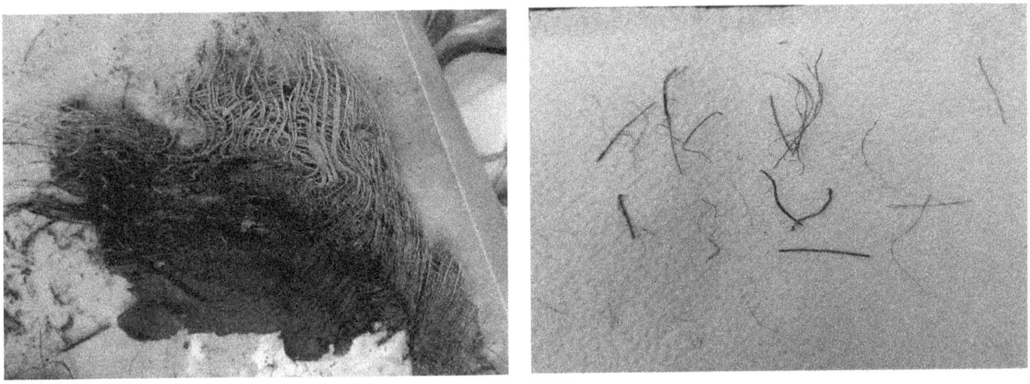

(**a**) Untreated fiber in woven form (**b**) Treated single fiber

Figure 4. Date palm fiber.

2.2. Mix Proportioning

The control concrete mix design was carried out adopting ACI 211.1R [30] based on the absolute volume method. The quantity of superplasticizer used throughout all the mixes was constant (1% by mass of binder) to mitigate the effects of water-to-cementitious materials ratio variations and to cut down the quantity of mixing water needed to obtain a standard consistency. The treated date palm fiber (DPF) was added in variations of 0%, 1%, 2%, and 3% by mass of binder. Silica fume was used as partial replacement to cement using the volume replacement method in different percentages by weight of binder materials at 0%, 5%, 10%, and 15%. A total of thirteen (13) mixes were generated as displayed in Table 3. Each of the mixes was named based on the quantities of the variables. Mix M1F0S was a mix comprising of 1% DPF and 0% silica fume, while mix M2F10S was a mix with 2% DPF and 10% silica fume and mix M3F15S was a mix with 3% DPF and 15% silica fume.

Table 3. Properties of DPF (2).

Mix	Variables (%)		Quantities in kg/m^3						
	Fiber (%)	Silica Fume (%)	Cement	Silica Fume	Fiber	Fine Agg	Coarse Agg	Water	S. P *
Control	0	0	490	0.0	0.0	750	905	185	4.9
M1F0S	1	0	490	0.0	4.9	750	905	185	4.9
M2F0S	2	0	490	0.0	9.8	750	905	185	4.9
M3F0S	3	0	490	0.0	14.7	750	905	185	4.9
M1F5S	1	5	465.5	17.9	4.83	750	905	185	4.8
M2F5S	2	5	465.5	17.9	9.67	750	905	185	4.8
M3F5S	3	5	465.5	17.9	14.50	750	905	185	4.8
M1F10S	1	10	441	35.8	4.77	750	905	185	4.8
M2F10S	2	10	441	35.8	9.54	750	905	185	4.8
M3F10S	3	10	441	35.8	14.30	750	905	185	4.8
M1F15S	1	15	416.5	53.7	4.70	750	905	185	4.7
M2F15S	2	15	416.5	53.7	9.40	750	905	185	4.7
M3F15S	3	15	416.5	53.7	14.11	750	905	185	4.7

* S.P = Superplasticizer, Agg = Aggregate.

2.3. Samples Preparation and Test Methods

The mixing, sampling, and curing of the concrete were carried out based on the standard procedure highlighted in ASTM C192/C192M [21]. The fresh concrete was mixed using a rotating drum concrete mixer in the laboratory as shown in Figure 5a. Prior to mixing, it was ensured that the aggregates were saturated and surface dried. The cement and silica fume were also devoid of any lumps prior to casting. The batching of the component ingredients was conducted using the weight approach. The water and superplasticizer were first mixed in a container. The mixing process was as follows: Firstly, the fine aggregate, cement, and silica fume were poured into the mixer, and the mixing began for at least 1 min. Then, the fiber, coarse aggregate, and part of the water were poured. Then, the mixing continued. The remaining water was poured gradually while mixing was ongoing. The mixing continued until a consistent mix was accomplished. The samples were then placed in the designed molds based on the testing conducted, and the fresh density was determined. The samples were cast into the recommended molds as shown in Figure 5b. The samples were air-dried for 24 h in the laboratory prior to demolding. Following removal, the hardened concretes were then cured for the recommended period in water, as shown in Figure 5c, before testing.

(**a**) Mixing

(**b**) Cast samples

(**c**) Curing of samples

(**d**) Compressive strength testing

(**e**) Splitting tensile strength testing

(**f**) Flexural strength testing

Figure 5. Experimental Set up.

2.4. Test Methods

The density of the freshly mixed concrete was determined instantly after mixing following BS-EN 12350-6 [31], while the density of the hardened concrete (unit weight) was determined regarding BS EN 12390-7 [32] using 100 mm cubes. The standard specifications defined in BS EN 12390-3 [33] were adopted for the determination of the compressive strength of the DPF-reinforced concrete using a universal testing machine (UTM). Cube specimens of 10 cm sizes were prepared and tested as shown in Figure 5d. For each mix, nine specimens tested after 3-, 7-, and 28-days curing duration. Three specimens were tested for each mix after each curing period to obtain the standard errors. The procedures outlined in BS EN 12390-6 [25] were utilized for the splitting tensile strength test. A UTM of 300 kN capacity was used for the testing. Cylinder-shaped specimens of 200 mm height and 100 mm diameter were tested after 3-, 7-, and 28-days curing duration as shown in Figure 5e. The flexural strength test was carried out using the beam with third-point load method as outlined in ASTM C78/C78M [34] using 100 mm × 100 mm × 500 mm prisms as shown in Figure 5f. The samples were tested after curing for a period of 7 and 28 days. The durability of the DPF-reinforced concrete was determined using a water absorption test, which was performed by adopting the procedure outlined in ASTM C642 [35]. Cube samples of 100 mm were made and cured for a period of 28 days before testing. To compute the standard error for each test and each mix, three samples were tested per test/mix/curing time.

The microstructure of the DPF concrete was examined with the aid of field emission scanning electron microscope (FESEM). Some mixes were randomly selected to represent both the DPF and silica fume proportions. After curing the samples for 28 days, the small fracture was extruded, dried, and used for the FESEM test. The sample was then cleaned using a high-pressure vacuum to remove any dust and impurities and to dry it completely. Prior to placing the sample in the FESEM machine, the sample was gold-coated with a thin film. Pictures of a high magnification up to 10,000 were taken from the computer screen attached to the machine. The pictures, with the help of energy dispersive spectroscopy (EDS) obtained from the FESEM machine, were used to analyze the morphology of the concrete.

3. Results and Discussions

3.1. Fresh Density and Unit Weight

The results of the fresh and hardened unit weights are presented in Figures 6 and 7. Adding DPF led to a slight reduction in both the fresh and hardened unit weight. The fresh unit weight of the concrete decreased by about 2.1%, 4%, and 5.9%, with inclusion of 1%, 2%, and 3% DPF, respectively. At the same time, the hardened unit weight decreased by 3.4%, 4.2%, and 5.4% by adding 1%, 2%, and 3% DPF, respectively. This can be illustrated by comparing the unit weight values of M1F0S, M2F0S, and M3F0S with that of the control mix in Figures 6 and 7. Benaniba, Driss [11] also reported a reduction in the unit weight of concrete with increase in addition of DPF. This gives DPF the advantage of being used as a good insulating material in concrete, as they reported that the lower the density of the DPF concrete, the lower its thermal conductivity. The slight reduction in the unit weight due to the addition of DPF was ascribed to the increased volume of voids in the concrete caused by fiber inclusion resulting from its hydrophilic nature and poor adhesion between the DPF and the concrete, thus increasing the porosity and causing the reduction in unit weight [1,36]. However, the decrease in the unit weight is one of the advantages of using DPF in cementitious composites, as higher thermal properties can be achieved with a lower unit weight. The concrete's fresh and hardened unit weights further decreased with the increment in percentage replacement of cement using silica fume as shown in Figures 6 and 7. For instance, comparing the unit weight values of the mixes comprising of 5% and 10% silica fume content with the same dosage of DPF, those with a higher silica fume content had lower fresh and hardened unit weight values. Mixes M1F5S, M2F5S, and M3F5S had higher fresh and hardened unit weights compared to mixes M1F10S, M2F10S, and M2F10S. Furthermore, mixes with 15% silica fume, i.e., M1F15S, M2F15S, and M3F15S, had the least fresh and hardened unit weights. Gencel, Nodehi [20] reported a slight reduction in the fresh density by about 4% in basalt-fiber-reinforced concrete when 15% cement was replaced with siica fume, while Saradar, Nemati [37] reported a decrease in the hardened density of basalt-fiber-reinforced concrete with the addition of silica fume as partial substitute to cement. The decrease in unit weight with the increment in silica fume can mainly be attributed to the inferior specific gravity of the silica fume compared to that of cement since the volume substitution method was used.

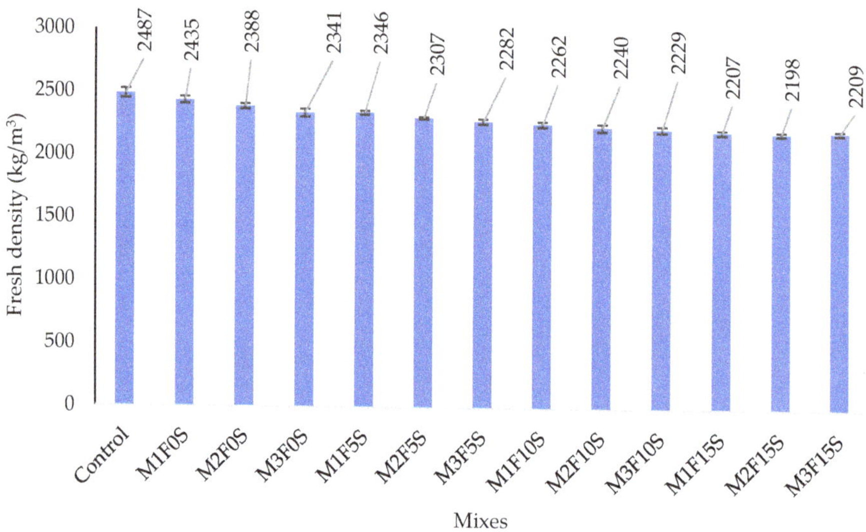

Figure 6. Results of fresh density.

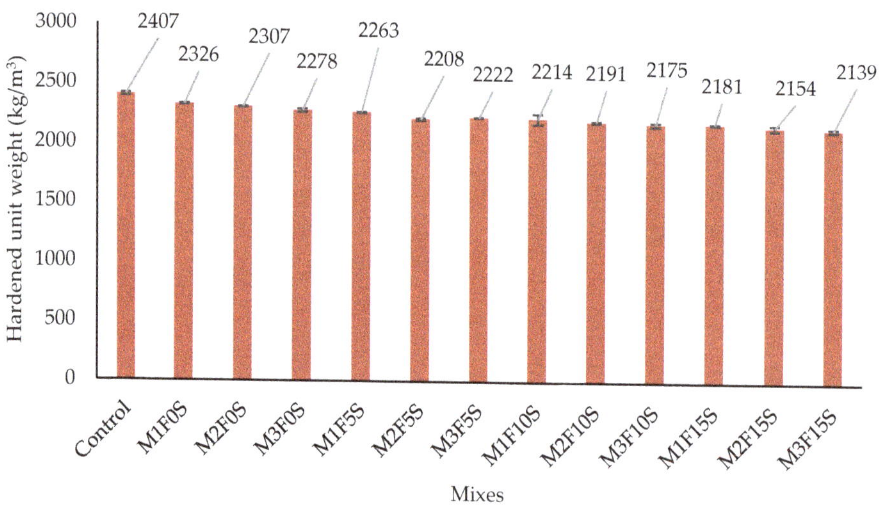

Figure 7. Results of hardened unit weight.

3.2. *Compressive Strength*

The results of the compressive strength of the DPF-reinforced concrete containing different proportions of silica fume as SCM are presented in Figure 8. The compressive strength increased with the curing time due to continuous hydration reactions. The strength declined with an increment in DPF content at all ages. The 3-day strength declined by 4.03%, 14.2%, and 20.1%, while the 28-day strength decreased by 4.3%, 21.2%, and 20.3% for 1%, 2%, and 3% DPF inclusion, respectively, as seen from mixes M1F0S, M2F0S, and M3F0S in Figure 8. These reductions can be ascribed to the following reasons: (i) the higher porosity in the concrete matrix resulting from air entrained on the fiber surface during mixing, difficult packing and compaction, and a balling effect due to the addition of a high fiber content. (ii) Weak adhesion between the DPF and cementitious paste resulting in a weak interfacial transition zone increasing the weak path for premature failure. (iii) The lower density and strength of the DPF added also affected the density and strength of the concrete [1,38,39]. Additionally, in the presence of the excess cement hydration product, i.e., calcium hydroxide, DPF loses its durability and strength, causing a reduction in the strength of DPF-reinforced concrete [40]. However, the poor bonding between the DPF and cement paste was minimized by treating the DPF with NaOH solution before usage, which eliminated impurities and enhanced the fiber's surface roughness leading to the enhancement of the bondage strength [29]. Benaimeche, Carpinteri [41] and Kriker, Debicki [42] also reported a reduction in the strength of concrete with the inclusion of DPF to cementitious composites.

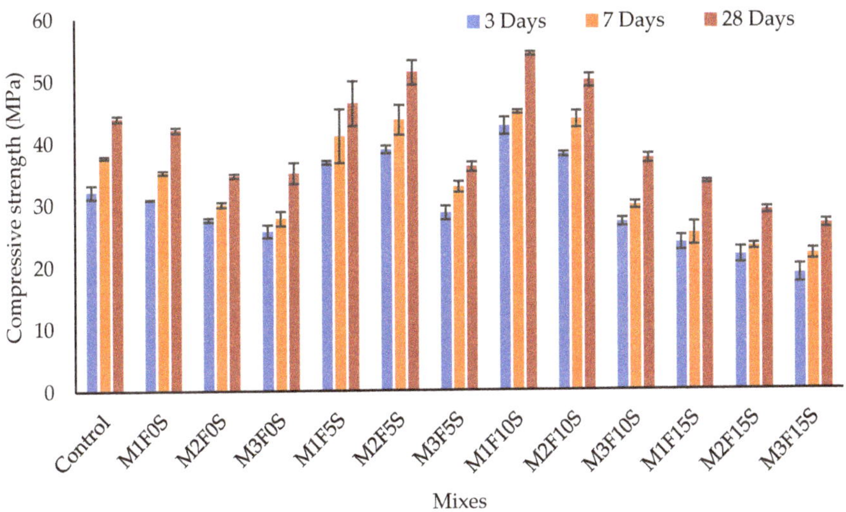

Figure 8. Compressive strength results.

The partial replacement of up to 10% cement with silica fume in the DPF-reinforced concrete enhanced the compressive strength and mitigated the loss in strength due to DPF's undesirable effects. For instance, the compressive strength of mix M1F5S, which contained 5% silica fume as a substitute to cement, was higher than the control mix by 14.6%, 8.8%, and 5.3% at 3, 7 and 28 days, respectively. Likewise, mix M2F5S had a higher compressive strength by 20.9%, 15.5%, and 16.5% at 3, 7 and 28 days, respectively. Similar result improvements were seen for the mixes containing 10% silica fume, i.e., mixes M1F10S and M2F10S at all ages of curing. Therefore, it can be said that the replacement of up to 10% cement using silica fume successfully mitigated the negative effects of up to 2% DPF addition on the compressive strength of the fiber-reinforced concrete. The enhancement in compressive strength with silica fume addition to the DPF concrete resulted from silica fume's higher surface area, thus acting as a filler in addition to it being a pozzolanic material, hence densifying the concrete microstructure and improving the strength. Silica fume contains a very high amount of reactive silica (SiO_2), reacting with cement hydration products such as $Ca(OH)_2$ from cement hydration to produce supplementary calcium aluminate silicate hydrate (C-AS-H) and calcium silicate hydrates (C-S-H), which are the major compounds for strength development in concrete [43,44]. Furthermore, the consumption of the excess calcium hydroxide by the silica fume prevented further deterioration of the durability and strength of the DPF. The improvement in strength was more noticeable at early ages up to 7 days compared to 28 days. This could be because silica fume is a highly reactive SCM, thereby its addition fastens the hydration reaction at early ages [44]. Gencel, Nodehi [20] carried out similar work by using silica fume as an SCM in basalt-fiber-reinforced concrete and found that the compressive strength was enhanced by up to 46% with the replacement of up to 15% cement using silica fume. Saradar, Nemati [37] also reported an increase in compressive strength by about 29% with the replacement of 10% cement with silica fume in basalt-fiber-reinforced concrete containing up to 0.5% fiber. On the contrary, the substitution of cement using a higher silica fume content of 15% reduced the compressive strength. This might have resulted from the fineness of the silica fume, which might have caused it to soak up the mixing water and decrease the consistency. This, in addition to the DPF, resulted to poor compaction, causing porosity in the cement matrix and hence a reduction in strength.

3.3. Flexural Strength

The bending resistance of the concrete was measured using a flexural strength test. This is very important, especially when the material will be used to construct flexural members such as beams and slabs. Figure 9 presents the results of the flexural strength of the DPF concrete. The incorporation of DPF resulted in a slight improvement in the flexural strength of the concrete at both 7 and 28 days. The flexural strength increased by about 3%, 9.7%, and 14.8% at 7 days and by 5.3%, 12.8%, and 17.3% at 28 days by incorporating 1%, 2%, and 3% DPF, respectively. As the DPF was treated using NaOH solution, this increased the DPF's surface roughness and improved the bonding between the cement matrix and fiber [29]. In addition to the fiber's bridging effect, this led to a delaying of the crack formation propagation, improving the post-cracking load resistance, hence enhancing the flexural strength. Similar improvements in flexural strength have been reported by Benaniba, Driss [11] with the addition of up to 6% DPF in mortar, and Ali-Boucetta, Ayat [29] with the addition of 2% DPF, and [10] found an improvement in flexural strength with the addition of up to 1% DPF to cement mortar.

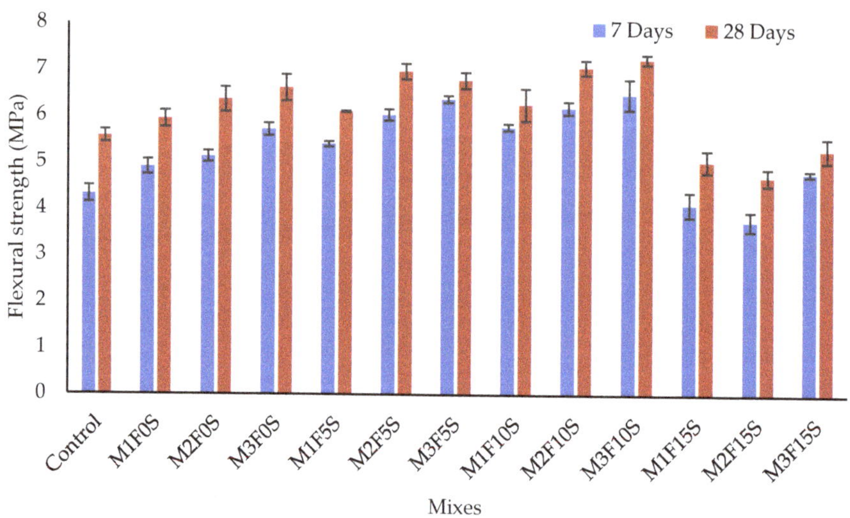

Figure 9. Flexural strength results.

Partially substituting up to 10% cement with silica fume enhanced the flexural strength of the DPF-reinforced concrete at both 7 and 28 days. This is further explained by comparing the flexural strength values of the mixes containing 5% and 10% silica fume with those without any silica fume addition with the same amount of DPF. For the mixes containing 1% DPF at 7 days, the flexural strength of the mixes M1F5S and M1F10S were greater by 16.3% and 25.4 %, respectively, and at 28 days by 5.81% and 7.83%, respectively, compared to mix M1F0S. Similar results were seen for the mixes containing 2% DPF and 3% DPF, as shown in Figure 9. This clearly explains the synergetic effects of DPF and silica fume. The enhancement in the bending resistance of the DPF-reinforced concrete with the addition of silica fume can be attributed to the high pozzolanic-hydration reactivity between the silica fume compounds, i.e., SiO_2 and cement hydration by-products, i.e., $Ca(OH)_2$, to produce surplus hydration compounds, which filled the pores created by the DPF and densified the concrete microstructure. This resulted in the improved adhesion and densification of the ITZ between the cement matrix and fiber, resulting in a higher modulus of rupture. Gencel et al. [16] reported an increase in flexural strength at 18 days by 6%, 11%, and 88% for basalt-fiber concrete containing 1%, 2%, and 3% fiber, respectively, with the replacement of 15% cement with silica fume. At an earlier age, i.e., 7 days, they recorded a higher increment in flexural strengths. Similarly, Saradar, Nemati [37] and Sadrmomtazi, et al. [17] reported a comparable improvement in flexural strength in natural-fiber-reinforced concrete when silica fume was used as a partial substitute to cement. On the contrary, when a higher content of silica fume (15%) was used, it decreased the flexural strength. This might be due to the larger surface area of silica fume, making it absorb more water. This significantly reduced the consistency of the fresh concrete and resulted in poor compaction, packing and fiber distribution within the cement matrix, leading to increased porosity and discontinuities within the microstructure, causing premature failure to occur and consequently lowering flexural strength.

3.4. *Splitting Tensile Strength*

The results of the splitting tensile strength of the DPF-reinforced concrete are presented in Figure 10. Adding up to 2% DPF improved the tensile strength of the concrete. This was noticed as the splitting tensile strength of M1F0S was higher than that of the control by 5.3%, 8.9%, and 12.6% at 3, 7, and 28 days, respectively. The tensile strength of M2F0S was superior by 11.4% and 16.5% at 7 and 28 days, respectively, compared to the control. This enhancement in the tensile strength was ascribed to the crack arrest through the fiber-bridging effect and fiber transfer energy. This resulted in an increased post-cracking load resistance and a higher tensile strength [45]. Other fibers such as jute, sisal, coconut, and sugarcane fibers were found to increase the tensile strength of concrete, as testified by Jamshaid, Mishra [46]. However, the addition of 3% DPF decreased the tensile strength of the composite, whereas for M3F0S the strength was less than the control by 15%, 7.2%, and 14.6% at 3, 7, and 28 days, respectively. This might be due to poor dispersion and the balling effect of the fiber, which, in addition to air being entrapped during mixing on the fiber surface, resulted in an increased porosity in the cement matrix. This caused a weak and discontinuous path leading to premature failure upon load application, thus causing a reduction in the tensile strength. Another reason might be ascribed to lack of proper adhesion between the DPF and cement matrix. The inclusion of silica fume resulted in a further enhancement in the splitting tensile strength of the concrete. The replacement of 5% and 10% cement with silica fume successfully mitigated the undesirable effects of the addition of 3% DPF on the tensile strength of the DPF-reinforced concrete. This can be explained by comparing the tensile strength values of mixes M3F5S and M3F10S with that of the control. For mix M3F5S, its tensile strength was greater at 3, 7, and 28 days by 18.9%, 21.5%, and 11.3%, respectively, compared to the control. Similarly, the tensile strength of M3F10S at 3, 7, and 28 days were greater by 14.6%, 21.1%, and 15.4%, respectively, compared to the control. Additionally, even for the mixes containing 1% and 2% DPF, adding silica fume further increased their tensile strengths. Saradar, Nemati [37] reported an increase in tensile strength by 10% with the replacement of 10% cement with silica fume in basalt-fiber-reinforced concrete made with up to 0.5% fiber, while Fallah and

Nematzadeh [47] recorded an improvement in tensile strength by 17.33%, 28.37%, and 25.17% with the replacement of 8%, 10%, and 12% cement with silica fume, respectively, in polypropylene-fiber-reinforced concrete. They attributed the increment to the enhancement of bonding between the cement matrix and fiber/aggregate. The improvement in the tensile strength of the DPF concrete with the addition of silica fume was due to the microstructural and fiber/aggregate–paste bond enhancement and the resulting filler effect and reactivity of the silica fume [44]. On the contrary, the substitution of 15% cement with silica fume also decreased the splitting tensile strength of the DPF-reinforced concrete containing any percentage of fiber addition. The reason for such a decrease was the same as compressive strength results.

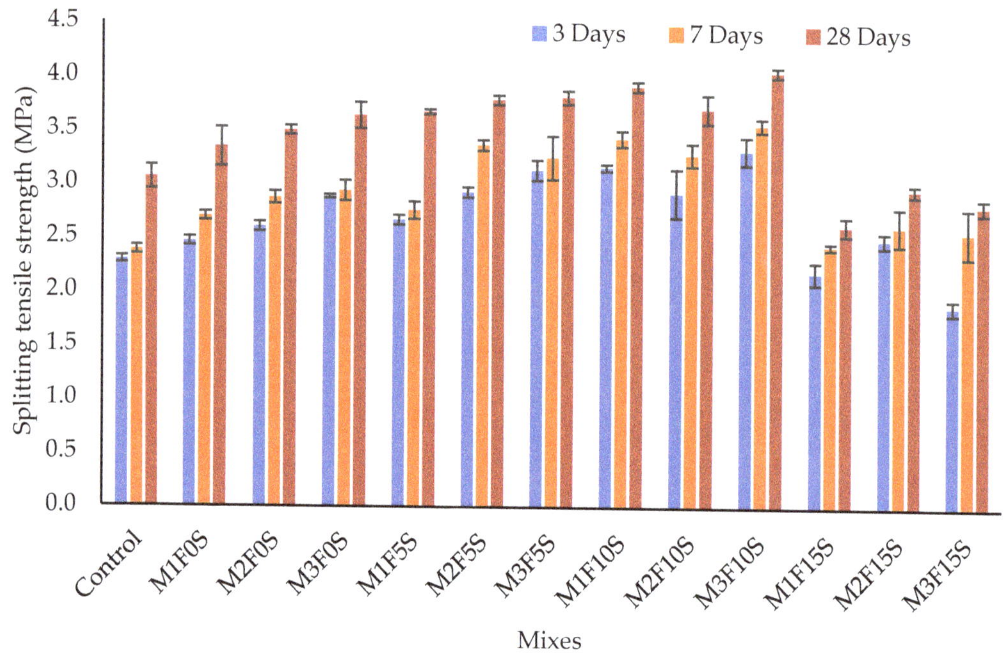

Figure 10. Splitting tensile strength results.

3.5. Water Absorption

The rate of water absorption is one of the methods of measuring porosity, which is a measuring factor influencing concrete's durability. The results of the water absorption for the DPF-reinforced concrete are presented in Figure 11. The water absorption was found to increase with the increment in the percentage addition of DPF. The water absorption for mixes M1F0S, M2F0S, and M3F0S was greater than the control by 4.9%, 21%, and 30.5%, respectively. The surge in water absorption was ascribed to the fiber's ligno-cellulose nature, making it hydrophilic and hence absorbing more water when mixing. This increased the porosity in the hardened cement matrix after drying, resulting in higher water absorption. Additionally, DPF can absorb water by up to 300% of its dry mass during mixing or curing, which can easily swell and promote micro-crack formation in the cement matrix, thereby causing a higher rate of water absorption [11,48]. A similar escalation in water absorption with the incorporation of DPF to the cement composite was reported by Bamaga [49] where they reported an increase of 27% and 74% with the inclusion of 1% and 3% DPF, correspondingly. However, the increase in water absorption for this study was lower compared to previous findings; this was due to the treatment of the fiber using NaOH solution, which significantly reduced the lignin and hemicellulose content of the fiber and created hydrophobicity in the fiber [10]. This resulted in a decline in the rate of water absorption in the concrete. Similar findings were reported by Ozerkan, Ahsan [10] where they found that the alkaline treatment of DPF removed hemicellulose and lignin from its surface and hence led to a reduction in the water absorption of the composite when the alkaline-treated DPF was added. The partial substitution of cement with silica fume caused a decline in the water absorption of the concrete. For the mixes containing 5% silica fume, the water absorption values of the concrete with 1% and 2% DPF were lesser than that of the control by 7.1% and 2.9%, respectively. Similarly, for the mixes containing 10% silica fume, their water absorptions of the concrete with 1%, 2%, and 3% were less by about 21.1%, 16.5%, and 6.1%, respectively, compared to the control mix. Similar findings were reported by Saradar, Nemati [37] where they reported a decline in water absorption by up to 38% with the replacement of 10% cement using silica fume in basalt-fiber-reinforced concrete containing up to 0.5% fiber, while Fallah and Nematzadeh [47] reported a reduction in water absorption by 18% with the replacement of 12% cement with silica fume in polypropylene-fiber reinforced composite. The decline in water absorption by the inclusion of silica fume in the DPF composite resulted from the lower porosity due to the formation of excess hydration products. The generated hydration products densified the pores generated by the DPF in the matrix and hence refined the microstructure of the DPF-reinforced concrete yielding a lower water absorption [44]. However, 15% silica content in the DPF composite caused an increase in water absorption, which became more severe with the increment in the addition of DPF. This might be due to the fineness of silica fume, which caused further absorption of mixing water, thereby reducing the consistency of the paste. This caused poor compaction and distribution of the fibers within the matrix and thus led to honeycombs and voids within the concrete microstructure and, consequently, a higher water absorption. Khan [50] also reported similar findings, where up to 10% silica fume addition as the SCM in conventional concrete decreased its water absorption, but 15% increased it.

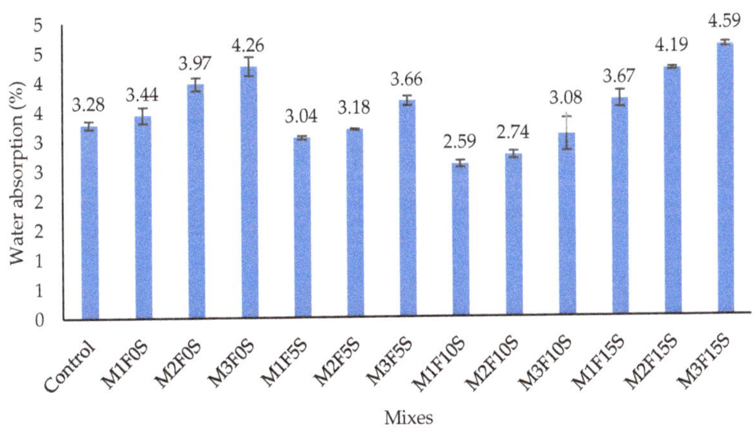

Figure 11. Water absorption results.

3.6. FESEM Analysis

FESEM analysis was used to obtain and study the microstructural morphology of some of the representative mixes of the DPF-reinforced concrete, where the effects of DPF addition and the partial replacement of cement with silica fume were analyzed. Figure 12a–e presents the morphology of some selected DPF concrete mixes. From the morphology of the control mix in Figure 12a, traces of $Ca(OH)_2$ and C-S-H were seen resulting from the cement hydration reaction. Additionally, some micro-cracks might have resulted from plastic or drying shrinkage. Furthermore, the interfacial transition zone (ITZ) between the cement paste and aggregate could be seen and was densified by the hydration products. For all other mixes containing DPF, i.e., Figure 12b–e, DPF could be seen on the microstructure, without been disintegrated due to applied loads. Around the surface of the DPF, traces of hydration compounds such as C-S-H and $Ca(OH)_2$ were also seen to be deposited on the fiber surfaces. Furthermore, it was observed that the ITZ between the DPF and cement matrix was porous, which was attributed to poor bonding between the cement matrix and the fiber. For example, considering Figure 12e, i.e., the morphology of M2F0S, its ITZ was the most porous, resulting in poor adhesion between the fiber and cement matrix. Furthermore, the hydrophilic nature of the DPF due to the presence of ligno-cellulose compounds increased the voids in the matrix through the absorption of more water and entrapment of air during mixing, which eventually dried up, leaving pores in the matrix [11,48]. This, consequently, is one of the reasons for the diminution in the mechanical strength and durability of the concrete with the inclusion of DPF. Similar results have been established by Kriker, Debicki [42]. Partial cement replacement with silica fume densified the microstructure and ITZ between the DPF and cement paste. The mixture containing 1% DPF and 10% silica fume in Figure 12c had the most densified microstructure, with the ITZ being filled and densified with the cement hydration and silica fume pozzolanic reaction products. This densification enhanced the bonding between the fiber and cement paste as displayed in Figure 12c. The improvement was ascribed to the reaction between SiO_2 from silica fume with $CaOH_2$ from cement hydration. This led to surplus C-S-H generation, which densified the concrete's ITZ and microstructure. Additionally, the consumption of $CaOH_2$ also decreased its leaching out with time, reducing the pore generation caused by the leachate.

Energy dispersive X-ray (EDX) analysis was used to obtain and analyze the elemental configurations of some of the randomly selected DPF-reinforced concrete containing silica fume. The results are summarized in Table 4 and Figure 12i–v. The main elements present in all the cement matrices included silicon (Si), carbon (C), calcium (Ca), aluminum (Al), oxygen (O_2), and iron (Fe). Other traces of elements and impurities in minor form were also present. Mixes containing silica fume such as M1F1S (Figure 12ii), M1F10S (Figure 12iii), and M1F15S (Figure 12iv) had abundant Si and O_2 with less Ca. This explained or indicated that pozzolanic reactions took place between the SiO_2 from the silica fume and $Ca(OH)_2$ from the cement to produce excess C-S-H [51]. This can be elaborated more by comparing their ratios of Ca/Si and Si/Al with the other mixes. It can be observed from Table 4 that the mix M1F10S had the smallest value of Ca/Si and the greatest value of Si/Al, which signified the highest pozzolanic reaction, while mix M2F0S had the greatest value of Ca/Si and smallest value of Si/Al, thus denoting the least pozzolanic reaction due to the non-presence of silica fume.

(**a**) FESEM of control mix

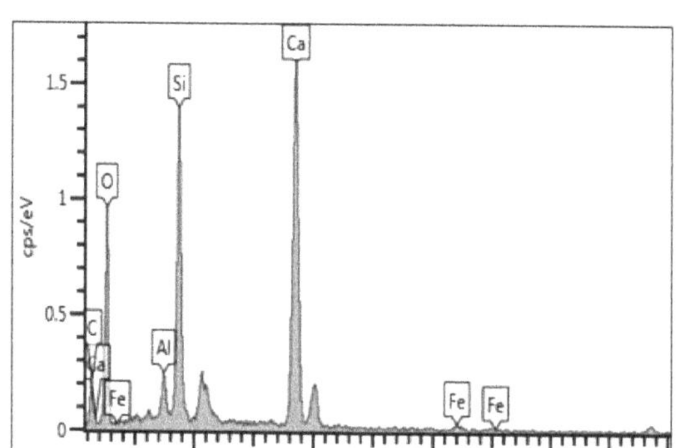

(i) EDX of control mix

Figure 12. *Cont.*

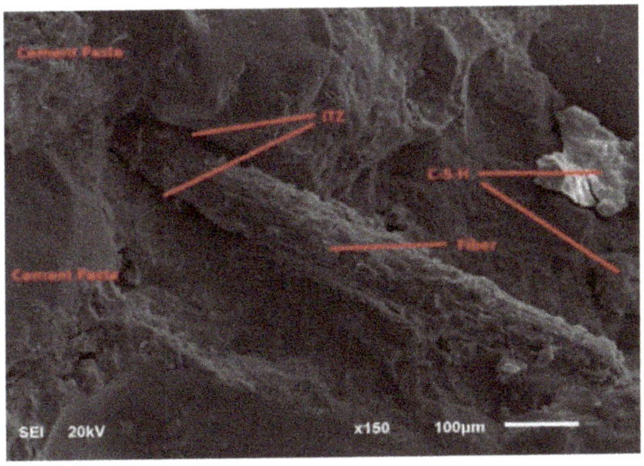

(**b**) FESEM of mix M1F5S

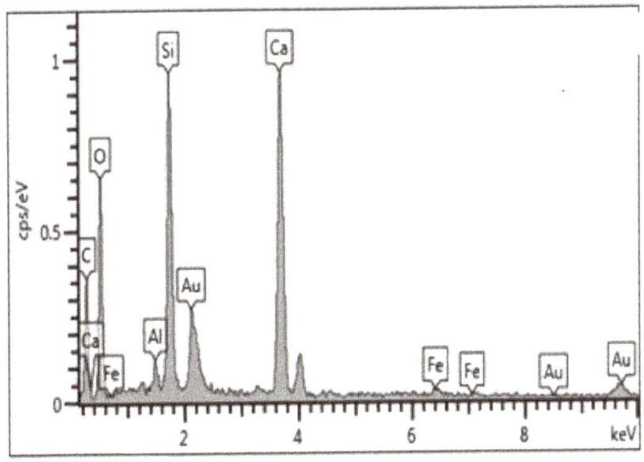

(**ii**) EDX of mix M1F5S

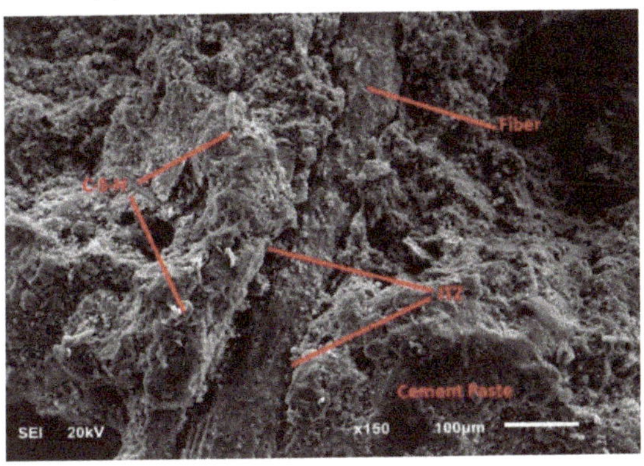

(**c**) FESEM of mix M1F10S

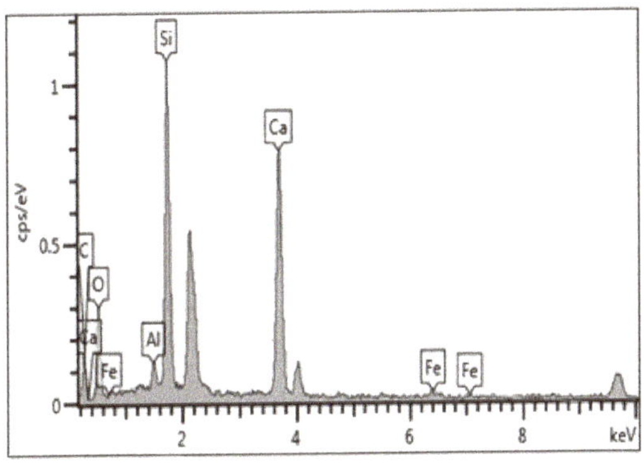

(**iii**) EDX of mix M1F10S

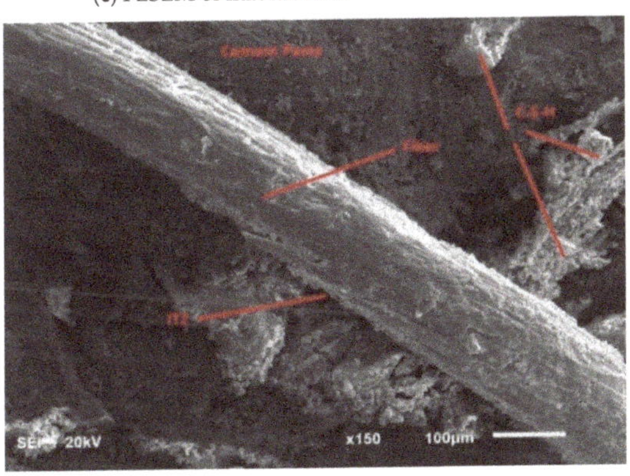

(**d**) FESEM of mix M1F15S

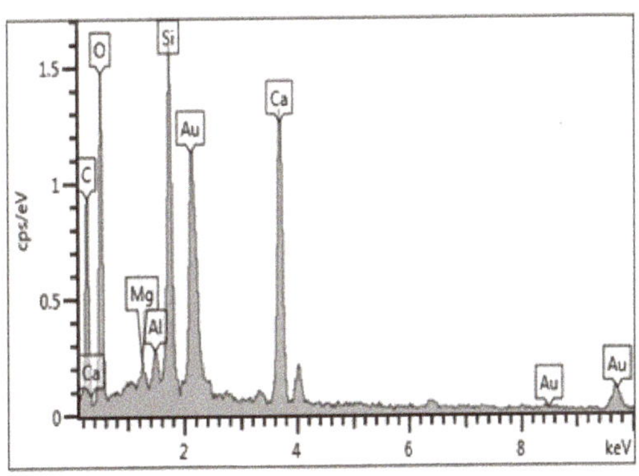

(**iv**) EDX of mix M1F15S

Figure 12. *Cont.*

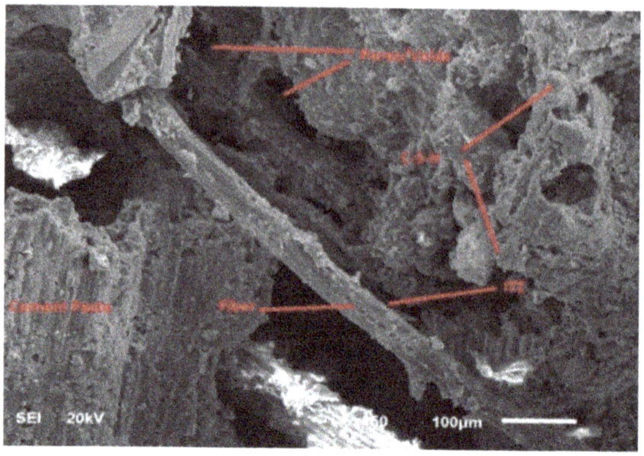

(**e**) FESEM of mix M2F0S

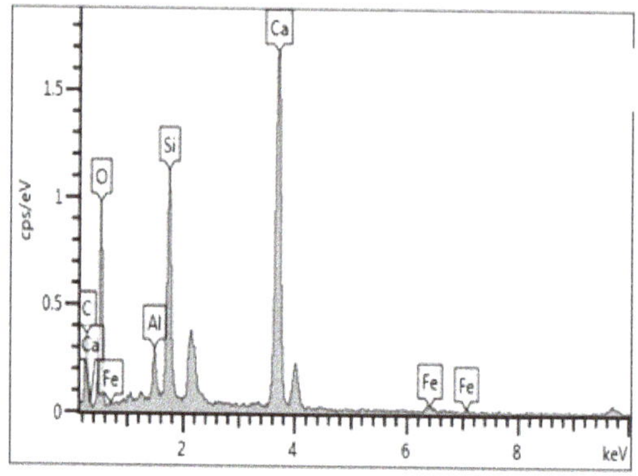

(**v**) EDX of mix M2F0S

Figure 12. Microstructural morphology of DPF-reinforced concrete.

Table 4. Elemental compositions obtained from EDX analysis.

Elements	Atomic Weight Concentration (%)				
	Control	M2F0S	M1F10S	M1F5S	M1F15S
C	18.15	23.91	28.87	31.61	37.01
O	48.98	46.01	33.84	43.38	44.73
Al	1.55	1.73	1.35	0.79	0.75
Si	9.63	7.04	14.4	8.08	6.49
Ca	20.98	20.22	20.6	15.32	10.52
Fe	0.71	1.09	0.95	0.82	-
Mg	-	-	-	-	0.5
Total	100	100	100	100	100
Si/Al	6.21	4.07	10.67	10.23	8.65
Ca/Si	2.18	2.87	1.43	1.90	1.62
Ca/(Si + Al)	1.88	2.31	1.31	1.73	1.45

4. Conclusions

This study investigated the effects of silica fume as an SCM on the mechanical properties of DPF-reinforced concrete. Some of the conclusions summarized include:

(1) The fresh and hardened densities of the DPF-reinforced concrete declined with the addition of DPF by weight of binder materials and partial substitution of cement with silica fume.

(2) The compressive strength of the DPF-reinforced concrete declined with the increase in DPF content as an additive by weight of binder materials

(3) The incorporation of up to 2% DPF by weight of binder constituents enhanced the splitting tensile and flexural strengths of the DPF-reinforced concrete

(4) The addition of DPF increased the porosity of the DPF-reinforced concrete as measured through the increment in water absorption with the addition of DPF.

(5) Partially substituting cement with silica fume significantly enhanced the mechanical strength of the DPF-reinforced concrete. The loss in compressive strength up to 2% DPF addition was mitigated by replacing up to 10% cement with silica fume.

(6) Silica fume densified the microstructure and refined the ITZ between the fiber and cement matrix, hence significantly reducing the porosity and enhancing the durability.

(7) A sustainable DPF-reinforced concrete mix was produced by adding up to 2% DPF by weight of binder materials and replacing up to 10% cement with silica fume.

Author Contributions: Conceptualization, Y.E.I. and M.A.; methodology, M.A., M.L.M., O.S.A. and Q.A.D.; software M.A. and M.L.M.; validation, M.A., Y.E.I. and M.A.M.; formal analysis, M.A. and Q.A.D.; investigation, Y.E.I., M.A., M.L.M. and M.A.M.; resources, O.S.A. and M.A.M.; data curation, M.A. and M.L.M.; writing—original draft preparation, M.A. and M.L.M.; writing—review and editing, Y.E.I. and M.A.M.; visualization, O.S.A. and M.A.M.; supervision, Y.E.I.; project administration, Y.E.I. and O.S.A.; funding acquisition, Y.E.I. All authors have read and agreed to the published version of the manuscript.

Funding: This research was supported by the Structures and Materials (S&M) Research Laboratory of Prince Sultan University, Saudi Arabia. Furthermore, the authors acknowledge the support of Prince Sultan University in paying the article processing charges (APC) of this publication.

Data Availability Statement: Not applicable.

Acknowledgments: The authors wish to acknowledge the support of the Structures and Materials laboratory (S&M Lab) of the College of Engineering, Prince Sultan University, Riyadh, Saudi Arabia, for their vital support.

Conflicts of Interest: The authors declare no conflict of interest.

References

1. Vantadori, S.; Carpinteri, A.; Zanichelli, A. Lightweight construction materials: Mortar reinforced with date-palm mesh fibres. *Theor. Appl. Fract. Mech.* **2019**, *100*, 39–45. [CrossRef]
2. Yousefi, M.; Khandestani, R.; Gharaei-Moghaddam, N. Flexural behavior of reinforced concrete beams made of normal and polypropylene fiber-reinforced concrete containing date palm leaf ash. *Structures* **2022**, *37*, 1053–1068. [CrossRef]
3. Zhang, P.; Han, S.; Golewski, G.L.; Wang, X. *Nanoparticle-Reinforced Building Materials with Applications in Civil Engineering*; SAGE Publications Sage UK: London, UK, 2020; p. 1687814020965438.
4. Bentur, A.; Mindess, S. *Fibre Reinforced Cementitious Composites*; CRC Press: Boca Raton, FL, USA, 2006.
5. Balea, A.; Fuente, E.; Monte, M.C.; Blanco, Á.; Negro, C. Fiber reinforced cement based composites. In *Fiber Reinforced Composites*; Elsevier: Amsterdam, The Netherlands, 2021; pp. 597–648.
6. Balaguru, P.N.; Shah, S.P. *Fiber-Reinforced Cement Composites*; McGraw-Hill: New York, NY, USA, 1992.
7. Banthia, N. Fiber reinforced concrete. In *ACI SP-142 Fiber Reinforced Concrete Developments and Innovations*; American Concrete Institute: Farmington Hills, MI, USA, 1994; Volume 91, p. 119.
8. Banthia, N. *ACI 544.5R-Report on the Physical Properties and Durability of Fiber-Reinforced Concrete*; American Concrete Institute: Farmington Hills, MI, USA, 2010.
9. Al-Oqla, F.M.; Alothman, O.Y.; Jawaid, M.; Sapuan, S.M.; Es-Saheb, M.H. Processing and Properties of Date Palm Fibers and its Composites. In *Biomass and Bioenergy*; Springer: Berlin/Heidelberg, Germany, 2014; pp. 1–25.
10. Ozerkan, N.G.; Ahsan, B.; Mansour, S.; Iyengar, S. Mechanical performance and durability of treated palm fiber reinforced mortars. *Int. J. Sustain. Built Environ.* **2013**, *2*, 131–142. [CrossRef]
11. Benaniba, S.; Driss, Z.; Djendel, M.; Raouache, E.; Boubaaya, R. Thermo-mechanical characterization of a bio-composite mortar reinforced with date palm fiber. *J. Eng. Fibers Fabr.* **2020**, *15*, 1558925020948234. [CrossRef]
12. Zanichelli, A.; Carpinteri, A.; Fortese, G.; Ronchei, C.; Scorza, D.; Vantadori, S. Contribution of date-palm fibres reinforcement to mortar fracture toughness. *Procedia Struct. Integr.* **2018**, *13*, 542–547. [CrossRef]
13. Kareche, A.; Agoudjil, B.; Haba, B.; Boudenne, A. Study on the Durability of New Construction Materials Based on Mortar Reinforced with Date Palm Fibers Wastes. *Waste Biomass Valorization* **2020**, *11*, 3801–3809. [CrossRef]
14. Alatshan, F.; Altlomate, A.M.; Mashiri, F.; Alamin, W. Effect of date palm fibers on the mechanical properties of concrete. *Int. J. Sustain. Build. Technol. Urban Dev.* **2017**, *8*, 68–80.
15. Hakkoum, S.; Kriker, A.; Mekhermeche, A. Thermal characteristics of Model houses Manufactured by date palm fiber reinforced earth bricks in desert regions of Ouargla Algeria. *Energy Procedia* **2017**, *119*, 662–669. [CrossRef]
16. Bamaga, S.O. A Review on the Utilization of Date Palm Fibers as Inclusion in Concrete and Mortar. *Fibers* **2022**, *10*, 35. [CrossRef]
17. Adamu, M.; Alanazi, F.; Ibrahim, Y.E.; Alanazi, H.; Khed, V.C. A Comprehensive Review on Sustainable Natural Fiber in Cementitious Composites: The Date Palm Fiber Case. *Sustainability* **2022**, *14*, 6691. [CrossRef]
18. Alawar, A.; Hamed, A.M.; Al-Kaabi, K. Characterization of treated date palm tree fiber as composite reinforcement. *Compos. Part B Eng.* **2009**, *40*, 601–606. [CrossRef]
19. Wazzan, A.A. Effect of fiber orientation on the mechanical properties and fracture characteristics of date palm fiber reinforced composites. *Int. J. Polym. Mater. Polym. Biomater.* **2005**, *54*, 213–225. [CrossRef]
20. Gencel, O.; Nodehi, M.; Bayraktar, O.Y.; Kaplan, G.; Benli, A.; Gholampour, A.; Ozbakkaloglu, T. Basalt fiber-reinforced foam concrete containing silica fume: An experimental study. *Constr. Build. Mater.* **2022**, *326*, 126861. [CrossRef]
21. Sadrmomtazi, A.; Tahmouresi, B.; Saradar, A. Effects of silica fume on mechanical strength and microstructure of basalt fiber reinforced cementitious composites (BFRCC). *Constr. Build. Mater.* **2018**, *162*, 321–333. [CrossRef]

22. Khan, M.; Rehman, A.; Ali, M. Efficiency of silica-fume content in plain and natural fiber reinforced concrete for concrete road. *Constr. Build. Mater.* **2020**, *244*, 118382. [CrossRef]
23. Zhang, P.; Wang, K.; Wang, J.; Guo, J.; Hu, S.; Ling, Y. Mechanical properties and prediction of fracture parameters of geopolymer/alkali-activated mortar modified with PVA fiber and nano-SiO2. *Ceram. Int.* **2020**, *46*, 20027–20037. [CrossRef]
24. Golewski, G. Green Concrete Based on Quaternary Binders with Significant Reduced of CO_2 Emissions. *Energies* **2021**, *14*, 4558. [CrossRef]
25. Golewski, G.L. An extensive investigations on fracture parameters of concretes based on quaternary binders (QBC) by means of the DIC technique. *Constr. Build. Mater.* **2022**, *351*, 128823. [CrossRef]
26. Gil, D.M.; Golewski, G.L. Potential of siliceous fly ash and silica fume as a substitute for binder in cementitious concretes. *E3S Web Conf.* **2018**, *49*, 00030. [CrossRef]
27. *ASTM C150/150M*; Standard Specification for Portland Cement. ASTM International: West Conshohocken, PA, USA, 2015.
28. *ASTM C33*; Standard Specification for Concrete Aggregates. ASTM International: West Conshohocken, PA, USA, 2018.
29. Ali-Boucetta, T.; Ayat, A.; Laifa, W.; Behim, M. Treatment of date palm fibres mesh: Influence on the rheological and mechanical properties of fibre-cement composites. *Constr. Build. Mater.* **2021**, *273*, 121056. [CrossRef]
30. *ACI 211.1R*; Standard Practice for Selecting Proportions for Normal, Heavyweight, and Mass Concrete. American Concrete Institute: Farmington Hills, MI, USA, 2002.
31. *BS-EN 12350-6*; Testing Fresh Concrete. Density. British Standard Institution: London, UK, 2019.
32. *BS EN 12390-7*; Testing Hardened Concrete. Density of Hardened Concrete. British Standards Institution: London, UK, 2019.
33. *BS EN 12390-3*; Testing Hardened Concrete. Compressive Strength of Test Specimens. British Standards Institution: London, UK, 2009.
34. *ASTM C78/C78M*; Standard Test Method for Flexural Strength of Concrete (Using Simple Beam with Third-Point Loading). ASTM International: West Conshohocken, PA, USA, 2015.
35. *ASTM C642*; Standard Test Method for Density, Absorption, and Voids in Hardened Concrete. ASTM International: West Conshohocken, PA, USA, 2001.
36. Rachedi, M.; Kriker, A. Thermal Properties of Plaster Reinforced with Date Palm Fibers. *Civ. Environ. Eng.* **2020**, *16*, 259–266. [CrossRef]
37. Saradar, A.; Nemati, P.; Paskiabi, A.S.; Moein, M.M.; Moez, H.; Vishki, E.H. Prediction of mechanical properties of lightweight basalt fiber reinforced concrete containing silica fume and fly ash: Experimental and numerical assessment. *J. Build. Eng.* **2020**, *32*, 101732. [CrossRef]
38. Boumhaout, M.; Boukhattem, L.; Hamdi, H.; Benhamou, B.; Nouh, F.A. Thermomechanical characterization of a bio-composite building material: Mortar reinforced with date palm fibers mesh. *Constr. Build. Mater.* **2017**, *135*, 241–250. [CrossRef]
39. Pickering, K.L.; Aruan Efendy, M.G.; Le, T.M. A review of recent developments in natural fibre composites and their mechanical performance. *Compos. Part A Appl. Sci. Manuf.* **2016**, *83*, 98–112. [CrossRef]
40. Kriker, A.; Bali, A.; Debicki, G.; Bouziane, M.; Chabannet, M. Durability of date palm fibres and their use as reinforcement in hot dry climates. *Cem. Concr. Compos.* **2008**, *30*, 639–648. [CrossRef]
41. Benaimeche, O.; Carpinteri, A.; Mellas, M.; Ronchei, C.; Scorza, D.; Vantadori, S. The influence of date palm mesh fibre reinforcement on flexural and fracture behaviour of a cement-based mortar. *Compos. Part B Eng.* **2018**, *152*, 292–299. [CrossRef]
42. Kriker, A.; Debicki, G.; Bali, A.; Khenfer, M.; Chabannet, M. Mechanical properties of date palm fibres and concrete reinforced with date palm fibres in hot-dry climate. *Cem. Concr. Compos.* **2005**, *27*, 554–564. [CrossRef]
43. Nochaiya, T.; Wongkeo, W.; Chaipanich, A. Utilization of fly ash with silica fume and properties of Portland cement–fly ash–silica fume concrete. *Fuel* **2010**, *89*, 768–774. [CrossRef]
44. Siddique, R. Utilization of silica fume in concrete: Review of hardened properties. *Resour. Conserv. Recycl.* **2011**, *55*, 923–932. [CrossRef]
45. Abbass, W.; Khan, M.I.; Mourad, S. Evaluation of mechanical properties of steel fiber reinforced concrete with different strengths of concrete. *Constr. Build. Mater.* **2018**, *168*, 556–569. [CrossRef]
46. Jamshaid, H.; Mishra, R.K.; Raza, A.; Hussain, U.; Rahman, L.; Nazari, S.; Chandan, V.; Muller, M.; Choteborsky, R. Natural Cellulosic Fiber Reinforced Concrete: Influence of Fiber Type and Loading Percentage on Mechanical and Water Absorption Performance. *Materials* **2022**, *15*, 874. [CrossRef] [PubMed]
47. Fallah, S.; Nematzadeh, M. Mechanical properties and durability of high-strength concrete containing macro-polymeric and polypropylene fibers with nano-silica and silica fume. *Constr. Build. Mater.* **2017**, *132*, 170–187. [CrossRef]
48. Benmansour, N.; Agoudjil, B.; Gherabli, A.; Kareche, A.; Boudenne, A. Thermal and mechanical performance of natural mortar reinforced with date palm fibers for use as insulating materials in building. *Energy Build.* **2014**, *81*, 98–104. [CrossRef]
49. Bamaga, O.S. Physical and mechanical properties of mortars containing date palm fibers. *Mater. Res. Express* **2022**, *9*, 015102. [CrossRef]
50. Khan, M. Isoresponses for strength, permeability and porosity of high performance mortar. *Build. Environ.* **2003**, *38*, 1051–1056. [CrossRef]
51. Kelechi, S.E.; Uche, O.A.U.; Adamu, M.; Alanazi, H.; Okokpujie, I.P.; Ibrahim, Y.E.; Obianyo, I.I. Modeling and Optimization of High-Volume Fly Ash Self-Compacting Concrete Containing Crumb Rubber and Calcium Carbide Residue Using Response Surface Methodology. *Arab. J. Sci. Eng.* **2022**, 1–20. [CrossRef]

buildings

MDPI

Article

Towards Sustainable Construction Materials: A Comparative Study of Prediction Models for Green Concrete with Metakaolin

Jiandong Huang [1,2], Mengmeng Zhou [1], Hongwei Yuan [1,*], Mohanad Muayad Sabri Sabri [2] and Xiang Li [1]

1 School of Mines, China University of Mining and Technology, Xuzhou 221116, China; huang@cumt.edu.cn (J.H.); ts21020232p21@cumt.edu.cn (M.Z.); ts21020224p21@cumt.edu.cn (X.L.)
2 Peter the Great St. Petersburg Polytechnic University, 195251 St. Petersburg, Russia; mohanad.m.sabri@gmail.com
* Correspondence: 3372@cumt.edu.cn

Abstract: Cement-based materials are widely used in transportation, construction, national defense, and other fields, due to their excellent properties. High performance, low energy consumption, and environmental protection are essential directions for the sustainable development of cement-based materials. To alleviate the environmental pressure caused by carbon emissions in cement production, this paper studies cement-based materials containing metakaolin by a comparison of prediction models for the compressive strength. To more accurately evaluate the compressive strength of metakaolin cement-based materials, this paper compares the prediction effects of four models, namely, support vector machine (SVM), decision tree (DT), k-nearest neighbor (KNN), and random forest (RF), with hyperparameters optimized by the Firefly Algorithm (FA) to study the compressive strength of cement-based materials containing metakaolin. The results demonstrated that the RF model showed the optimized prediction effect considering the lowest RSME value and the highest R value among the hybrid models for predicting metakaolin cement-based materials' compressive strength. The importance test showed that the cement grade and the water-to-binder ratio greatly influence the compressive strength of cement-based materials with metakaolin compared to the other design parameters.

Keywords: metakaolin; green concrete; machine learning; compressive strength; hyperparameters

check for updates

Citation: Huang, J.; Zhou, M.; Yuan, H.; Sabri, M.M.S.; Li, X. Towards Sustainable Construction Materials: A Comparative Study of Prediction Models for Green Concrete with Metakaolin. *Buildings* **2022**, *12*, 772. https://doi.org/10.3390/buildings12060772

Academic Editor: Tomáš Dvorský

Received: 30 March 2022
Accepted: 9 May 2022
Published: 6 June 2022

1. Introduction

Cement-based materials refer to the materials with new properties obtained by combining cement with other materials [1–3]. Common cement base materials have concrete, mortar, and grouting materials [4–12]. Cement-based materials are widely used in construction, transportation, national defense, and other fields because of their advantages: high performance, convenient production, convenient construction, high compressive strength, good corrosion resistance, good durability, and so on [13–17]. In cement production, when $CaCO_3$, as cement raw material, decomposes into CO_2, the combustion of combustion materials also releases a certain amount of CO_2 [18]. Carbon emission has caused the greenhouse effect and brought a certain burden to the environment [4]. It is of great significance to improve the production of cement-based materials, effectively utilize resources, and alleviate the environmental load for promoting the sustainable development of cement-based materials [19]. Using mineral materials such as auxiliary gel materials to improve the performance of cement-based materials has become an important development direction in the manufacture of high-performance cement-based materials. The commonly used mineral materials are fly ash, silica fume, ore powder, metakaolin, and so on [20].

Mineral admixture can reduce the use of cement and reduce the cost and the carbon emission caused by cement production. In addition, mineral admixture can improve the performance and internal structure of cement-based materials and improve the strength and durability of cement-based materials [21,22]. Metakaolin is a kind of active mineral admixture formed by calcination of kaolin as raw material at high temperatures [23,24]. Metakaolin can be hydrated with cement to generate the corresponding cementing substance and improve cement-based materials' strength and corrosion resistance [25,26].

In recent years, there are many types of research on improving cement-based materials' performance by adding the appropriate amount of metakaolin into cement-based materials, mainly focusing on the compressive strength and durability of cement-based materials [27]. Metakaolin was used in the cement composite because it can significantly improve the properties of cement-based materials because of its pozzolanic activity, nucleation, and macroaggregates filling effect. Batis et al. studied the influence of metakaolin on the compressive strength and corrosion of cement mortar and found that the compressive strength of cement mortar could be improved by replacing cement with partial metakaolin. When the content of metakaolin is 10%, its contribution to the compressive strength of cement mortar is the largest. As a substitute for sand or cement, Metakaolin plays a positive role in improving the corrosion resistance of cement mortar [1,2]. Khater et al. studied the resistance of cement mortar with different amounts of metakaolin to magnesium chloride solution. The research results showed that the higher the metakaolin content within a certain range, the stronger the resistance of cement mortar to magnesium chloride solution [28]. Wang et al. evaluated various properties of metakaolin mixed concrete with the comprehensive hydration force durability model [4]. Kirsanova et al. studied the influence of metakaolin on the structural change of modified cement set to evaluate the influence of metakaolin on the phase composition of cement set. The results showed that the use of metakaolin promoted the composition of metastable calcium hydroaluminate and thus promoted the structure composition of cement set [29]. Kadri et al. studied the influence of metakaolin on hydration heat and compressive strength of mortar, and the results showed that adding a certain amount of metakaolin could bring mortar to a higher temperature. The contribution of metakaolin to mortar strength is mainly determined by packing effect, hydration acceleration of ordinary Portland cement, and pozzolans reaction of metakaolin [30]. Vu et al. studied the influence of metakaolin on the durability and compressive strength of cement mortar and concrete, and the research results showed that the replacement level of metakaolin for cement mortar or concrete was within 30%, and it had a positive effect on the compressive strength and durability of cement mortar and concrete [31]. Courard et al. found that metakaolin can inhibit chloride diffusion and sulfate erosion to a certain extent when the dosage of metakaolin is 5–20% of the mass of cement foundation by studying the durability of metakaolin on modified mortar [13].

Machine learning is an interdisciplinary subject involving many fields, such as probability theory, statistics, and algorithm complexity theory [32–38]. Machine learning mainly studies how to simulate and implement human learning behavior, acquire new skills, and constantly improve their skills [39,40]. The essence of machine learning is to process and analyze data in large quantities through the computer's powerful data processing and analysis ability [41–45]. In recent years, machine learning has been applied in many fields, such as finance, medicine, education, and architecture, due to its superior performance [35–37,46,47]. Many scholars at home and abroad have studied the application of machine learning in cement-based materials and made some progress [48–57]. Shamsabadi et al. studied the influence of waste marble powder (WMP) on the compressive strength of concrete by using the extreme gradient enhancement model (XGB) and artificial neural network model (ANN), and the results showed that the influence of WMP on the compressive strength of concrete was limited to 10–20% due to the inertia of WMP. Asteris et al. established an optimization model using a machine learning method to evaluate the impact of six input parameters on the compressive strength of metakaolin concrete: the specimen age, the metakaolin-to-glue ratio, the water–glue ratio, the superplastic fortifier ratio, and the thick–thin aggregate ratio [58]. Ahmad et al., based on Artificial Neural Network (ANN), Boosting, and AdaBoost machine learning, studied the prediction of compressive strength of high calcium fly ash base (GPC) based on Python coding. In addition, the comparison of the prediction performance of the two technologies showed that the ensemble ML method, AdaBoost, and Boosting method are better than the single machine learning technology in predicting the compressive strength of GPC [59]. Barnat-Hunek et al. proposed a new method based on image texture analysis and machine learning technology to evaluate Lightweight Cement composites with hydrophobic coatings modified by the durability of nanocellulose (LLC) materials [60]. Asteris et al. proposed to evaluate the applicability of cement mortar compressive strength based on traditional machine learning models, such as support vector machine, random forest, decision tree, and AdaBoost. The above machine learning methods have achieved good results in the performance evaluation of cement-based materials, proving that machine learning has a certain development prospect in the performance evaluation of cement-based materials.

Although some researchers used machine learning methods to evaluate the compressive strength of cement-based materials before, it should be pointed out that due to the characteristics and application of different machine learning models, their prediction effects for different application scenarios will be different to some extent [61,62]. However, in previous studies, few researchers have analyzed the influence of different machine learning methods on the compressive strength of cement-based materials with metakaolin. To more accurately study the influence of the cement grade, the water-to-binder ratio, the binder-to-sand ratio, the metakaolin-to-binder ratio, and the superplasticizer on the compressive strength of cement-based materials with metakaolin, this study will use FA to optimize the hyperparameters of SVM, DT, KNN, and RF. The prediction effect of these four models on the compressive strength of cement-based materials with metakaolin was compared, and the model with the best prediction effect was selected.

2. Methodology

2.1. Dataset Collection

A reliable and huge database is significant for predicting the compressive strength of cement-based materials containing metakaolin with machine learning models. In this study, data on the effect of the cement grade, the water-to-binder ratio, the binder-to-sand ratio, the metakaolin-to-binder ratio, and the superplasticizer on the compressive strength of cement-based materials with metakaolin were collected from published articles, and a reliable database containing 361 data sets was established. The influence of these five parameters on the compressive strength of cement-based materials with metakaolin has been confirmed in previous studies. Therefore, they were selected as the input variables in the present study; considering the fact that compressive strength has been regarded as one of the most important parameters to evaluate the performance of cement-based materials, it has been selected as the output variable. The database is randomly divided into the training set and testing set. The training set contains about 80% of the data in the database, whereas the testing set contains about 20% of the data. The frequency distribution histogram of the six variables is shown in Figure 1.

By comparing the predicted value and the actual value of the training set and the testing set of different machine learning models, the prediction effect of different machine learning models on the compressive strength of metakaolin cement-based materials was analyzed. The five machine learning models are introduced below.

Correlation analysis is an analysis method to verify the linear relationship between two variables through a correlation coefficient, which can analyze the correlation between two variables and the strength of the correlation. Before using the machine learning model to predict the compressive strength of cement-based materials containing metakaolin, it is necessary to analyze the correlation between the input variables. The correlation coefficient between the two variables is high positive or high negative, that is, there is a high positive or negative correlation between the two variables, which will affect the prediction effect of the machine learning model. The correlation analysis of the cement grade, the water-to-binder ratio, the binder-to-sand ratio, the metakaolin-to-binder ratio, and the superplasticizer is shown in Figure 2. As shown in Figure 2, the correlation between the same variables is 1, and that between different variables is 0.2–0.6. Although the correlation coefficients between the water-to-binder ratio and the superplasticizer, the water-to-binder ratio, the binder-to-sand ratio, the cement grade, and the binder-to-sand ratio are high, about 0.5, the correlation coefficients between them are all lower than 0.6, that is, these five input variables are independent of each other. Therefore, using these five variables as input variables to predict the compressive strength of cement-based materials with metakaolin will achieve good results.

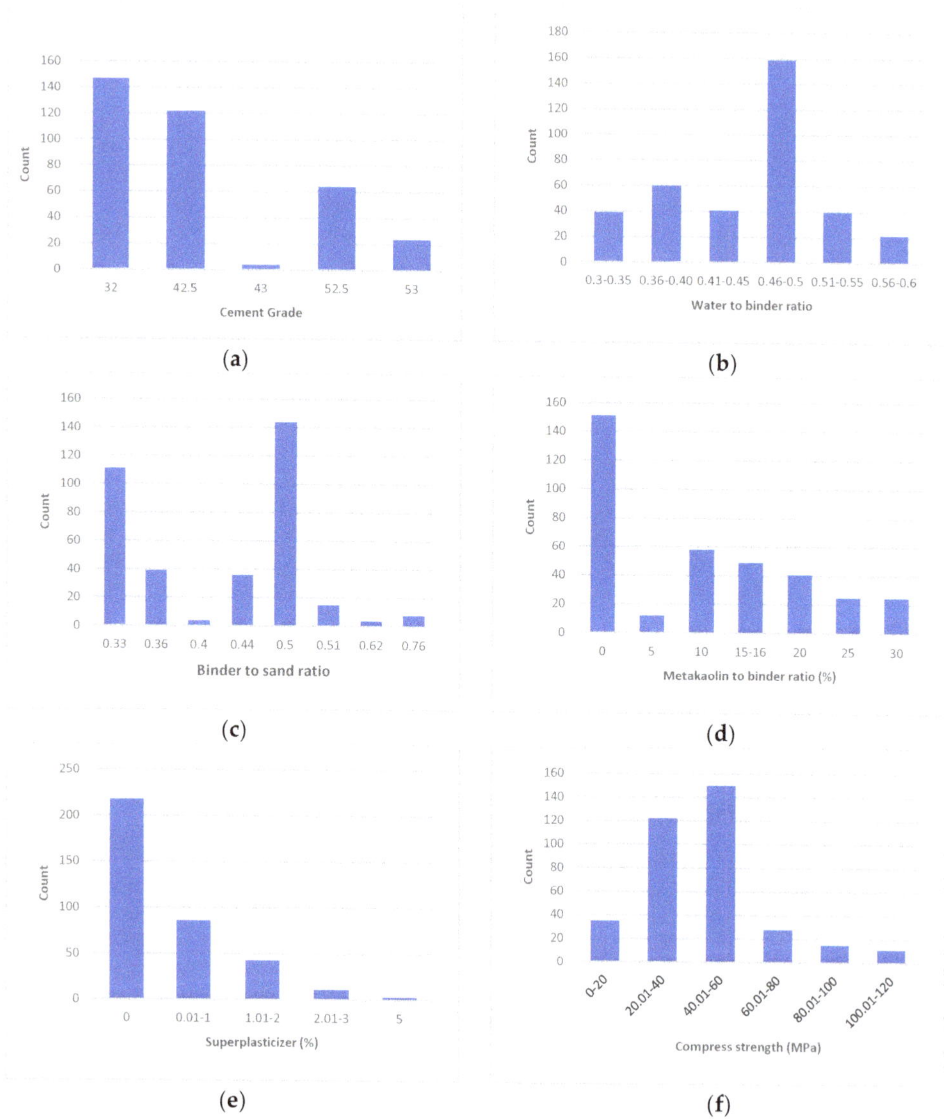

Figure 1. Frequency distribution histogram of variables. (**a**) Cement grade. (**b**) Water-to-binder ratio. (**c**) Binder-to-sand ratio. (**d**) Metakaolin-to-binder ratio. (**e**) Superplasticizer. (**f**) Compress strength.

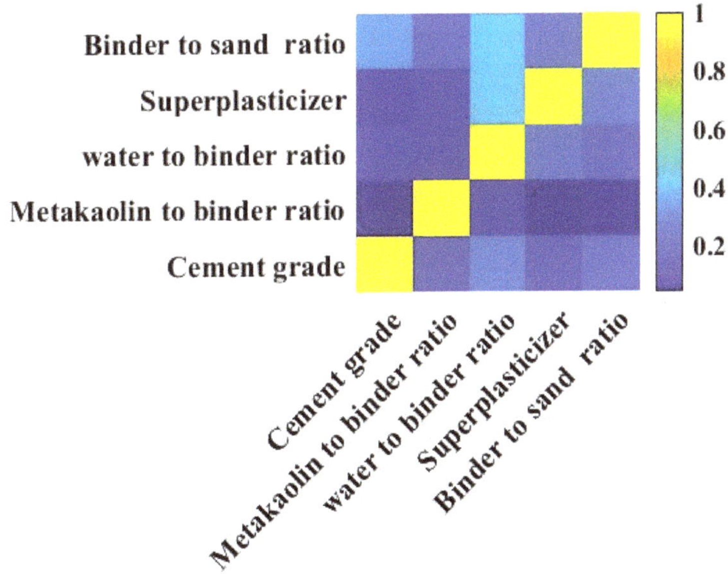

Figure 2. Correlation coefficients matrix diagram.

2.2. Applied Machine Learning Models

Five machine learning models, FA, SVM, DT, KNN, and RF, are used in this study, among which FA is used for hyperparameters tuning of the other four machine learning models. Four machine learning models including SVM, DT, KNN, and RF were used to predict the compressive strength of cement-based materials containing metakaolin with the cement grade, the water-to-binder ratio, and the binder-to-sand ratio, the metakaolin -to-binder ratio, and the superplasticizer as input variables.

FA is a random algorithm for global optimization problems based on firefly luminous behavior [63,64]. In the FA, it is assumed that all fireflies are hermaphroditic and are attracted to each other regardless of sex. FA is based on two key concepts: the intensity of the light emitted by the firefly and the degree of attraction between the two fireflies. The algorithm solves the optimization problem by simulating the living habits of fireflies attracted by light. The flow chart of FA can be found in Figure 3.

SVM is a method to classify data according to supervised learning. The decision boundary of the classification method is the maximum margin hyperplane of learning samples. SVM is suitable for small and medium-sized data samples, high latitude, and nonlinear classification problems [65]. High latitude means that SVM can efficiently deal with the classification of high-latitude feature space, which is of far-reaching significance in practical application. At the same time, although there are many training samples, SVM can only rely on limited samples for decision-making, so the computer only needs to store the limited samples used in decision-making, which saves the computer memory to a great extent [66]. DT is a predictive model in machine learning, representing the mapping between object attributes and object values. DT achieves the ability to predict target variables by input predictive variables by analyzing the relationship between input variables and target variables. A decision tree is similar to a flowchart in that each internal node is a test attribute and each branch is a test result. KNN is a data mining classification algorithm. Each sample in KNN can be represented by its nearest K neighboring values. This algorithm can classify every record in geometry. The idea of classification is that K samples are closest to each other in space, so the sample and the K samples belong to the same category. This method is a classification method to determine the classification of the samples to be classified by the category of one or several adjacent samples. RF is a common classification algorithm in machine learning. RF's basic idea is to establish a forest composed of many decision trees randomly, and there is no correlation between these decision trees. RF is mainly used to solve the classification problem. It uses multiple classification trees to classify data. In the process of classification, the algorithm obtains the importance of each variable classification by scoring the importance of each variable.

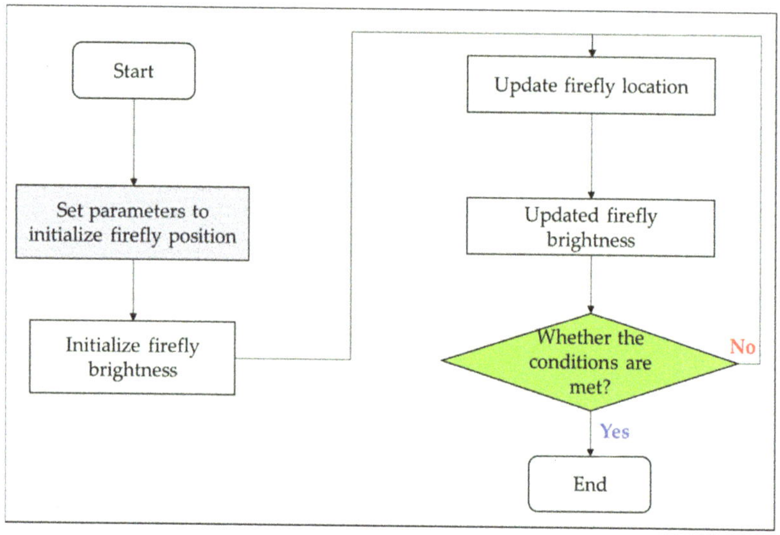

Figure 3. Flow chart of FA.

3. Results and Discussion

3.1. Hyperparameter Tuning

In this study, FA was used for hyperparameter tuning of SVM, DT, KNN, and RF models. The results of hyperparameter tuning are shown in Figure 4. As can be seen from Figure 4, with the increase in iteration times, RMSE values of these four models all decline sharply at first and then tend to be stable. The RMSE values of the SVM, DT, and RF models have similar rates of change, whereas the RMSE values of the KNN model are significantly different from those of the other three models, and its RMSE values are generally higher than those of the other three models, that is, FA has a poor tuning effect on this model. It can be seen from Figure 4 that after the hyperparameter tuning of SVM, DT, KNN, and RF by FA, the minimum RMSE values of SVM, DT, and RF models are all low, that is, the FA model has a better hyperparameters tuning effect on these three models. In this study, the prediction effects of four models, including SVM, DT, KNN, and RF on compressive strength of cement-based materials containing metakaolin after FA hyperparameter tuning were compared, and the model with the best prediction effect was selected.

The 10-fold cross-validation is a common test method to test the accuracy of the algorithm. This method needs to divide the data set into ten parts, selecting one of them as the testing data in turn and the remaining nine as the training data for training. The results of hyperparameter tuning for the above four models by 10-fold cross-validation are shown in Figure 5. Figure 5 shows that the minimum RSME of SVM, DT, and RF are all about 8, which are obtained at the fourth, ninth, and ninth iterations, respectively. The minimum RMSE value of KNN is 12, which is obtained at the sixth iteration. Therefore, after FA hyperparameter tuning, RSME values of SVM, DT, and RF models are lower, that is, the prediction effect of compressive strength of cement-based materials with metakaolin is better.

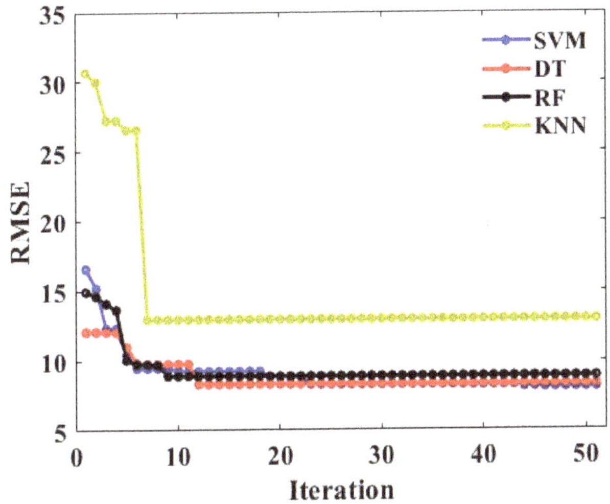

Figure 4. Root-mean-square error (RMSE) versus iterations.

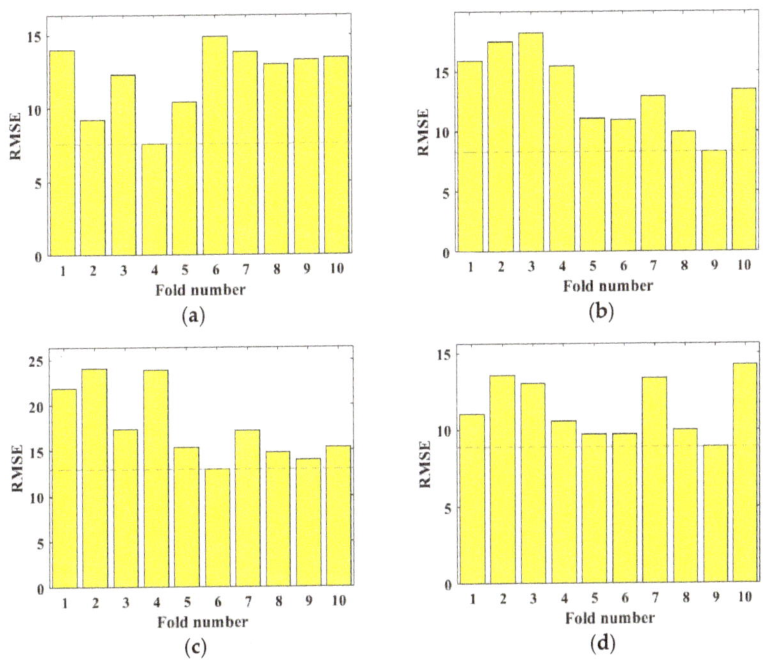

Figure 5. RMSE values of different folds. (**a**) RMSE values of different folds (SVM model). (**b**) RMSE values of different folds (DT model). (**c**) RMSE values of different folds (KNN model). (**d**) RMSE values of different folds (RF model).

3.2. Model Evaluation

In this study, the accuracy of the prediction of compressive strength of cement-based materials with metakaolin was evaluated by comparing the predicted values with the actual values of the training set and the testing set of SVM, DT, KNN, and RF. The comparison results between the predicted value and the actual value of the four models are shown in Figure 6, where the horizontal line represents the error between the predicted value and the actual value of the dataset. As shown from Figure 6, the predicted values and the actual values of SVM, DT, and RF are consistent, with only a few points with large errors, while there is a large error between the predicted values and the actual values of KNN.

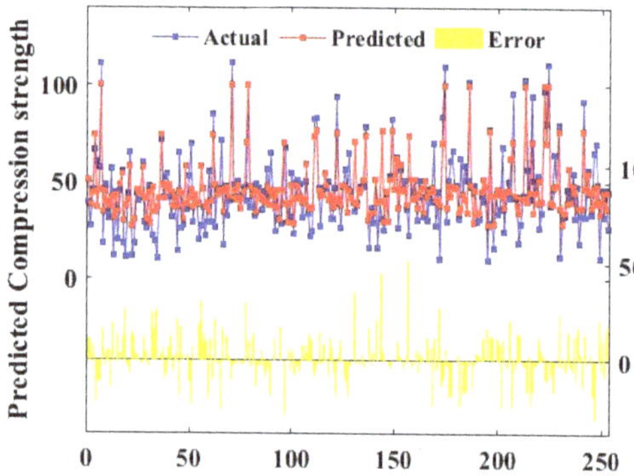

(a)

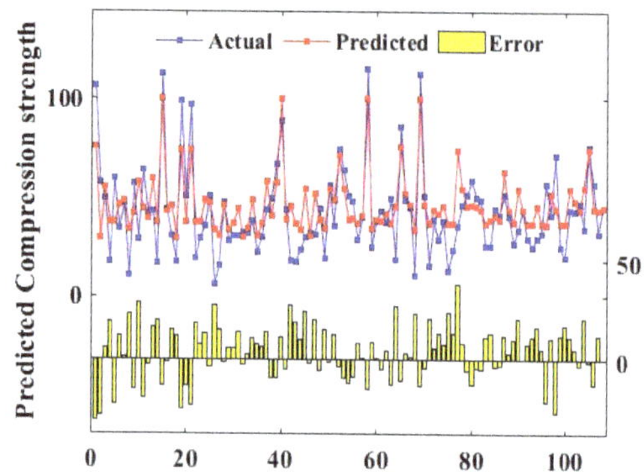

(b)

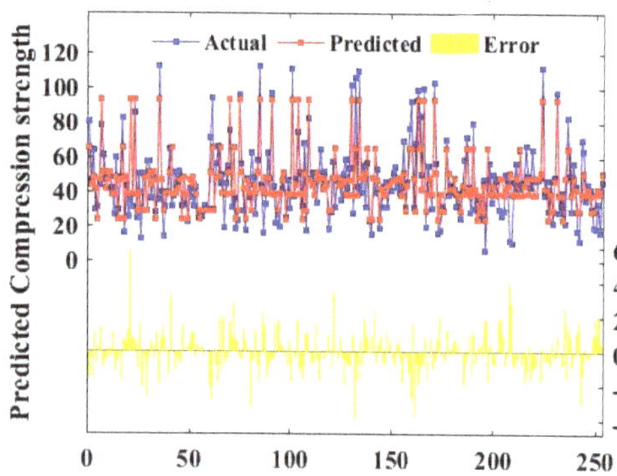

(c)

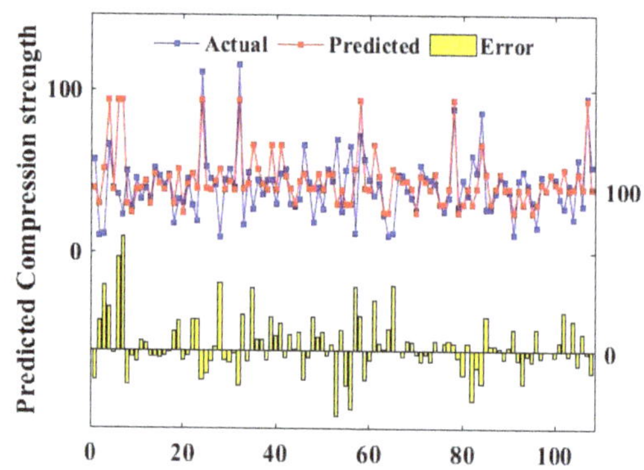

(d)

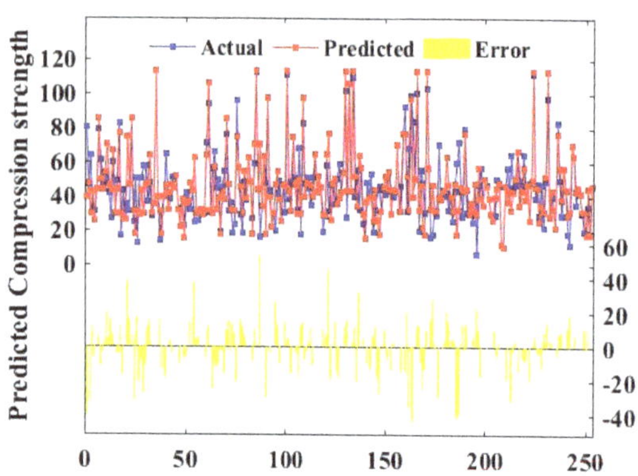

(e)

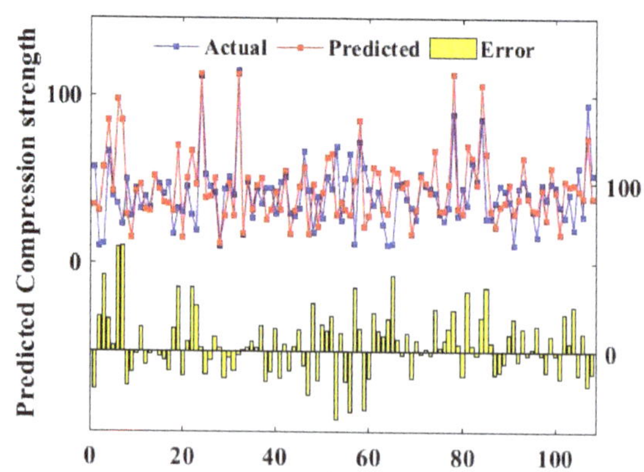

(f)

Figure 6. *Cont.*

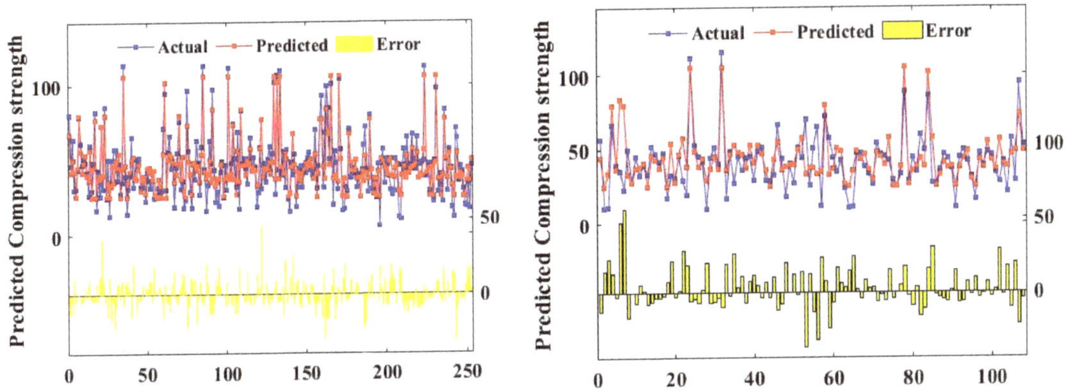

(g) (h)

Figure 6. Comparison of the actual compressive strength and predicted compressive strength. (**a**) SVM model (training dataset). (**b**) SVM model (testing dataset). (**c**) DT model (training dataset). (**d**) DT model (testing dataset). (**e**) KNN model (training dataset). (**f**) KNN model (testing dataset). (**g**) RF model (training dataset). (**h**) RF model (testing dataset).

Figure 7 compares the predicted and actual values of the training set and the testing set of SVM, DT, KNN, and RF. It can be seen from Figure 7 that most of the predicted values of the training set and the testing set of SVM, DT, and RF models are concentrated in 20–60 MPa, most of the actual values are concentrated in 0–60 MPa, and a few of the predicted values and the actual values of the training set and the testing set are concentrated in 60–120 MPa. Most of the predicted values and the actual values of the training set and the testing set of the KNN model are concentrated in 0–60 MPa and a few are concentrated in 60–120 MPa. The RMSE values corresponding to the SVM, DT, KNN, and RF training sets are 12.4494, 12.942, 13.6014, and 10.4863, respectively, and the R values are 0.7856, 0.7933, 0.7924, and 0.8709, respectively. The RMSE values corresponding to the testing sets of these four models are 14.3837, 17.6971, 19.2549, and 14.2904, respectively, and the R values are 0.7928, 0.5246, 0.5492, and 0.696, respectively. Among the four models, RF has the lowest RMSE value and the highest R value on both the training set and the testing set, whereas KNN has the highest RMSE value and the lowest R value on both the training set and the testing set. That is, after the hyperparameters of SVM, DT, KNN, and RF models are adjusted by the FA model, RF has the best prediction effect on the compressive strength of cement-based materials with metakaolin, whereas KNN has the worst prediction effect.

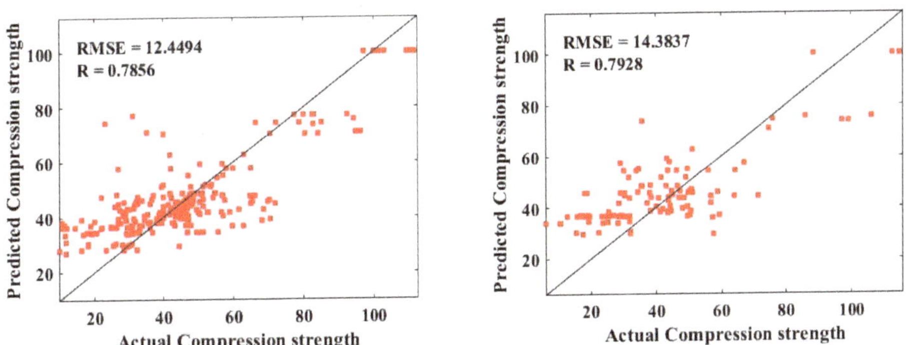

(a) (b)

Figure 7. *Cont.*

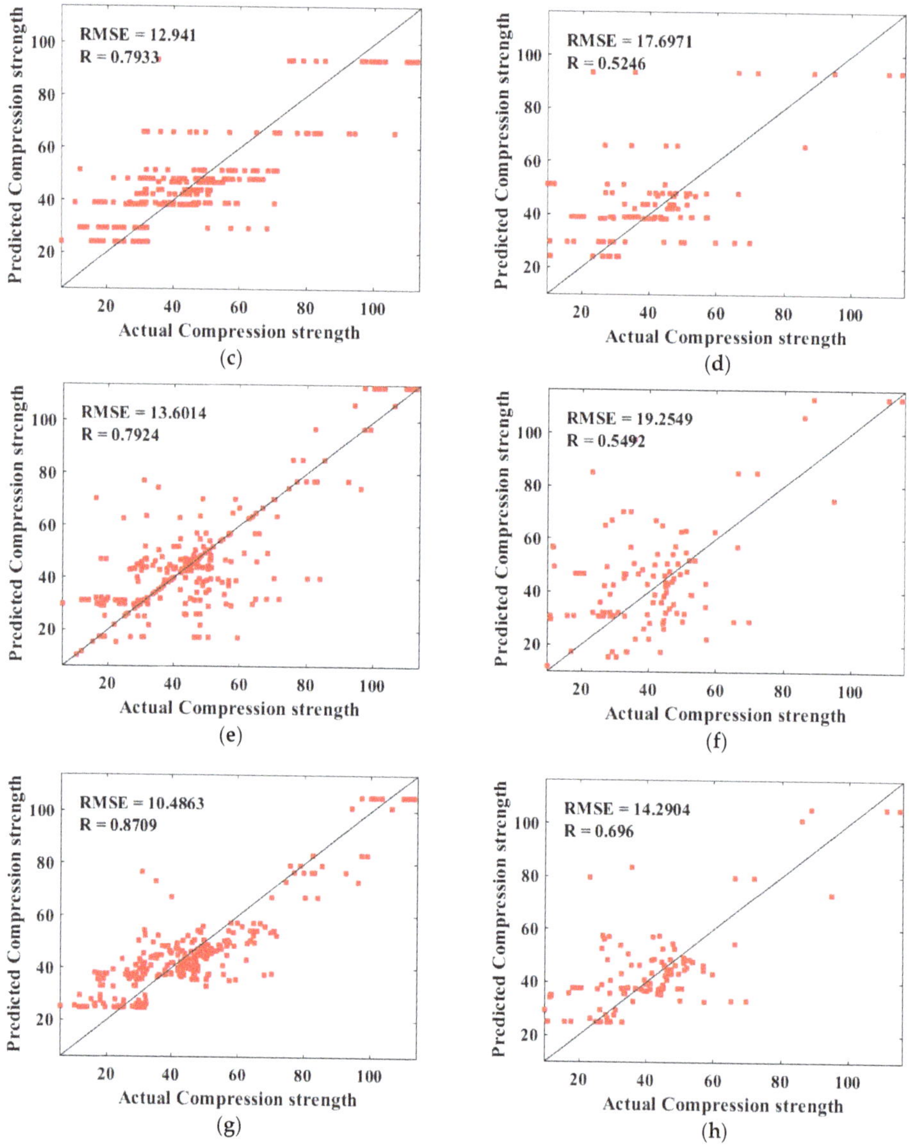

Figure 7. Comparison of the actual compressive strength and predicted compressive strength. (**a**) SVM model. (**b**) SVM model. (**c**) DT model. (**d**) DT model. (**e**) KNN model. (**f**) KNN model. (**g**) RF model. (**h**) RF model.

Figure 8 compares RMSE values and R values of SVM, DT, KNN, and RF with hyperparameter tuning by FA. It can be clearly seen from Figure 8 that after hyperparameter tuning by FA, RMSE values of these four models decrease one by one in the order of KNN, DT, SVM, and RF, while R values increase one by one. In other words, RF has the lowest RSME value and the highest R value among the four models, which confirms again that RF has the best prediction effect on the compressive strength of cement-based materials with metakaolin among the four models after FA hyperparameter tuning. Such prediction results can provide certain technical support in the production process of green concrete with Metakaolin in the future. However, it should be noted that the proposed model should be clarified before practical application, and the raw materials used should be consistent with the ones used to build the prediction model.

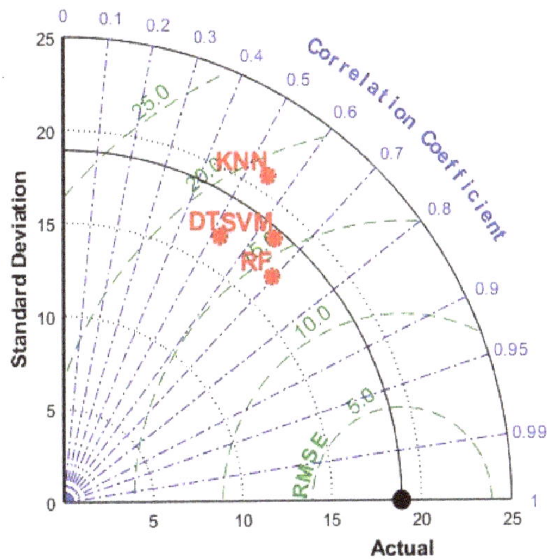

Figure 8. Comparison of RSME values and R values of different models.

3.3. Variable Importance Evaluation

Based on the analysis of the importance of the five input variables (the cement grade, the water-to-binder ratio, the binder-to-sand ratio, the metakaolin-to-binder ratio, and the superplasticizer) affecting the compressive strength of cement-based materials with metakaolin studied in this paper, we can understand the importance of these five variables to the compressive strength of cement-based materials with metakaolin and then provide some guidance for engineers to configure high-strength cement-based materials with metakaolin. Machine learning models can be used to predict the compressive strength of cement-based materials with metakaolin according to five input parameters: the cement grade, the water-to-binder ratio, the binder-to-sand ratio, the metakaolin-to-binder ratio, and the superplasticizer. Similarly, machine learning models can also be used to evaluate the importance of these five variables to the compressive strength of cement-based materials with metakaolin. According to the above analysis, RF is the best prediction model of SVM, DT, KNN, and RF after hyperparametric optimization with FA. Therefore, this paper will use the RF model to analyze the importance of the five input variables. It can be seen from Figure 9 that the importance of the cement grade, the water-to-binder ratio, the binder-to-sand ratio, the metakaolin-to-binder ratio, and the superplasticizer are all positive, that is, these five variables are directly proportional to the compressive strength of cement-based materials with metakaolin. Therefore, the compressive strength of cement-based materials with metakaolin can be improved by increasing any one of these five parameters. Among the five variables of the cement grade, the water-to-binder ratio, the binder-to-sand ratio, the metakaolin-to-binder ratio, and the superplasticizer, the importance scores of the cement grade and the water-to-binder ratio are the highest, which are 1.4400 and 1.4155, respectively, that is, these two parameters are the two most important variables affecting the compressive strength of cement-based materials with metakaolin. The importance score of the superplasticizer on the compressive strength of cement-based materials with metakaolin is 0.5981. Among the five input variables, it is the variable with the lowest importance score on the compressive strength of cement-based materials with metakaolin. Therefore, when configuring high-strength cement-based materials with metakaolin, engineers should focus on the impact of the cement grade and the water-to-binder ratio on the compressive strength of cement-based materials with metakaolin and do not need to focus too much on the dosage of the superplasticizer.

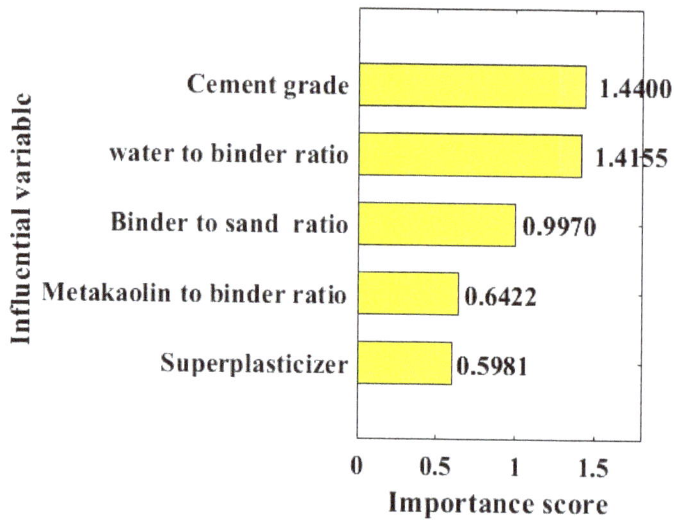

Figure 9. Variable importance of the compressive strength.

4. Conclusions

According to the published literature, this study established a database. The database takes the cement grade, the water-to-border ratio, the binder-to-sand ratio, the metakaolin-to-border ratio, and the superplasticizer as input variables, and the compressive strength of cement-based material with metakaolin as output variable. The data set in the database is randomly divided into the training set and the testing set. The training set contains about 80% of the data in the database, and the testing set contains about 20% of the data in the database. Based on using FA to tune the hyperparameters of SVM, DT, KNN, and RF, the author analyzed the model with the highest prediction accuracy of the compressive strength of cement-based materials with metakaolin by comparing the predicted values and the actual values of the training set and the testing set of these four models. The corresponding conclusions are as follows:

1. The FA model can achieve good results in the hyperparameter tuning of SVM, DT, and RF, whereas the FA model is relatively poor in the hyperparameter tuning of KNN. In other words, SVM, DT, and RF models can accurately predict the compressive strength of cement-based materials with metakaolin, of which the RF model has the best prediction effect, whereas the KNN model has the worst prediction effect.
2. The compressive strength of cement-based materials with metakaolin is directly proportional to the five variables of the cement grade, the water-to-binder ratio, the binder-to-sand ratio, the metakaolin-to-binder ratio, and the superplasticizer, that is, the compressive strength of cement-based materials with metakaolin increases with the increase of any one of these five variables. It decreases with the decrease of any one of these five variables.
3. The five variables of the importance of the compressive strength of cement-based material with metakaolin decrease one by one according to the order of the cement grade, the water-to-binder ratio, the binder-to-sand ratio, the metakaolin-to-binder ratio, and the superplasticizer. The cement grade and the water-to-binder ratio have a significant influence on the compressive strength of cement-based materials with

metakaolin. In contrast, the superplasticizer has less influence on the importance of the compressive strength of cement-based materials with metakaolin.

It should be noted that the predictive solutions may not be directly applied to the construction industry at present. This is because the 361 databases come from different studies, and there are significant differences regarding the raw materials in morphology characteristics, chemical composition, and other factors and the environmental parameters (such as the temperature). One goal of future research should be to establish a gene database of raw materials and take the genetic characteristics of raw materials (for example, morphology characteristics and chemical composition) as input variables to conduct in-depth data mining in the case of big data to establish reliable and effective prediction methods for actual production.

Author Contributions: Conceptualization, X.L., J.H. and M.Z.; methodology, J.H. and M.M.S.S.; writing—original draft preparation, J.H. and M.Z.; writing—review and editing, M.M.S.S. and M.Z.; visualization, M.Z.; supervision, J.H. and H.Y.; funding acquisition, M.M.S.S. All authors have read and agreed to the published version of the manuscript.

Funding: The research is partially funded by the Ministry of Science and Higher Education of the Russian Federation under the strategic academic leadership program 'Priority 2030' (Agreement 075-15-2021-1333 dated 30 September 2021).

Data Availability Statement: The data presented in this study are available on request from the corresponding author.

Acknowledgments: The research is partially funded by the Ministry of Science and Higher Education of the Russian Federation under the strategic academic leadership program 'Priority 2030' (Agreement 075-15-2021-1333 dated 30 September 2021).

Conflicts of Interest: The authors declare no conflict of interest.

References

1. Tang, S.W.; Huang, J.S.; Duan, L.; Yu, P.; Chen, E. A review on fractal footprint of cement-based materials. *Powder Technol.* **2020**, *370*, 237–250. [CrossRef]
2. Batis, G.; Pantazopoulou, P.; Tsivilis, S.; Badogiannis, E. The effect of metakaolin on the corrosion behavior of cement mortars. *Cem. Concr. Compos.* **2005**, *27*, 125–130. [CrossRef]
3. Huang, J.; Zhang, J.; Ren, J.; Chen, H. Anti-rutting performance of the damping asphalt mixtures (dams) made with a high content of asphalt rubber (ar). *Constr. Build. Mater.* **2021**, *271*, 121878. [CrossRef]
4. Wang, X.-Y. Analysis of hydration-mechanical-durability properties of metakaolin blended concrete. *Appl. Sci.* **2017**, *7*, 1087. [CrossRef]
5. Xu, W.; Huang, X.; Huang, J.; Yang, Z. Structural analysis of backfill highway subgrade on the lower bearing capacity foundation using the finite element method. *Adv. Civ. Eng.* **2021**, *2021*, 1690168. [CrossRef]
6. Ren, J.; Xu, Y.; Zhao, Z.; Chen, J.; Cheng, Y.; Huang, J.; Yang, C.; Wang, J. Fatigue prediction of semi-flexible composite mixture based on damage evolution. *Constr. Build. Mater.* **2022**, *318*, 126004. [CrossRef]
7. Ren, J.; Xu, Y.; Huang, J.; Wang, Y.; Jia, Z. Gradation optimization and strength mechanism of aggregate structure considering macroscopic and mesoscopic aggregate mechanical behaviour in porous asphalt mixture. *Constr. Build. Mater.* **2021**, *300*, 124262. [CrossRef]
8. Ren, J.; Li, D.; Xu, Y.; Huang, J.; Liu, W. Fatigue behaviour of rock asphalt concrete considering moisture, high-temperature, and stress level. *Int. J. Pavement Eng.* **2021**, 1–11. [CrossRef]
9. Liang, X.; Yu, X.; Chen, C.; Ding, G.; Huang, J. Towards the low-energy usage of high viscosity asphalt in porous asphalt pavements: A case study of warm-mix asphalt additives. *Case Stud. Constr. Mater.* **2022**, *16*, e00914. [CrossRef]
10. Cao, D.; Malakooti, S.; Kulkarni, V.N.; Ren, Y.; Lu, H. Nanoindentation measurement of core–skin interphase viscoelastic properties in a sandwich glass composite. *Mech. Time-Depend. Mater.* **2021**, *25*, 353–363. [CrossRef]
11. Cao, D.; Malakooti, S.; Kulkarni, V.N.; Ren, Y.; Liu, Y.; Nie, X.; Qian, D.; Griffith, D.T.; Lu, H. The effect of resin uptake on the flexural properties of compression molded sandwich composites. *Wind Energy* **2022**, *25*, 71–93. [CrossRef]
12. Wang, X.; Xu, T.; Andrade, M.J.D.; Rampalli, I.; Cao, D.; Haque, M.; Roy, S.; Baughman, R.H.; Lu, H. The interfacial shear strength of carbon nanotube sheet modified carbon fiber composites. In *Challenges in Mechanics of Time Dependent Materials*; Springer: Cham, Switzerland, 2021; Volume 2, pp. 25–32.
13. Courard, L.; Darimont, A.; Schouterden, M.; Ferauche, F.; Willem, X.; Degeimbre, R. Durability of mortars modified with metakaolin. *Cem. Concr. Res.* **2003**, *33*, 1473–1479. [CrossRef]

14. Wang, Q.-A.; Zhang, C.; Ma, Z.-G.; Huang, J.; Ni, Y.-Q.; Zhang, C. Shm deformation monitoring for high-speed rail track slabs and bayesian change point detection for the measurements. *Constr. Build. Mater.* **2021**, *300*, 124337. [CrossRef]
15. Gao, Y.; Huang, J.; Li, M.; Dai, Z.; Jiang, R.; Zhang, J. Chemical modification of combusted coal gangue for u(vi) adsorption: Towards a waste control by waste strategy. *Sustainability* **2021**, *13*, 8421. [CrossRef]
16. Ahmad, M.; Tang, X.-W.; Ahmad, F.; Pirhadi, N.; Wan, X.; Cheng, K. Probabilistic Evaluation of Cpt-Based Seismic Soil Liquefaction Potential: Towards the Integration of Interpretive Structural Modeling and Bayesian Belief Network. *Math. Biosci. Eng.* **2021**, *18*, 9233–9252. [CrossRef] [PubMed]
17. Ma, H.X.; Liu, J.D.; Zhang, J.; Huang, J.D. Estimating the compressive strength of cement-based materials with mining waste using support vector machine, decision tree, and random forest models. *Adv. Civ. Eng.* **2021**, *2021*, 6629466. [CrossRef]
18. Shi, L.; Liu, J.Z.; Liu, J.P. Influence and mechanism of polymer coating on shrinkage of cement-based materials. In Proceedings of the 1st International Congress on Advanced Materials, Jinan, China, 13–16 May 2011; pp. 923–926.
19. Huang, H.; Yuan, Q.; Deng, D.H.; Peng, J.W.; Huang, Y.L. Effects of chemical and mineral admixtures on the foam indexes of cement-based materials. *Case Stud. Constr. Mater.* **2019**, *11*, e00232. [CrossRef]
20. Duan, P.; Yan, C.J.; Zhou, W.; Ren, D.M. Fresh properties, compressive strength and microstructure of fly ash geopolymer paste blended with iron ore tailing under thermal cycle. *Constr. Build. Mater.* **2016**, *118*, 76–88. [CrossRef]
21. Cheng, A.; Chao, S.J.; Lin, W.T. Effects of leaching behavior of calcium ions on compression and durability of cement-based materials with mineral admixtures. *Materials* **2013**, *6*, 1851–1872. [CrossRef]
22. Huang, J.; Li, X.; Kumar, G.S.; Deng, Y.; Gong, M.; Dong, N. Rheological properties of bituminous binder modified with recycled waste toner. *J. Clean. Prod.* **2021**, *317*, 128415. [CrossRef]
23. Qian, X.Q.; Li, Z.J. The relationships between stress and strain for high-performance concrete with metakaolin. *Cem. Concr. Res.* **2001**, *31*, 1607–1611. [CrossRef]
24. Zhang, S.; Fan, Y.; Huang, J.; Shah, S.P. Effect of nano-metakaolinite clay on hydration behavior of cement-based materials at early curing age. *Constr. Build. Mater.* **2021**, *291*, 123107. [CrossRef]
25. Hu, L.L.; He, Z. A fresh perspective on effect of metakaolin and limestone powder on sulfate resistance of cement-based materials. *Constr. Build. Mater.* **2020**, *262*, 119847. [CrossRef]
26. Mobili, A.; Giosue, C.; Bellezze, T.; Tittaretti, F. Corrosion behavior of galvanized and bare steel reinforcements embedded in carbonated alkali-activated metakaolin mortar. *Metall. Ital.* **2020**, *4*, 22–26.
27. Bhat, A.H.; Naqash, J.A. Experimental studies of sustainable concrete modified with colloidal nanosilica and metakaolin. *J. Build. Pathol. Rehabil.* **2021**, *7*, 18. [CrossRef]
28. Khater, H.M. Influence of metakaolin on resistivity of cement mortar to magnesium chloride solution. *Ceramics-Silikáty* **2010**, *54*, 325–333. [CrossRef]
29. Kirsanova, A.A.; Kirsanova, M.A. Investigation into effect of metakaolin-based additives incorporation on formation of cement stone structure. In Proceedings of the International Conference on Construction, Architecture and Technosphere Safety (ICCATS), Ural State University, Yekaterinburg, Russia, 26–28 September 2018.
30. Kadri, E.-H.; Kenai, S.; Ezziane, K.; Siddique, R.; De Schutter, G. Influence of metakaolin and silica fume on the heat of hydration and compressive strength development of mortar. *Appl. Clay Sci.* **2011**, *53*, 704–708. [CrossRef]
31. Vu, D.D.; Stroeven, P.; Bui, V.B. Strength and durability aspects of calcined kaolin-blended portland cement mortar and concrete. *Cem. Concr. Compos.* **2001**, *23*, 471–478. [CrossRef]
32. Benton, W.C. Machine learning systems and intelligent applications. *IEEE Softw.* **2020**, *37*, 43–49. [CrossRef]
33. Xu, W.; Huang, X.; Yang, Z.; Zhou, M.; Huang, J. Developing hybrid machine learning models to determine the dynamic modulus (e*) of asphalt mixtures using parameters in witczak 1-40d model: A comparative study. *Materials* **2022**, *15*, 1791. [CrossRef]
34. Koopialipoor, M.; Tootoonchi, H.; Armaghani, D.J.; Mohamad, E.T.; Hedayat, A. Application of deep neural networks in predicting the penetration rate of tunnel boring machines. *Bull. Eng. Geol. Environ.* **2019**, *78*, 6347–6360. [CrossRef]
35. Armaghani, D.J.; Koopialipoor, M.; Marto, A.; Yagiz, S. Application of several optimization techniques for estimating tbm advance rate in granitic rocks. *J. Rock Mech. Geotech. Eng.* **2019**, *11*, 779–789. [CrossRef]
36. Koopialipoor, M.; Armaghani, D.J.; Hedayat, A.; Marto, A.; Gordan, B. Applying various hybrid intelligent systems to evaluate and predict slope stability under static and dynamic conditions. *Soft Comput.* **2019**, *23*, 5913–5929. [CrossRef]
37. Armaghani, D.J.; Hasanipanah, M.; Mohamad, E.T. A combination of the ica-ann model to predict air-overpressure resulting from blasting. *Eng. Comput.* **2016**, *32*, 155–171. [CrossRef]
38. Armaghani, D.J.; Asteris, P.G. A comparative study of ann and anfis models for the prediction of cement-based mortar materials compressive strength. *Neural Comput. Appl.* **2021**, *33*, 4501–4532. [CrossRef]
39. Wang, Q.-A.; Zhang, J.; Huang, J. Simulation of the compressive strength of cemented tailing backfill through the use of firefly algorithm and random forest model. *Shock Vib.* **2021**, *2021*, 5536998. [CrossRef]
40. Huang, J.; Zhang, J.; Gao, Y. Intelligently predict the rock joint shear strength using the support vector regression and firefly algorithm. *Lithosphere* **2021**, *2021*, 2467126. [CrossRef]
41. Zelinka, J.; Romportl, J.; Muller, L. A priori and a posteriori machine learning and nonlinear artificial neural networks. In Proceedings of the 13th International Conference on Text, Speech and Dialogue, Brno, Czech Republic, 6–10 September 2010; pp. 472–479.

42. Huang, J.; Koopialipoor, M.; Armaghani, D.J. A combination of fuzzy delphi method and hybrid ann-based systems to forecast ground vibration resulting from blasting. *Sci. Rep.* **2020**, *10*, 1939. [CrossRef]
43. Huang, J.; Duan, T.; Zhang, Y.; Liu, J.; Zhang, J.; Lei, Y. Predicting the permeability of pervious concrete based on the beetle antennae search algorithm and random forest model. *Adv. Civ. Eng.* **2020**, *2020*, 8863181. [CrossRef]
44. Huang, J.; Duan, T.; Lei, Y.; Hasanipanah, M. Finite element modeling for the antivibration pavement used to improve the slope stability of the open-pit mine. *Shock Vib.* **2020**, *2020*, 6650780. [CrossRef]
45. Huang, J.; Asteris, P.G.; Pasha, S.M.K.; Mohammed, A.S.; Hasanipanah, M. A new auto-tuning model for predicting the rock fragmentation: A cat swarm optimization algorithm. *Eng. Comput.* **2020**, *38*, 2209–2220. [CrossRef]
46. Hasanipanah, M.; Armaghani, D.J.; Amnieh, H.B.; Abd Majid, M.Z.; Tahir, M.M. Application of pso to develop a powerful equation for prediction of flyrock due to blasting. *Neural Comput. Appl.* **2017**, *28*, 1043–1050. [CrossRef]
47. Armaghani, D.J.; Raja, R.S.N.S.B.; Faizi, K.; Rashid, A.S.A. Developing a hybrid pso–ann model for estimating the ultimate bearing capacity of rock-socketed piles. *Neural Comput. Appl.* **2017**, *28*, 391–405. [CrossRef]
48. Koopialipoor, M.; Ghaleini, E.N.; Tootoonchi, H.; Armaghani, D.J.; Haghighi, M.; Hedayat, A. Developing a new intelligent technique to predict overbreak in tunnels using an artificial bee colony-based ann. *Environ. Earth Sci.* **2019**, *78*, 165. [CrossRef]
49. Armaghani, D.J.; Mohamad, E.T.; Narayanasamy, M.S.; Narita, N.; Yagiz, S. Development of hybrid intelligent models for predicting tbm penetration rate in hard rock condition. *Tunn. Undergr. Space Technol.* **2017**, *63*, 29–43. [CrossRef]
50. Hajihassani, M.; Armaghani, D.J.; Marto, A.; Mohamad, E.T. Ground vibration prediction in quarry blasting through an artificial neural network optimized by imperialist competitive algorithm. *Bull. Eng. Geol. Environ.* **2015**, *74*, 873–886. [CrossRef]
51. Koopialipoor, M.; Armaghani, D.J.; Haghighi, M.; Ghaleini, E.N. A neuro-genetic predictive model to approximate overbreak induced by drilling and blasting operation in tunnels. *Bull. Eng. Geol. Environ.* **2019**, *78*, 981–990. [CrossRef]
52. Chen, W.; Hasanipanah, M.; Rad, H.N.; Armaghani, D.J.; Tahir, M. A new design of evolutionary hybrid optimization of svr model in predicting the blast-induced ground vibration. *Eng. Comput.* **2019**, *37*, 1455–1471. [CrossRef]
53. Lu, S.; Koopialipoor, M.; Asteris, P.G.; Bahri, M.; Armaghani, D.J. A novel feature selection approach based on tree models for evaluating the punching shear capacity of steel fiber-reinforced concrete flat slabs. *Materials* **2020**, *13*, 3902. [CrossRef]
54. Mahdiyar, A.; Jahed Armaghani, D.; Koopialipoor, M.; Hedayat, A.; Abdullah, A.; Yahya, K. Practical risk assessment of ground vibrations resulting from blasting, using gene expression programming and monte carlo simulation techniques. *Appl. Sci.* **2020**, *10*, 472. [CrossRef]
55. Koopialipoor, M.; Nikouei, S.S.; Marto, A.; Fahimifar, A.; Armaghani, D.J.; Mohamad, E.T. Predicting tunnel boring machine performance through a new model based on the group method of data handling. *Bull. Eng. Geol. Environ.* **2019**, *78*, 3799–3813. [CrossRef]
56. Asteris, P.G.; Koopialipoor, M.; Armaghani, D.J.; Kotsonis, E.A.; Lourenço, P.B. Prediction of cement-based mortars compressive strength using machine learning techniques. *Neural Comput. Appl.* **2021**, *33*, 13089–13121. [CrossRef]
57. Mohamad, E.T.; Armaghani, D.J.; Momeni, E.; Yazdavar, A.H.; Ebrahimi, M. Rock strength estimation: A pso-based bp approach. *Neural Comput. Appl.* **2018**, *30*, 1635–1646. [CrossRef]
58. Asteris, P.G.; Lourenco, P.B.; Roussis, P.C.; Adami, C.E.; Armaghani, D.J.; Cavaleri, L.; Chalioris, C.E.; Hajihassani, M.; Lemonis, M.E.; Mohammed, A.S.; et al. Revealing the nature of metakaolin-based concrete materials using artificial intelligence techniques. *Constr. Build. Mater.* **2022**, *322*, 126500. [CrossRef]
59. Ahmad, A.; Ahmad, W.; Chaiyasarn, K.; Ostrowski, K.A.; Aslam, F.; Zajdel, P.; Joyklad, P. Prediction of geopolymer concrete compressive strength using novel machine learning algorithms. *Polymers* **2021**, *13*, 3389. [CrossRef]
60. Barnat-Hunek, D.; Omiotek, Z.; Szafraniec, M.; Dzierzak, R. An integrated texture analysis and machine learning approach for durability assessment of lightweight cement composites with hydrophobic coatings modified by nanocellulose. *Measurement* **2021**, *179*, 109538. [CrossRef]
61. Cheng, M.Y.; Hoang, N.D. Groutability prediction of microfine cement based soil improvement using evolutionary ls-svm inference model. *J. Civ. Eng. Manag.* **2014**, *20*, 839–848. [CrossRef]
62. Tong, Z.; Huo, J.Y.; Wang, Z.J. High-throughput design of fiber reinforced cement-based composites using deep learning. *Cem. Concr. Compos.* **2020**, *113*, 103716. [CrossRef]
63. Sarangi, S.K.; Panda, R.; Priyadarshini, S.; Sarangi, A. A new modified firefly algorithm for function optimization. In Proceedings of the International Conference on Electrical, Electronics, and Optimization Techniques (ICEEOT), Palanchur, India, 3–5 March 2016; pp. 2944–2949.
64. Ma, L.Y.; Cao, P.R. Comparative study of several improved firefly algorithms. In Proceedings of the IEEE International Conference on Information and Automation (ICIA), Ningbo, China, 1–3 August 2016; pp. 910–914.
65. Gong, W.; Wang, W.S. Application research of support vector machine in e-learning for personality. In Proceedings of the IEEE International Conference on Cloud Computing and Intelligence Systems (CCIS20111), Beijing, China, 15–17 September 2011; pp. 638–642.
66. Hu, Y.H.; Fang, Q.Q.; Hu, Y.M.; Xiao, L.; Atlantis, P. Electronic commerce talent requirement forecasting model by genetic-support vector machine. In Proceedings of the International Conference on Intelligence Science and Information Engineering (ISIE 2012), Lushan, China, 25–26 August 2012; pp. 116–118.

buildings

MDPI

Article

Modelling the Strength and Fracture Parameters of Dam Gallery Concrete Considering Ambient Temperature and Humidity

Xiaofeng Gao [1,2], Qingbin Li [2], Zhihong Liu [3], Jiangnan Zheng [1], Kai Wei [1], Yaosheng Tan [4,5], Ning Yang [4,5], Chunfeng Liu [4,5], Yongjiu Lu [6] and Yu Hu [2,*]

1 College of Civil Engineering, Zhejiang University of Technology, Hangzhou 310023, China; gaoxf@zjut.edu.cn (X.G.); 2112006155@zjut.edu.cn (J.Z.); 2112106070@zjut.edu.cn (K.W.)
2 State Key Laboratory of Hydroscience and Engineering, Tsinghua University, Beijing 100084, China; qingbinli@tsinghua.edu.cn
3 Faculty of Electric Power Engineering, Kunming University of Science and Technology, Kunming 650500, China; liuzhihong@stu.kust.edu.cn
4 China Three Gorges Construction Engineering Corporation, Chengdu 610041, China; tan_yaosheng@ctg.com.cn (Y.T.); yang_ning1@ctg.com.cn (N.Y.); liu_chunfeng@ctg.com.cn (C.L.)
5 China Three Gorges Group Corporation, Beijing 100038, China
6 China Three Gorges Corporation Test Center, Yichang 443133, China; lu_yongjiu@ctg.com.cn
* Correspondence: yu-hu@tsinghua.edu.cn; Tel.: +86-10-62781161

Abstract: Due to the complex structure and stress distribution of dam galleries, cracks often appear during the construction, operation and maintenance periods of dams. This paper proposes a method to determine the real strength and fracture parameters of gallery concrete, considering environmental temperature and humidity. The strength and fracture tests were carried out for gallery concrete at various ages and under different curing temperature and humidity conditions. The influence of curing conditions on the mechanical properties of gallery concrete was quantitatively analyzed. The prediction equations of the strength and fracture parameters of gallery concrete under arbitrary temperature and humidity were established. Based on the measured temperature and humidity data, the real mechanical parameters of gallery concrete were predicted. The research results show that the influence of environmental conditions on mechanical parameters cannot be neglected, as this can result in a strength reduction of up to 33.81%. The proposed equations can be used to predict the mechanical parameters of gallery concrete, subject to real environmental conditions, which can help to correct a maximum deviation of 54.62% on parameters calculated using actual ages. The proposed method can provide a scientific basis for the cracking risk analysis and safety assessment of the gallery structure under actual conditions.

Keywords: dam; gallery concrete; strength; fracture parameters; maturity model; cracking risk; ambient temperature; ambient humidity

Citation: Gao, X.; Li, Q.; Liu, Z.; Zheng, J.; Wei, K.; Tan, Y.; Yang, N.; Liu, C.; Lu, Y.; Hu, Y. Modelling the Strength and Fracture Parameters of Dam Gallery Concrete Considering Ambient Temperature and Humidity. *Buildings* **2022**, *12*, 168. https://doi.org/10.3390/buildings12020168

Academic Editor: Tomáš Dvorský

Received: 6 January 2022
Accepted: 1 February 2022
Published: 3 February 2022

1. Introduction

In recent years, with the continuous improvement of China's dam intelligent construction technology [1–3], the dam engineering community has put forward higher requirements for the risk prevention of dam projects. The cracking of dam surface and dam body concrete is one of the important factors affecting the safe operation of dam structures [4–6]. Among them, dam galleries are openings or passageways, constructed in the dam structure, for the purpose of drainage, inspection, foundation grouting, gate operation and equipment control and, due to the complex structure and stress distribution of dam galleries, cracks often appear during the construction, operation and maintenance periods of dams [7–15]. For example, the Punt dal Gall double curvature arch dam, with a height of 130 m, on the border between Switzerland and Italy, has formed multiple cracks in the top arch of the gallery during its operation period, and the maximum crack opening in one year reached

1.5 mm [7]. On another gravity arch dam, with a height of 45 m, a horizontal crack has been observed in the upper gallery after 25 years of operation, which has propagated along almost the full length of the dam and the crack opening continued to increase at a rate of 0.1 mm per year [8]. The Dagangshan high arch dam in China experienced different degrees of gallery cracking in the top arch of the traffic gallery and foundation gallery, within two months of the completion of the concrete pouring, with the maximum crack depth reaching 3.7 m [9]. In addition, the Dworshak gravity dam in the western United States [10], Konar gravity dam in India [11], Sefid-Rud buttress dam in northern Iran [12], Xiaowan super-high arch dam [13] and Shaxikou gravity dam in China [14], all experienced varying degrees of gallery cracking problems during the construction or operation period. In order to accurately analyze the working behavior and cracking risk of the dam gallery structure, one of the prerequisites is the acquisition of real strength and fracture parameters for dam gallery concrete. In the traditional gallery cracking analysis process, the strength and fracture parameters of gallery concrete, at different ages, provided by different inspection units, are generally taken directly [15]. However, existing studies have shown [16–22] that the environmental temperature and humidity of concrete have a greater impact on the development of its strength and fracture parameters, hence, the gallery structure cracking analysis carried out directly, using the strength and fracture parameters at a specific age, may lead to the large deviations between the analysis results and the real anti-cracking safety factor of the gallery concrete. Therefore, it is necessary to propose a method for determining the real strength and fracture parameters of dam gallery concrete, considering the environmental temperature and humidity, which can provide a scientific basis for the cracking risk analysis and safety assessment of the dam gallery structure under actual conditions.

In order to quantitatively analyze the influence of environmental temperature on the strength of concrete, various scholars have carried out a large number of experimental studies [23–26]. For example, Kim et al. [23] carried out strength tests at ages of 1 day, 3 days, 7 days and 28 days, and under curing temperatures of 10 °C, 23 °C, 35 °C and 50 °C, for five types of concretes with different mixture proportions, which were Type I cement concrete with water–binder ratios of 0.4 and 0.5, Type V cement concrete with water–binder ratios of 0.4 and 0.5, and Type V cement concrete with water–binder ratios of 0.4 and a fly ash replacement rate of 15%. The research results show that high curing temperature conditions, such as 35 °C or 50 °C, are conducive to the development of early-age concrete strength. However, when the age is greater than 7 days, the strength of concrete subjected to high temperature conditions may be less than the strength of concrete under normal temperature conditions, such as 10 °C or 23 °C; the effect of the curing temperature on the strength of concrete varies with the variation of concrete mixture proportions. Castellano et al. [24] carried out compressive strength tests, at ages of 2 day, 7 days, 28 days, 90 days and 365 days, and with curing temperatures of 20 °C, 40 °C, 60 °C for cement pastes with water–binder ratios of 0.3, 0.4 and 0.5. It was found that when the water–binder ratio is 0.3, the compressive strength of cement paste at 2 days and 7 days increases with the increase in the curing temperature, while it decreases as the increase in curing temperature for cement paste at 28, 90 and 365 days. For cement pastes with water–binder ratios of 0.4 and 0.5, the effect of the curing temperature on cement strength is basically the same, but the compressive strength started to decrease with the increase in curing temperature, at the age of 7 days. Regarding the effect of curing temperature on concrete fracture parameters, Li et al. [27] cast and cured the dam concrete wedge-splitting specimens on the construction site, during the high temperature condition, in summer, and the low temperature condition, in winter, and carried out the fracture tests at different ages. The results show that the fracture parameters of concrete poured in summer are greater than those of the same age in winter. Yu and Ansari [28] performed fracture tests on three-point bending concrete beams, cured at 14 °C, 23 °C and 35 °C. It was established that for concrete of the same age, the higher the curing temperature, the greater the fracture energy of concrete. Mi et al. [29,30] carried out concrete wedge-splitting fracture tests under curing temperatures of 5 °C, 20 °C, 40 °C, and 60 °C and found that at early ages, the fracture parameters of concrete increase with the increase in the curing temperature, but from 14 days to 60 days of age, the fracture energy and equivalent fracture toughness of concrete under the curing condition of 60 °C are smaller than the fracture parameters of the concrete at the same age when the curing temperature is 40 °C; hence, the high curing temperature condition may have an adverse effect on the development of concrete fracture parameters. From the above research results, it can be seen that a higher curing temperature can promote the development of early-age concrete strength and fracture parameters, but an excessive curing temperature may have an adverse effect on the growth of later-age concrete's mechanical properties. Furthermore, for different concretes, the curing temperature has different effects on the development of strength and fracture parameters. Therefore, for important projects, such as dams, it is necessary to carry out the tests under different curing temperature conditions, on the construction site, to clarify the impact of the curing temperature on concrete strength and fracture parameters.

Similar to curing temperature, curing humidity also has a significant impact on the development of concrete strength [31–36] and fracture parameters [18,37,38]. Powers [31] found that when the relative humidity inside the concrete is lower than 80%, the hydration rate of the cementitious material will gradually decrease or even stop, which will slow down the development of the macro mechanical properties of the concrete. Liao et al. [32] carried out the compressive strength test of concrete, under curing temperatures of 12 °C, 27 °C and 40 °C, and curing humidity of 70%, 80%, 90% and 100%. The results show that under certain curing temperature conditions, the difference in the compressive strength of concrete at an early age, due to changes in curing humidity, is small. With the increase in age, the reduction proportion of compressive strength, due to the decrease in curing humidity, gradually increases. For example, when the curing temperature is 12 °C, the age is 17.33 days, and the curing humidity is 75%, the compressive strength is about 34% lower than the strength at the same age when the curing humidity is 100%. Liang et al. [33] performed the flexural tensile strength test of pavement cement concrete, when the curing humidity was 40%, 60%, and 100%, and found that the lower the relative humidity, the lower the flexural tensile strength of concrete at different ages, and the difference in flexural tensile strength grows with the increase in age. Compared with the experimental research carried out by scholars, on the influence of curing humidity on the strength of concrete, there are still few studies on the effect of environmental humidity on the fracture parameters of concrete. Mi et al. [18] carried out concrete wedge splitting fracture tests, with curing temperatures of 5 °C, 20 °C, 40 °C and 60 °C, and curing humidity of 30%, 50%, 70% and 98%. The results show that curing humidity has little effect on the fracture parameters of early-age concrete, but as the age increases, the lower the relative humidity, the lower the fracture toughness and fracture energy of concrete at each age, and the difference in fracture parameters becomes larger with the increase in age; that is, the detrimental effect of the decrease in curing humidity on the fracture parameters of concrete becomes more and more obvious. Other scholars [37,38] paid more attention to the influence of water content on concrete fracture parameters. For example, Bažant [37] carried out an experimental study on the effect of concrete internal water content on fracture energy, at different curing temperatures, and found that under normal temperature conditions, concrete internal water content has little effect on concrete fracture energy. As the curing temperature increases, the influence of the internal water content on the fracture energy of the concrete gradually increases. Lau et al. [38] also studied the influence of the internal water content of concrete on fracture toughness. The test results show that the fracture toughness of concrete decreases with the increase in its internal water content. In summary, the increase in curing humidity is conducive to the development of concrete strength and fracture parameters and its influencing effects vary, depending on the concrete material and curing temperature. For actual projects, concrete strength and fracture tests under different curing temperature and humidity conditions should be carried out at the same time, so as to provide a basis for the determination of concrete mechanical parameters under real environmental conditions.

This study used the gallery concrete produced by the on-site mixing plant of a dam under construction in southwest China as the research object, then designed and carried out the strength and fracture tests of gallery concrete, at various ages and under different curing temperature and humidity conditions. Based on the test results, theoretical study was carried out to establish the prediction formula of the strength and fracture parameters of the gallery concrete, under arbitrary temperature and humidity history conditions. Finally, combined with the measured temperature and humidity data inside and outside the dam gallery, as well as the temperature of the gallery concrete, the real strength and fracture parameters of the gallery concrete at any age were predicted.

This research aims at establishing the strength and fracture properties of gallery concrete under real environmental conditions. Both temperature and humidity effects on the development of mechanical parameters of gallery concrete have been studied experimentally and theoretically, and the prediction equations of strength and fracture parameters have been established accordingly, which can help to correct the deviations of mechanical parameters of gallery concrete determined by traditional methods, using the actual age. This research provides a basis for all parties involved in dam engineering to better analyze and control the gallery cracks, which often appear during the construction, operation and maintenance periods of dams, thereby ensuring the safety of the dam gallery structure, as well as the whole dam structure.

2. Materials and Methods

China's dam construction has entered the era of intelligence, and a complete set of technologies, such as intelligent water-cooling, intelligent grouting, and intelligent spraying, based on extensive monitoring measurements, real analyses, and intelligent decision making, as well as closed-loop feedback control theory, have been gradually formed [3]. As an integral part of this set of technologies, this study proposes a method for determining and applying the real strength and fracture parameters of dam gallery concrete considering the ambient temperature and humidity. The specific process is shown in Figure 1. This method can determine the reliable material parameters of the gallery concrete at any time under real environmental conditions, which can be used for real-time cracking risk analysis of the dam gallery structure, and can then provide a basis for the optimization of the dam structure's operating status, the formulation of the construction schedule, or the control of gallery's temperature and humidity, etc.

This article focuses on the method of determining the real strength and fracture parameters of dam gallery concrete considering the environmental temperature and humidity. This method directly uses the gallery concrete produced by the on-site mixing plant of the dam project to pour concrete specimens, then carries out tests on the strength and fracture performance of gallery concrete under different ages and curing temperature and humidity conditions. Furthermore, a temperature–humidity maturity method is used to analyze the test results and determine the prediction equations of the gallery concrete strength and fracture parameters. Finally, based on the measured temperature and humidity data inside and outside the dam gallery, and the measured temperature data of the gallery concrete, the real strength and fracture parameter development curves of the gallery concrete at any age are established, which can be used to evaluate the cracking risk of the gallery structure under different working conditions.

2.1. Strength and Fracture Tests

2.1.1. Raw Materials and Mixture Proportions

The test directly used the gallery concrete with a maximum aggregate size of 40 mm and a slump of 70–90 mm produced by the concrete mixing system at the construction site of a dam project in southwest China. The water–binder ratio of concrete is 0.42, and the compressive strength for 180 days is 40 MPa. The mixture proportion of dam gallery concrete is shown in Table 1. The cement is P·LH42.5 low-heat Portland cement with a density of 1450 kg/m^3, which is produced by Sichuan Jiahua Jinping Special Cement Co., Ltd. located in Mianning County, China. The fly ash is produced in Qujing City, China, by Xuanwei Zhonghecheng Building Materials Co., Ltd. The fly ash conformed to Type F Class I, according to the standard of Chinese hydraulic engineering industry DL/T 5055-2007 [39]. The replacement rate of fly ash was 35% of the cementitious material, which was used to reduce the hydration heat release of large volume concrete and save cement. The chemical composition of the cement and fly ash is shown in Table 2. Both coarse aggregate and fine aggregate were limestone produced by a Hangudi sand and aggregate production system located in Qiaojia County, China, with a sand ratio of 34%. Air-entraining agent and water reducing agents were GYQ-I air-entraining agent and SBTJM-II retarder type II high-range water reducing admixture, respectively, both of which are produced by Sobute New Materials Co., Ltd. in Nanjing City, China.

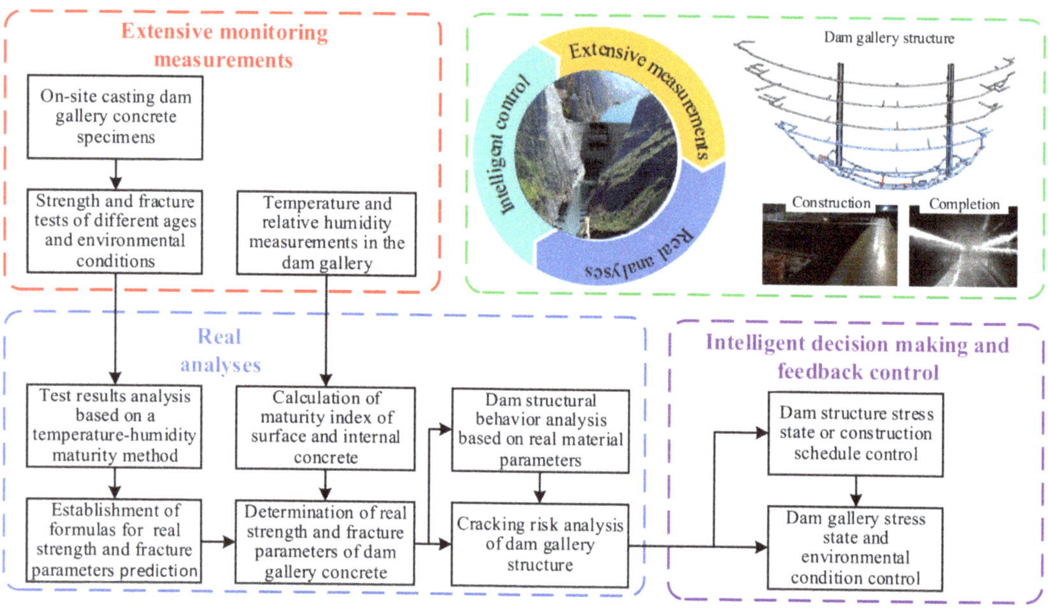

Figure 1. Determination and application method in intelligent dam construction platform (IDAM 2.0) of real strength and fracture parameters of dam gallery concrete considering ambient temperature and humidity.

Table 1. Mixture proportion of dam gallery concrete.

Water (kg/m^3)	Cement (kg/m^3)	Fly Ash (kg/m^3)	Sand (kg/m^3)	Aggregate (kg/m^3)		High-Range Water Reducing Admixture (kg/m^3)	Air-Entraining Agent (kg/m^3)
				5–20 (mm)	20–40 (mm)		
112	173	93	687	669	669	1.33	0.093

Table 2. Chemical composition of cementitious material (% by mass).

Composition	CaO	SiO_2	Al_2O_3	Fe_2O_3	MgO	SO_3	K_2O	Na_2O	R_2O	Others
Cement	60.66	22.92	3.27	4.58	4.02	2.50	0.46	0.17	0.47	0.95
Fly ash	3.25	55.58	21.68	8.99	1.30	0.35	1.16	0.20	0.96	6.53

2.1.2. Test Scheme

To study the effects of curing temperature and humidity on the strength and fracture parameters of the gallery concrete, this research designed the strength and fracture tests of gallery concrete for 7 days, 14 days, 28 days, 60 days and 90 days and under the following 5 different kinds of curing conditions: curing humidity of 98%, curing temperature of 10 °C, 20 °C, 40 °C, and curing temperature of 20 °C, curing humidity of 30%, 60%. The detailed test scheme is shown in Table 3. The test used the large-scale temperature and humidity environmental chambers manufactured by Chongqing Haoyuan Environmental Test Equipment Co., Ltd. located in Chongqing City, China, to accurately control the curing temperature and humidity of the specimens. It should be noted that the temperature and humidity range designed by the experiment is basically consistent with the temperature and humidity variation range of the construction site of dam projects over the years and the actual temperature and humidity history of the gallery concrete. Therefore, the test results can better reflect the influence of actual environmental conditions on the strength and fracture parameters of the gallery concrete.

Table 3. Test scheme for the strength and fracture parameters of the environmental chamber-cured dam gallery concrete.

Curing Conditions	Temperature (°C)	Humidity (%)	Specimen Size (mm × mm × mm)	Number of Specimens	Number of Groups	Measured Parameters	Test Ages (Days)
T10RH98	10	98	150 × 150 × 150 φ150 × 300 330 × 300 × 120	30 30 15	10 5 5	f_c, f_t f_{acs}, E_c K_{IC}, G_F	7, 14, 28, 60, 90
T20RH98	20	98	150 × 150 × 150 φ150 × 300 330 × 300 × 120	30 30 15	10 5 5	f_c, f_t f_{acs}, E_c K_{IC}, G_F	7, 14, 28, 60, 90
T40RH98	40	98	150 × 150 × 150 φ150 × 300 330 × 300 × 120	30 30 15	10 5 5	f_c, f_t f_{acs}, E_c K_{IC}, G_F	7, 14, 28, 60, 90
T20RH60	20	60	150 × 150 × 150 φ150 × 300 330 × 300 × 120	30 30 15	10 5 5	f_c, f_t f_{acs}, E_c K_{IC}, G_F	7, 14, 28, 60, 90
T20RH30	20	30	150 × 150 × 150 φ150 × 300 330 × 300 × 120	30 30 15	10 5 5	f_c, f_t f_{acs}, E_c K_{IC}, G_F	7, 14, 28, 60, 90

φ is the diameter of the specimen.

In this study, a total of 300 gallery concrete strength specimens and 75 fracture specimens were cast. There are 60 strength specimens and 15 fracture specimens of 5 different test ages for each temperature and humidity condition. The three different types of specimens cast in the experiment are shown in Figure 2. Among them, the cube specimens with dimensions of 150 mm × 150 mm × 150 mm are used to determine the compressive strength f_c and splitting tensile strength f_t of concrete. The cylinder specimens with a size of φ150 mm × 300 mm are used to measure the axial compressive strength f_{acs} and static compressive elastic modulus E_c of concrete. The wedge-splitting specimen with a size of 330 mm × 300 mm × 120 mm and initial crack length to height ratio of 0.4 is used to determine the equivalent fracture toughness K_{IC} and fracture energy G_F of the concrete.

2.1.3. Specimen Preparation

The production, transportation and pouring of concrete is shown in Figure 3. To ensure the test results obtained in this research can be directly applied to the cracking risk analysis of the dam gallery structure, the concrete used in this test and in the construction of the dam gallery was produced at the same time and under the same conditions. After picking up the concrete material at the outlet of the mixing plant on the construction site, the concrete was transported to the pouring site by a thermal insulation dump truck and unloaded on the floor which had been sprinkled with water in advance, and then started to cast the 3 types of specimens at the same time, namely 150 cube specimens with a size of 150 mm × 150 mm × 150 mm, 150 cylinder specimens with a size of φ150 mm × 300 mm, and 75 wedge-splitting specimens with a size of 330 mm × 300 mm × 120 mm.

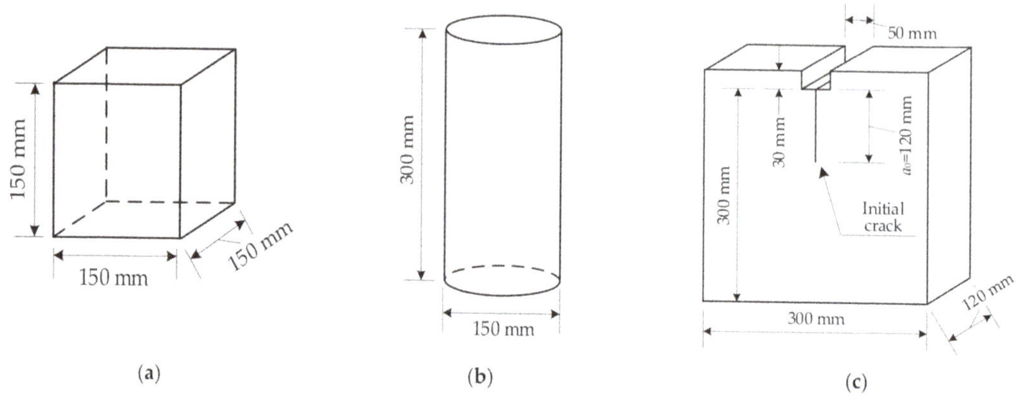

Figure 2. Three types of specimens cast on-site: (**a**) cube specimen; (**b**) cylinder specimen; (**c**) wedge-splitting specimen.

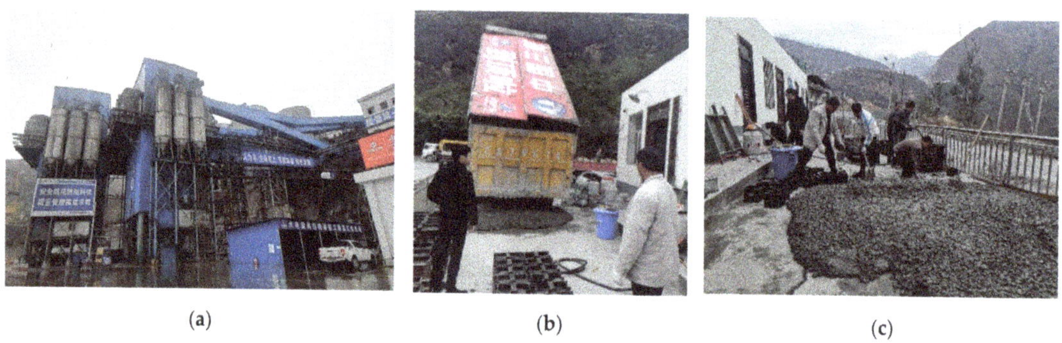

Figure 3. Production, transportation and pouring of concrete: (**a**) concrete mixing system at the construction site; (**b**) concrete transportation with thermal insulation dump truck; (**c**) pouring of specimens.

The cube and cylinder specimens were cast using high-quality engineering plastic molds, and the wedge-splitting specimens were cast using steel molds. A groove was reserved at the top of the steel mold for fixing a square tube with a size of 30 mm × 50 mm, so that the groove (as shown in Figure 2c) could be formed on the top of the specimen. The two side plates of the steel mold were designed with a seam with a height of 120 mm and width of 3 mm, which was used to embed a 3-mm thick steel plate with V-shaped bottom, so an initial crack (as shown in Figure 2c) could be formed in the molded specimen. Before casting the specimens, each mold needed to be brushed with oil in advance for later demolding. To minimize the impact of the casting batch and the environmental conditions during pouring on the test results, all the specimens in this test were cast at one time, that is, the casting was completed within 2 h of the concrete being transported to the pouring site.

To reduce the possible influence of the loss of moisture on the test results, all specimens were covered with plastic wrap after pouring, as shown in Figure 4a. After curing for 24 h, the specimens were removed from the mold and placed in the large environmental chamber shown in Figure 4b with the curing temperature and humidity set in advance. During the operation of the environmental chamber, the temperature and humidity in the environmental chamber were regularly monitored, to ensure that the curing conditions always met the design requirements. Figure 4c shows an example of three different types of specimens cured in an environmental chamber.

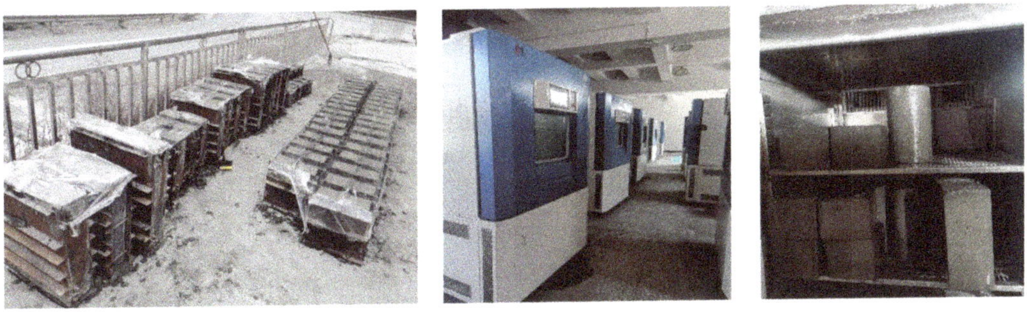

(**a**) (**b**) (**c**)

Figure 4. Curing of concrete specimens: (**a**) the surface of the specimen was covered with plastic wrap to prevent the loss of concrete moisture; (**b**) large temperature and humidity environmental chamber; (**c**) an example of three different types of specimens cured in the environmental chamber.

2.1.4. Test Methods

In this study, the strength tests were carried out on a Servo Hydraulic Universal Testing Machine with a range of 1000 kN, manufactured by Shanghai Hualong Test Instruments Corporation in Shanghai City, China, and the fracture tests were conducted on a computer-controlled electronic universal testing machine with a range of 300 kN, produced by Jiangsu Tianyuan Test Equipment Co., Ltd. in Yangzhou City, China, as shown in Figure 5a,b, respectively. During the fracture test, the data acquisition system shown in Figure 5c was used to record the applied vertical load F and crack mouth opening displacement ($CMOD$) by a 10 kN capacity force transducer (Shanghai East China Electronic Instrument Factory, Shanghai City, China) placed on the stiff steel profile and a clip gauge (Central Iron Steel & Research Institute, Beijing City, China) with a range of 4 mm installed ahead of the initial crack.

(**a**) (**b**) (**c**)

Figure 5. Test equipment: (**a**) Servo Hydraulic Universal Testing Machine; (**b**) computer-controlled electronic universal testing machine; (**c**) data acquisition system.

The strength tests were carried out in accordance with the relevant test standard DL/T 5150-2017 [40] of Chinese hydraulic engineering industry. According to this standard, the average value of the test results of 3 specimens in the same group for a certain test age and curing condition was taken as the final strength result.

The compressive strength tests adopted the stress loading method with a loading rate of 0.3 MPa/s. Before the test, the specimen was placed on the pressure plate of the testing machine, ensuring that the pressure surface of the specimen was not the pouring surface. Then, the load was applied with the preset rate until the specimen was broken, the failure load was recorded, the specimen was taken out, then all the above steps were repeated, and all the compressive strength tests for certain groups were completed. After the tests were completed, the failure loads were substituted into Equation (1) to calculate the compressive strength.

$$f_c = \frac{P}{A} \tag{1}$$

where f_c is the compressive strength, MPa; P is the failure load, kN; A is the pressure bearing area of the specimen, mm^2.

The splitting tensile strength tests also adopted the stress loading method with a loading rate of 0.04 MPa/s. Before the test, the center lines of the specimen were drawn to ensure that the splitting surface was parallel to the pouring surface. We then put the specimen in the center of the pressure plate of the testing machine and placed two rectangular strips (with cross-sectional dimensions of 5 mm × 5 mm) between the upper and lower pressure plates and the specimen. The test was conducted until the specimen was broken, the test data was then recorded, taken out the test piece, the above steps repeated, and all splitting tensile tests were completed. Substituting the failure load obtained from the tests into Equation (2) allows us to calculate the splitting tensile strength of the gallery concrete.

$$f_t = \frac{2P}{\pi A} = 0.637\frac{P}{A} \tag{2}$$

where f_t is splitting tensile strength of concrete; P is failure load; A is cross-sectional area of the cube.

The test method of the axial compressive strength test was the same as the compressive strength test, and the loading rate was also 0.3 MPa/s. Substituting the failure load obtained from the test into Equation (3), we can calculate the axial compressive strength.

$$f_{acs} = \frac{P}{A} \tag{3}$$

where f_{acs} is the axial compressive strength, MPa; P is the failure load, kN; A is the pressure bearing area of the specimen, mm^2.

For the test of static compressive elastic modulus, the strain gauge method was adopted. After starting the test, we preloaded the specimen, then applied the load step by step at 0.5 MPa intervals, recorded the strain value of each step until the pressure value exceeded 40% of the axial compressive strength. Afterwards, the load remained at a constant rate of 0.3 MPa/s until the specimen was broken. The static compressive elastic modulus E_c can be calculated according to the stress–strain relationship obtained during the step-by-step loading process.

The wedge-splitting method was adopted for fracture tests and a diagram of the loading device is shown in Figure 6a. The test data was collected by a data acquisition system, as shown in Figure 5c. The vertical loading was applied with a constant displacement rate of 0.05 mm/min. In order to eliminate the influences of the self-weight of the specimen and the vertical component of the applied load on the fracture test results, two hinge roller supports were placed at the quarter points of the bottom of the specimen, which were also aligned with the centers of the roller bearing, as shown in Figure 6a. Figure 6b shows a schematic diagram of the F_H-$CMOD$ curve, where F_H was obtained by transforming the vertical load F collected by the force transducer after considering the wedge angle of 15°, and the crack mouth opening displacement $CMOD$ was directly recorded by the clip gauge.

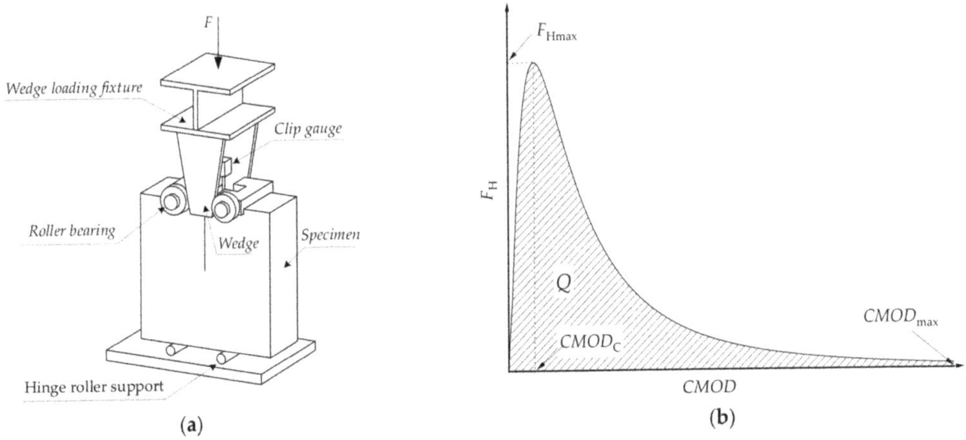

Figure 6. Wedge-splitting fracture test: (**a**) diagram of the loading device and (**b**) schematic diagram of the F_H-*CMOD* curve.

The fracture parameter of equivalent fracture toughness was calculated by the expressions proposed by Xu et al. [41], as shown in Equations (4)–(8), as follows:

$$K_{IC} = \frac{F_{Hmax} \times 10^{-3}}{th^{1/2}} f(\alpha) \quad (4)$$

$$F_{Hmax} = \frac{(F_{max} + mg \times 10^{-3})}{2 \tan 15^{\circ}} \quad (5)$$

$$f(\alpha) = \frac{3.675[1 - 0.12(\alpha - 0.45)]}{(1-\alpha)^{3/2}}, \alpha = \frac{a_c}{h} \quad (6)$$

$$a_c = (h + h_0)\left(1 - \sqrt{\frac{13.18}{CMOD_c \times E \times t / F_{Hmax} + 9.16}}\right) - h_0 \quad (7)$$

$$E = \frac{1}{tc_i}\left[13.18\left(1 - \frac{a_0 + h_0}{h + h_0}\right)^{-2} - 9.16\right] \quad (8)$$

where K_{IC} in Equation (4) is the equivalent fracture toughness, MPa·m$^{0.5}$; F_{Hmax} is the maximum horizontal load, kN; t and h are the thickness and effective height of the specimen, which equal 0.12 m and 0.3 m, respectively, in this study; $f(\alpha)$ is a dimensionless factor that can be calculated by Equation (6); F_{max} in Equation (5) is the maximum applied vertical load, kN; mg is the weight of the wedge loading fixture, N; a_c in Equation (6) is effective crack length, m; $CMOD_c$ is critical crack mouth opening displacement, μm; E is the modules of elasticity calculated by Equation (8), GPa; h_0 = 0.006 m, which is the thickness of the thin steel plate where the clip gauge is installed; c_i in Equation (8) is the initial compliance of the ascending branch of the F_H-*CMOD* curve (shown in Figure 6b); a_0 equals 0.12 m, which is the initial crack length.

In addition to the effective fracture toughness, another parameter to evaluate the fracture performance of concrete is the fracture energy G_F, which is the average energy consumed per unit area during the crack propagation process. According to its definition, the fracture energy calculation formula is as follows:

$$G_F = \frac{Q}{A_0} = \frac{\int_0^{CMOD_{max}} F_H dCMOD + mgCMOD_{max}}{t(h - a_0)} \quad (9)$$

where G_F is the fracture energy of concrete, N/m; Q is the work of fracture, J, which is the area enclosed by the F_H-$CMOD$ curve shown in Figure 6b; A_0 is the ligament area of the specimen, m^2; $CMOD_{max}$ is the maximum crack mouth opening displacement, m. The meanings of the remaining variables are consistent with the former equations.

2.2. Temperature–Humidity Maturity Method

2.2.1. Maturity Functions

The maturity of concrete is the extent of the development of its properties. Based on this concept, the maturity method which establishes the relationship between the concrete properties and maturity index has been proposed by scholars [42–46]. The maturity method can be used to predict the strength parameters of concrete or other properties related to the chemical reactions of the cementitious material. With the development of the maturity method, various maturity functions [46] have been proposed to calculate the maturity index. Among the different maturity functions, the equivalent age model proposed by Freiesleben Hansen and Pedersen [45] based on the Arrhenius equation has been widely used in academia and engineering. The calculation formula of the maturity index, that is, the equivalent age is as follows:

$$t_e = \sum \exp\left[\frac{E_a}{R}\left(\frac{1}{273+T_r} - \frac{1}{273+T}\right)\right]\Delta t \tag{10}$$

where t_e is the equivalent age at a specified temperature, hours or days; T is the average curing temperature of the concrete during time interval Δt, °C; T_r is the specified temperature, °C, generally T_r = 20 °C is adopted; E_a is the activation energy, J/mol, when $T \geq 20$ °C, E_a = 33,500 − T) J/mol; R is the gas constant, 8.3144 J/mol/K; Δt is the time interval, hours or days.

The above maturity function assumes that the cementitious material has sufficient water supply during the hydration process, so the effect of curing humidity on the strength of concrete can be neglected. However, existing studies [18,31–38] have shown that curing humidity also has a great influence on the development of concrete properties. In practice, the curing humidity of concrete is often difficult to reach a state where the relative humidity equals to 98% in a laboratory. Therefore, if the traditional maturity function is directly used to calculate the maturity index and then predict the concrete properties for a real structure, it may cause a large deviation between the predicted results and the actual concrete properties. Especially for surface concrete, the development of its properties is often greatly affected by the environmental humidity. Therefore, when using the maturity method to predict the properties of concrete, the effect of curing humidity should better be considered. Some scholars have carried out the humidity correction work based on the traditional temperature maturity functions. However, the parameter calibration of those modified functions is often complicated, hence, the direct application of these functions to the actual structure can be difficult. In recent years, in order to characterize the influence of environmental temperature and humidity on concrete fracture parameters, based on a humidity rate function, Mi et al. [30] established the following temperature–humidity maturity function, which reads:

$$t_e = \sum \frac{1}{\alpha_u - \alpha_0}\left(\frac{T}{T_r}\right)^{-m} \exp\left[\frac{E_a}{R}\left(\frac{1}{273+T_r} - \frac{1}{273+T}\right)\right] \times \left[H\Delta t + \frac{(1-H)}{\gamma}\ln(1+\gamma\Delta t)\right] \tag{11}$$

where α_u is the final degree of hydration; α_0 is the initial degree of hydration; m is defined as a microscopic coefficient; γ is the water diffusion coefficient; H is the average value of the curing humidity of the concrete during time interval Δt. The remaining variables are the same as the ones in Equation (10). To ensure the consistency of Equations (10) and (11), and for simplicity, the constants in Equation (11) are chosen as follows in this study: $\alpha_u = 1$, $\alpha_0 = 0$, $m = 0$, $\gamma = 1$. It should be noted that normally these constants can be calibrated by the best fit of the test results. However, these parameters only change the calculation results of equivalent age and will not affect the prediction results of concrete properties, hence, they can be pre-set according to the relevant research results [30].

2.2.2. Strength–Maturity Relationships

Once the maturity index is determined, the appropriate strength–maturity relationship should be selected to establish the equation for the prediction of the concrete properties. At present, the commonly used strength–maturity relationship equations mainly have the following three forms: logarithmic, exponential and hyperbolic functions. The logarithmic relationship reads [47–49], as follows:

$$S = a + b\log(M) \tag{12}$$

where S represents the concrete properties such as strength or fracture parameters; M is the maturity index such as temperature–time factor or equivalent age; a and b are the coefficients determined by best fitting of the test results. The prediction result given by the logarithmic function would monotonically increase with the increase in the maturity index, which is a shortcoming of the form of the logarithmic function for concrete with long curing age, while the exponential and hyperbolic functions could give the ultimate stable predictions. The expressions [46,50] are as follows:

$$S = S_{\infty}e^{-\left(\frac{\tau}{M}\right)^{d}} \tag{13}$$

where S_{∞} represents the ultimate material properties of concrete; τ is a time characteristic parameter and d is a shape factor. They can be determined by the test results.

$$S = \frac{M}{1/A_r + M/S_{\infty}} \tag{14}$$

where A_r is the initial slope of strength–maturity curve and other variables share the same definitions as the former equations.

In this study, the above three functions were adopted to fit the test results, and it was established that for the equivalent age range considered in this paper, the logarithmic function provides the best overall fitting accuracy. Hence, for the sake of simplicity, only the predictions based on the logarithmic function are given in this article.

3. Results and Discussion

3.1. Test Results

3.1.1. Strength Test Results

Table 4 shows the test results for compressive strength f_c, splitting tensile strength f_t, axial compressive strength f_{acs} and static compressive elastic modulus E_c of environmental chamber-cured dam gallery concrete, with different ages and different curing conditions.

Figure 7 shows the variation of strength parameters of low-heat cement gallery concrete, with age, when the curing humidity is 98% and the curing temperature is 10 °C, 20 °C and 40 °C. It can be seen from Figure 7 that as the age increases, the strength parameters of concrete at different curing temperatures continue to increase; not all of the strength parameters reached their final stable values when the test age reached the designed maximum age of 90 days. This is because the low-heat cement has a relatively high content of dicalcium silicate and a slow hydration rate, so the strength can continue to increase. Taking the compressive strength at a curing temperature of 20 °C (in Figure 7a as an example), the strength of the specimen increased to 17.61 MPa within 7 days of the completion of pouring, and the growth rate was 2.52 MPa/day. With the increase in age, the growth rate of compressive strength decreased to 0.92 MPa/day, 0.43 MPa/day, 0.28 MPa/day and 0.11 MPa/day, in the several designed test age intervals of 8–14 days, 15–28 days, 29–60 days and 61–90 days. This shows that as the age increases, the growth rate of strength gradually decreases, so the concrete strength parameters will eventually reach their stable values.

Table 4. Strength test results of environmental chamber-cured concrete at different temperatures and humidity.

Curing Conditions	Test Age (Days)	f_c (MPa) [Mean (Standard Deviation)]	f_t (MPa) [Mean (Standard Deviation)]	f_{acs} (MPa) [Mean (Standard Deviation)]	E_c (GPa) [Mean (Standard Deviation)]
T10RH98	7	11.92 (0.53)	0.98 (0.05)	7.71 (0.23)	22.27 (0.59)
	14	16.58 (0.16)	1.29 (0.07)	13.37 (0.44)	26.93 (1.67)
	28	23.93 (1.52)	1.93 (0.11)	17.77 (0.65)	31.05 (1.15)
	60	30.88 (1.50)	2.25 (0.14)	24.89 (1.77)	35.23 (1.27)
	90	33.92 (1.88)	2.51 (0.13)	27.99 (1.93)	38.35 (1.24)
T20RH98	7	17.61 (0.76)	1.26 (0.05)	11.61 (0.65)	25.50 (0.37)
	14	24.03 (1.10)	1.86 (0.05)	19.03 (1.29)	30.25 (1.10)
	28	30.05 (0.95)	2.26 (0.14)	25.69 (0.84)	35.42 (0.47)
	60	38.86 (1.93)	2.63 (0.09)	30.40 (0.77)	39.73 (0.70)
	90	42.03 (1.15)	2.93 (0.12)	35.12 (1.58)	41.15 (0.56)
T40RH98	7	24.41 (1.58)	1.76 (0.09)	18.77 (1.08)	30.02 (1.57)
	14	33.70 (1.45)	2.34 (0.11)	27.28 (0.64)	35.62 (1.48)
	28	37.43 (1.46)	2.73 (0.16)	30.68 (0.70)	39.20 (0.64)
	60	46.68 (2.27)	3.16 (0.13)	37.46 (1.18)	44.40 (1.25)
	90	50.79 (1.31)	3.31 (0.13)	39.67 (1.48)	46.77 (2.03)
T20RH60	7	15.49 (0.81)	1.27 (0.05)	10.71 (0.63)	24.25 (0.61)
	14	18.57 (0.94)	1.55 (0.04)	14.90 (0.44)	27.12 (0.14)
	28	25.05 (0.89)	1.86 (0.06)	20.77 (1.02)	31.43 (0.94)
	60	32.60 (1.50)	2.28 (0.12)	26.37 (1.29)	37.08 (1.07)
	90	36.30 (1.94)	2.51 (0.08)	29.06 (1.01)	38.82 (1.13)
T20RH30	7	14.32 (0.89)	1.24 (0.08)	10.78 (0.54)	24.00 (0.58)
	14	16.97 (0.76)	1.45 (0.08)	12.56 (0.77)	25.98 (0.57)
	28	19.89 (0.97)	1.62 (0.11)	16.29 (0.74)	28.12 (0.91)
	60	25.20 (0.81)	1.91 (0.08)	21.32 (1.41)	33.08 (0.14)
	90	30.73 (1.13)	2.23 (0.10)	24.83 (1.38)	35.03 (1.38)

On the other hand, when comparing the compressive strength of concrete under different curing temperature conditions (as in Figure 7a), we find that the higher the curing temperature, the greater the strength of the concrete at the same age, and the faster the strength growth rate. This is because a higher curing temperature speeds up the rate of hydration of the cementitious material. It should be noted that excessive curing temperatures may cause the deterioration of the internal microstructure of concrete, which will adversely affect the final strength and even have the opposite effect, a reverse growth phenomenon of concrete strength decreasing with age. It is generally believed that only when the curing temperature reaches 60 °C, will it have an adverse effect on the development of concrete strength. Therefore, in the curing temperature range of 10 °C to 40 °C, there will be no reduction in strength. It can be inferred that if the differences in early-stage pore characteristics and microcracks, that may be introduced by the difference in curing temperature, are ignored, as the age increases, the compressive strength of concrete under different curing temperatures will gradually approach its final compressive strength.

Figure 8 shows the variation of compressive strength, splitting tensile strength, axial compressive strength and elastic modulus of low-heat cement gallery concrete, with age, when the curing temperature is 20 °C and the curing humidity is 30%, 60% and 98%. It can be seen from Figure 8 that the strength development of low-heat cement concrete is closely related to its curing humidity. Under different curing humidity conditions, the growth law of each strength parameter with age is basically the same, which means that the greater the curing humidity, the faster the growth rate of strength. From the previous analysis, it is known that when the curing humidity is 98%, the compressive strength growth rate, within the interval of each test age designed in this study, would gradually decrease from

2.52 MPa/day to 0.92 MPa/day, 0.43 MPa/day, 0.28 MPa/day and 0.11 MPa/day. And when the curing humidity drops to 30%, the compressive strength growth rate in each age period decreases from 2.05 MPa/day to 0.38 MPa/day, 0.21 MPa/day, 0.17 MPa/day and 0.18 MPa/day. Comparing the above two groups, only in the last 30 days of age, from 61 to 90 days, the strength growth rate under a curing humidity of 30% is greater than one of 98% humidity. The main reason for this phenomenon is that, at the beginning, the water required for the hydration reaction of concrete cementitious materials, under low humidity conditions, may be insufficient, thus, delaying the growth of concrete strength. In the later stage, the hydration reaction of the cementitious material, under high humidity conditions, is near completion, so the strength growth rate is greatly reduced. However, although water is insufficient under low humidity conditions, there are still more cementitious materials that can participate in hydration, under conditions of insufficient water, resulting in its strength growth rate exceeding that of high humidity conditions. Within 7 days of the age of 7 to 14 days, the low-humidity curing condition, with a humidity of 30%, reduces the growth rate of compressive strength by 58.72%; hence, the influence of humidity on the strength of concrete cannot be ignored.

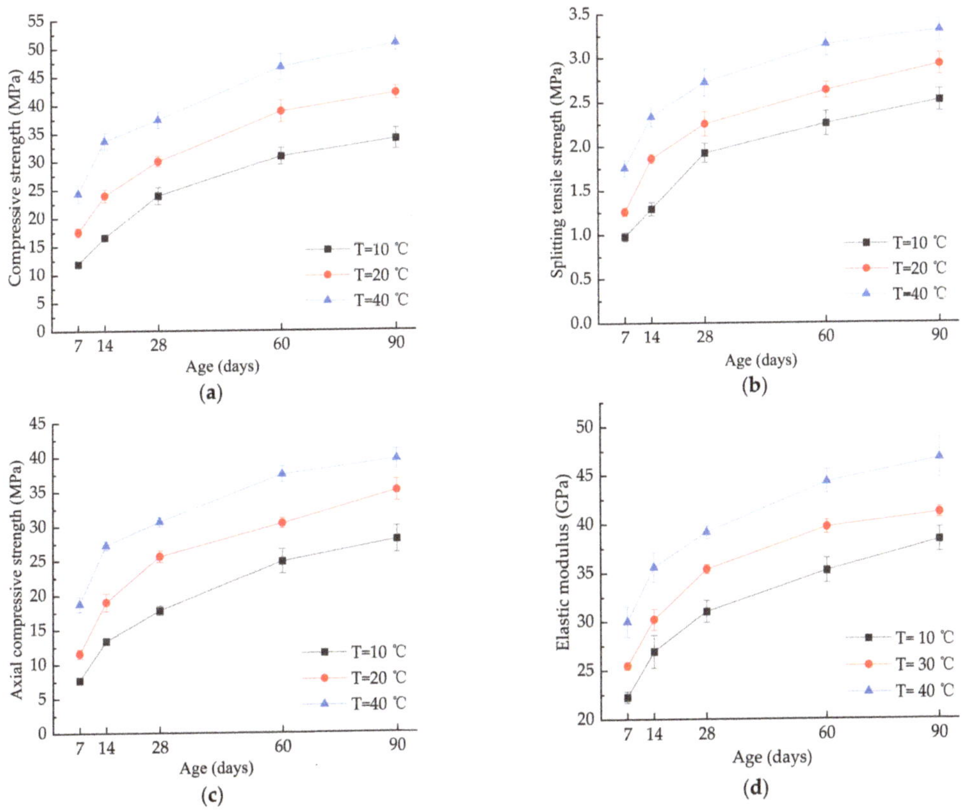

Figure 7. The variation of strength parameters of low-heat cement gallery concrete with age when the curing humidity is 98% and curing temperature is 10 °C, 20 °C and 40 °C: (**a**) compressive strength; (**b**) splitting tensile strength; (**c**) axial compressive strength and (**d**) static compressive elastic modulus.

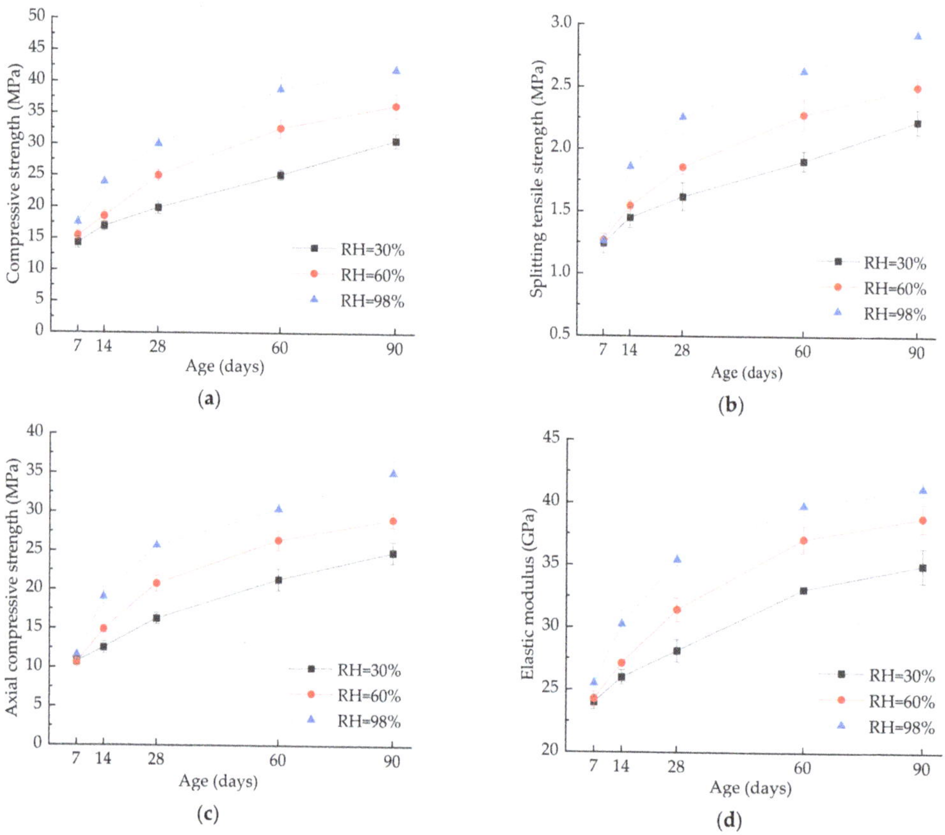

Figure 8. The variation of strength parameters of low-heat cement gallery concrete with age when the curing temperature is 20 °C and curing humidity is 30%, 60% and 98%: (**a**) compressive strength; (**b**) splitting tensile strength; (**c**) axial compressive strength and (**d**) static compressive elastic modulus.

When comparing the various strength parameters of concrete under different curing humidity conditions (as in Figure 8), at the age of 7 days, the difference in concrete strength parameters, caused by the variation in curing humidity, is almost negligible. This is mainly because at an early age, the moisture inside the concrete can already meet the hydration reaction demand of the cementitious material. However, as the age increases, the adverse effects of insufficient maintenance humidity on the strength parameters gradually appear. As the age increases, the difference between the strength parameters, under different humidity conditions, shows a law that first expands and then reduces. Taking Figure 8a, the compressive strength test results of differing curing humidity, as an example, when the curing humidity is 30% and 60%, the compressive strength at the age of 7 days is reduced by 18.68% and 12.04%, respectively, than when the humidity is 98%. The reduction at 28 days expanded to 33.81% and 16.64%, while the reduction in compressive strength at 90 days, due to insufficient humidity, was reduced to 26.89% and 13.63%.

3.1.2. Fracture Test Results

Figure 9 shows the failure modes of the fracture surface of the low-heat cement concrete fracture test specimens, under different ages and curing conditions. Figure 9a–e are the fracture surfaces of low-heat cement gallery concrete at 7 days, 14 days, 28 days, 60 days, and 90 days, when the curing temperature is 20 °C and the curing humidity is 98%. Figure 9f,c,g are the fracture surfaces at 28 days, when the curing humidity is 98%

and the curing temperature is 10 °C, 20 °C and 40 °C. Figure 9c,h,i are the fracture surfaces of the low-heat cement gallery concrete, when the curing temperature is 20 °C and the curing humidity is 98%, 60% and 30%, respectively. Comparing the fracture surfaces of different ages under the same curing conditions (as in Figure 9a–e), the aggregates on the fracture surface of the wedge–stretch specimen, at the early age, are basically undamaged, and the fracture surface is relatively rough. This means that the early-age wedge-splitting specimens are mainly damaged by the interface between the aggregate and the mortar. As the age increases, the bond strength between mortar and aggregate gradually increases, and the failure mode of the fracture surface gradually changes, from interface failure to aggregate splitting. As shown in Figure 9e, the fracture surface of the 90 days specimen is relatively flat, and it can be clearly observed that there are more aggregate splits on the fracture surface.

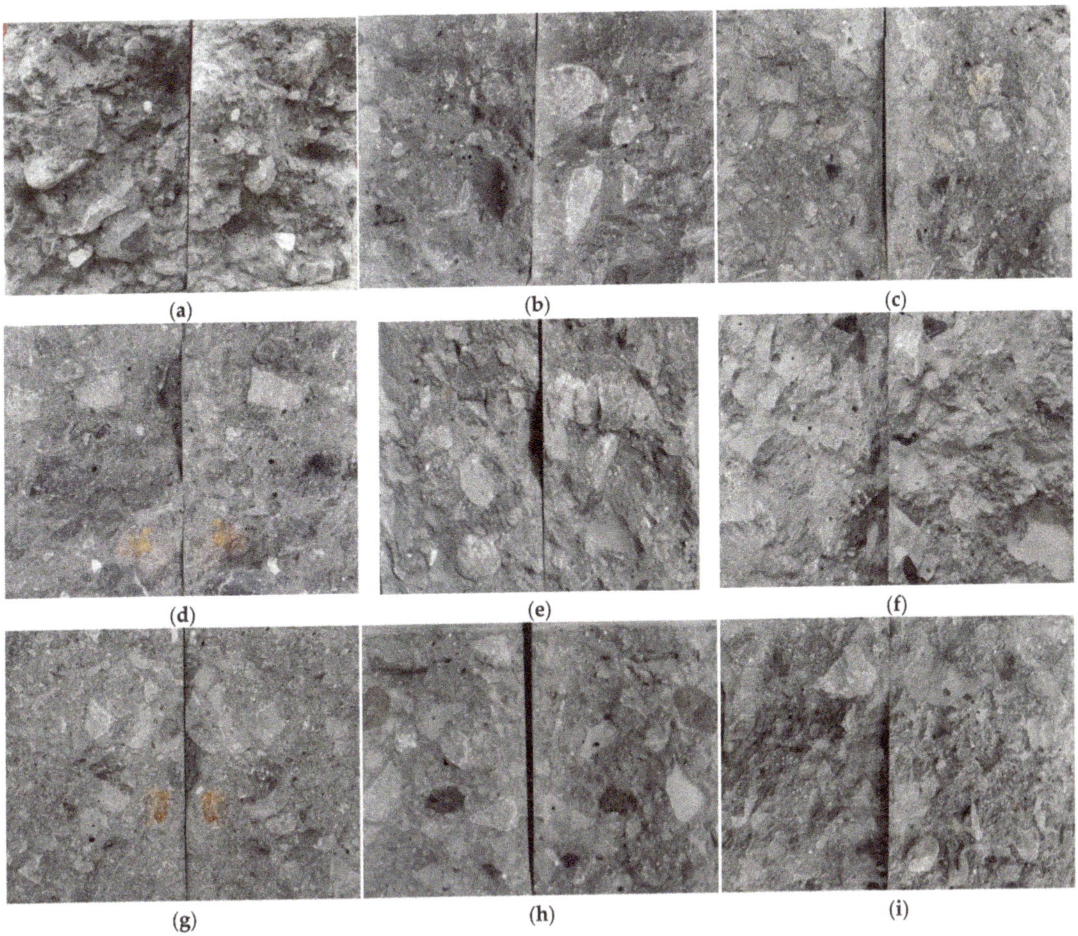

Figure 9. Fracture surfaces of wedge-splitting specimens at different ages and under different curing conditions: (**a**) T = 20 °C, RH = 98%, Age = 7 days; (**b**) T = 20 °C, RH = 98%, Age = 14 days; (**c**) T = 20 °C, RH = 98%, Age = 28 days; (**d**) T = 20 °C, RH = 98%, Age = 60 days; (**e**) T = 20 °C, RH = 98%, Age = 90 days; (**f**) T = 10 °C, RH = 98%, Age = 28 days; (**g**) T = 40 °C, RH = 98%, Age = 28 days; (**h**) T = 20 °C, RH = 60%, Age = 28 days; (**i**) T = 20 °C, RH = 60%, Age = 28 days.

In addition to age factor, curing temperature and humidity would also affect the failure modes of the fracture surfaces of the specimens. Compare Figure 9c,f,g, the fracture surface under different curing temperature conditions at 28 days, and Figure 9c,h,i, the fracture surface under different curing humidity conditions at 28 days; with the increase in temperature and humidity, the failure mode of the fracture surface gradually changes, from the interface failure between the aggregate and the mortar to the splits of the aggregates. This is mainly because higher a curing temperature and humidity contribute to the growth of the bond strength of aggregate and mortar, which leads to a macroscopic change in the failure mode of the fracture surface.

Table 5 shows the fracture test results of low-heat cement concrete, at different ages and under different curing conditions. The equivalent fracture toughness K_{IC} and fracture energy G_F are calculated by Equations (4) and (9), respectively. The test results listed in Table 5 are the average of the test results of three specimens in the same group, and the numbers in parentheses are the standard deviations of the test results.

Table 5. Fracture parameters of gallery concrete at different ages and under different curing conditions.

Curing Conditions	Age (Days)	F_{Hmax} (kN) [Mean (Standard Deviation)]	$CMOD_C$ (μm) [Mean (Standard Deviation)]	a_c (mm) [Mean (Standard Deviation)]	K_{IC} (MPa·m$^{0.5}$) [Mean (Standard Deviation)]	G_F (N/m) [Mean (Standard Deviation)]
T = 10 °C, RH = 98%	7	5.65 (0.26)	87.45 (4.63)	161.21 (6.53)	1.03 (0.03)	148.18 (4.29)
	14	6.99 (0.15)	95.20 (0.85)	160.09 (2.72)	1.22 (0.02)	165.49 (3.99)
	28	7.66 (0.34)	125.67 (16.81)	155.98 (5.29)	1.29 (0.03)	195.44 (3.98)
	60	9.14 (0.33)	122.85 (6.24)	157.64 (3.87)	1.55 (0.04)	226.46 (6.89)
	90	10.19 (0.21)	132.31 (16.87)	153.46 (3.00)	1.66 (0.06)	259.27 (5.88)
T = 20 °C, RH = 98%	7	6.70 (0.20)	108.36 (5.04)	157.20 (3.98)	1.18 (0.02)	167.50 (6.69)
	14	8.43 (0.46)	125.40 (9.95)	153.00 (9.10)	1.37 (0.05)	192.22 (4.29)
	28	9.48 (0.24)	88.80 (14.06)	145.63 (5.52)	1.43 (0.04)	240.82 (9.25)
	60	10.53 (0.48)	131.99 (16.95)	155.18 (6.33)	1.74 (0.03)	262.67 (3.75)
	90	11.53 (0.48)	130.49 (14.88)	154.08 (5.95)	1.86 (0.08)	282.20 (9.58)
T = 40 °C, RH = 98%	7	7.46 (0.27)	123.01 (13.98)	157.68 (3.53)	1.28 (0.03)	185.92 (9.02)
	14	9.42 (0.28)	115.53 (10.41)	151.6 (5.94)	1.51 (0.05)	224.62 (2.26)
	28	10.25 (0.40)	136.50 (6.01)	155.24 (5.01)	1.70 (0.04)	257.95 (5.53)
	60	11.72 (0.54)	125.55 (9.90)	153.32 (8.12)	1.91 (0.07)	288.67 (6.76)
	90	12.57 (0.44)	115.59 (10.50)	154.12 (5.36)	2.06 (0.06)	295.30 (6.04)
T = 20 °C, RH = 60%	7	6.20 (0.28)	110.26 (14.34)	158.59 (6.77)	1.08 (0.06)	158.53 (11.43)
	14	7.05 (0.12)	126.67 (29.80)	165.51 (4.85)	1.27 (0.02)	169.69 (5.60)
	28	7.79 (0.13)	141.92 (9.76)	160.29 (1.26)	1.36 (0.01)	204.82 (11.34)
	60	9.29 (0.07)	125.47 (37.99)	156.59 (2.30)	1.56 (0.05)	240.95 (0.67)
	90	9.71 (0.13)	141.17 (20.20)	159.07 (4.91)	1.67 (0.07)	249.95 (15.52)
T = 20 °C, RH = 30%	7	5.95 (0.14)	86.95 (11.54)	150.43 (5.48)	0.98 (0.05)	153.92 (0.59)
	14	6.34 (0.25)	125.21 (18.87)	163.37 (5.10)	1.14 (0.02)	157.68 (13.01)
	28	7.30 (0.01)	106.82 (14.66)	154.08 (3.17)	1.19 (0.04)	174.26 (6.55)
	60	8.50 (0.45)	111.18 (20.77)	151.79 (9.67)	1.36 (0.06)	210.31 (12.15)
	90	8.73 (0.25)	108.65 (21.09)	157.29 (4.85)	1.48 (0.04)	212.25 (8.03)

Figure 10a–e is the load–crack opening displacement (*CMOD*) curves at 7 days, 14 days, 28 days, 60 days and 90 days, under different curing temperature and humidity conditions. Each curve in the figure is selected from the same group of three accompanying specimens. It can be seen from Figure 10 that for different curing conditions, the variation of the load–*CMOD* curves are relatively similar. At the initial stage of loading, no macroscopic cracks were observed in the ligament zone of the specimen, and the load increased linearly with the increase in the crack mouth opening displacement, indicating that the concrete mainly undergoes linear elastic deformation. When *CMOD* increases to a certain value, the load no longer increases linearly with the increase in the *CMOD*; that is, the relationship between the load and *CMOD* changes from linear to nonlinear. In the double-*K* fracture theory, proposed by Xu et al. [41], this turning point is the crack initiation point, and the

corresponding load is the crack initiation load. With the further increase in the load, the crack tip enters a stable expansion stage. With the further increase in the load, the crack tip enters a stable propagation stage. At this time, the load increases nonlinearly with the increase in the opening displacement of the crack opening, until it reaches the peak load. Then, the load continues to decrease with the further increase in *CMOD*, which means that the crack has entered the instability propagation stage, and the testing machine begins to unload.

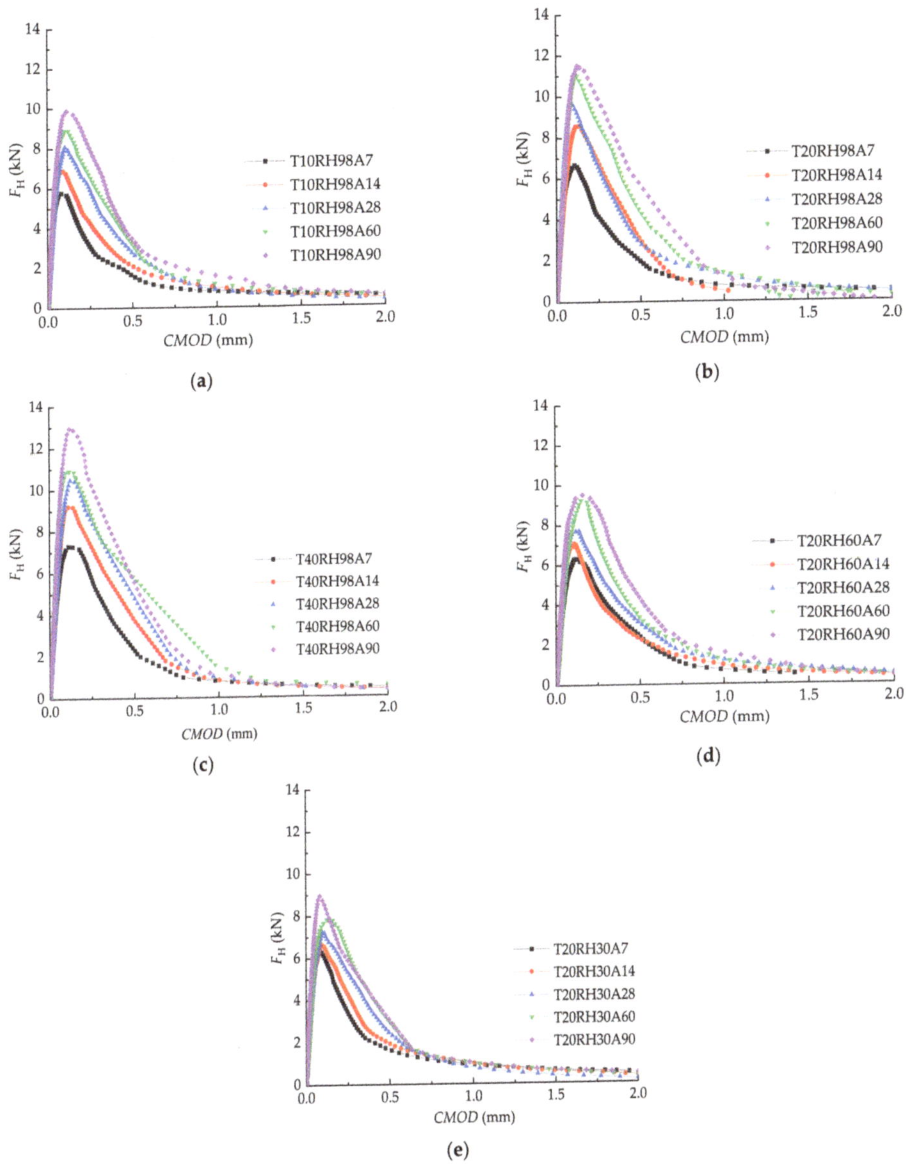

Figure 10. Load–crack mouth opening displacement curves of wedge-splitting specimens under different curing conditions: (**a**) T = 10 °C, RH = 98%; (**b**) T = 20 °C, RH = 98%; (**c**) T = 40 °C, RH = 98%; (**d**) T = 20 °C, RH = 60%; (**e**) T = 20 °C, RH = 30%.

From Figure 10a–c, it can be seen that, at a given age, the higher the curing temperature, the greater the peak load of the specimen. Taking the peak load at different curing temperatures, 90 days as an example, the curing humidity is 98% and the peak load, at a curing temperature of 40°C, is 23.36% and 9.02% larger than the peak loads at 10 °C and 20 °C, respectively. Similarly, looking at Figure 10b,d,e, it can be seen that, at a given age, the higher the curing humidity, the greater the peak load of the specimen. When the curing temperature is 20 °C and the curing humidity is 98%, the peak load is 32.07% and 18.74% larger than the peak load at 30% and 60%, respectively. In addition, it can be seen from Figure 10 that, under a given curing temperature and humidity, the initial slope and peak load of the load–*CMOD* curves both increase with the increase in age. This is mainly because the higher the age, the more sufficient the hydration degree of the cement under the same curing conditions, which increases the mechanical properties and bearing capacity of the concrete.

Figure 11 shows the variation of equivalent fracture toughness and fracture energy of low-heat cement gallery concrete, with test age, when the curing humidity is 98% and the curing temperature is 10 °C, 20 °C and 40 °C. Similar to the strength test results, with the increase in age, the fracture parameters of concrete under different curing temperatures continued to increase, and none reached their final stable value at the maximum test age of 90 days, designed in this study. Taking the fracture energy at a curing temperature of 20 °C, using Figure 11b as an example, the growth rate of fracture energy is gradually reduced, from 23.93 (N/m)/day to 3.53 (N/m)/day, 3.47 (N/m)/day, 0.68 (N/m)/day and 0.65 (N/m)/day, in the designed test age intervals of 1–7 days, 8–14 days, 15–28 days, 29–60 days and 61–90 days. This indicates that the growth rate of the fracture parameter decreases greatly with the increase in age, so that the fracture parameters will gradually reach their final stable values.

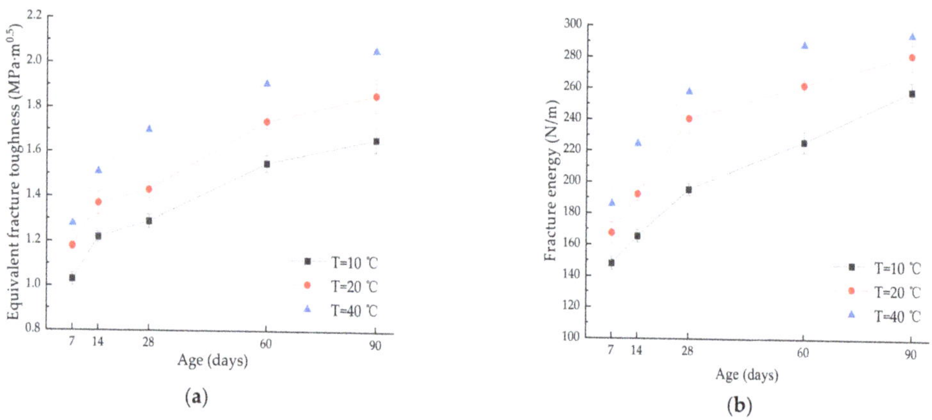

Figure 11. Fracture parameters of gallery concrete under different curing temperatures: (**a**) equivalent fracture toughness; (**b**) fracture energy.

In addition, comparing the fracture parameters of concrete under different curing temperature conditions, in Figure 11, it can be seen that the higher the curing temperature, the greater the fracture parameters of concrete at the same age. This is mainly because the higher curing temperature accelerates the rate of hydration of the cementitious material, thereby promoting the development of concrete material properties. However, with the increase in age, the growth rate of fracture parameters gradually decreases, and the higher the curing temperature, the more obvious the decline in the growth rate of fracture parameters. Therefore, the difference in the fracture parameters of the concrete, at the same age, caused by the curing temperature, gradually decreases with the increase in age. If the differences in early-stage pore characteristics and microcracks, that may be introduced by the difference in curing temperature, are neglected, as the age increases, the fracture parameters of concrete under different curing temperatures will gradually approach their final stable values.

Figure 12 shows the variation of the equivalent fracture toughness and fracture energy of gallery concrete, with age, when the curing temperature is 20 °C and the curing humidity is 30%, 60%, and 98%. It can be seen from Figure 12 that the development of the fracture parameters of gallery concrete is closely related to its curing humidity. For different curing humidity conditions, the growth trend of equivalent fracture toughness and fracture energy, with age, is basically the same. Similar to the test results of strength parameters, at the age of 7 days, since the moisture inside the concrete can already meet the hydration reaction demand of the cementitious material, the curing humidity condition has little effect on the fracture parameters of the concrete. With the increase in age, the adverse effects of insufficient curing humidity on fracture parameters gradually appear. Taking the test results of fracture energy, under different curing humidity, Figure 12b as an example, the fracture energy at the 7-day age, when the curing humidity is 30% and 60%, is reduced by 8.11% and 5.36%, respectively, compared with a humidity of 98%. At the 28-day age, this reduction rate is increased, to 27.64% and 14.95%, and the reduction rate at the 90-day age is 24.79% and 11.43%, respectively.

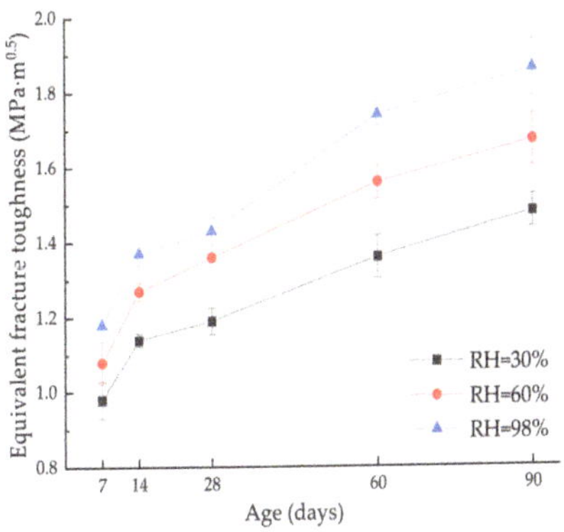

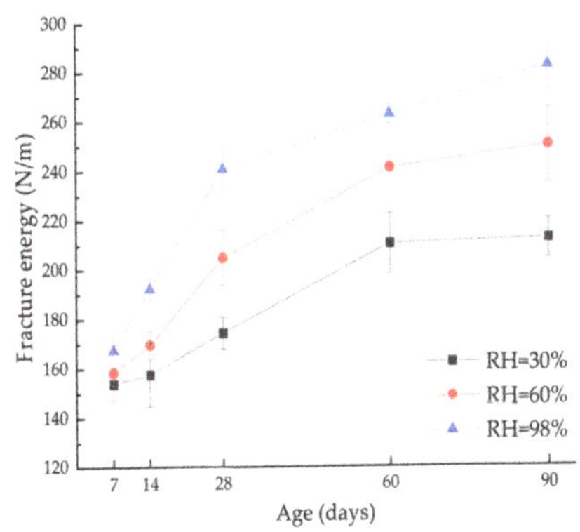

Figure 12. Fracture parameters of gallery concrete under different curing humidity: (**a**) equivalent fracture toughness; (**b**) fracture energy.

3.2. *Maturity Method Analysis*

3.2.1. Maturity Index Calculation

To carry out a unified analysis of the strength and fracture parameters of gallery concrete, under different curing temperature and humidity conditions, the temperature–humidity maturity function (Equation (11)) was adopted in this study, to calculate the equivalent age of gallery concrete at different ages, from 7 days to 90 days. The calculation results are shown in Table 6. The results show that, under the same curing humidity condition, the higher the curing temperature, the higher the equivalent age of the concrete at the same age. Furthermore, under the same curing temperature condition, when the curing humidity is insufficient, the equivalent age of the concrete of the same age is smaller.

3.2.2. Strength Parameters

Figure 13a–d shows the relationship between the test results of compressive strength, splitting-tensile strength, axial compressive strength and elastic modulus of gallery concrete, and the equivalent age, under different curing temperature and humidity conditions. The figure also shows the best fit curves and the corresponding determination coefficients. It should be noted that this section only provides the logarithmic relationship between the strength parameter and the maturity index, because the logarithmic, exponential, and hyperbolic function fitting results of equation 12 to 14 were compared in this study and it was found that the fitting result of the logarithmic function was the best. It can be seen from Figure 13 that the experimental data points are all located near the fitting curve, and the determination coefficients are all larger than 0.97, indicating that the temperature–humidity maturity method, considering curing temperature, humidity and age, and the logarithmic relationship of strength–maturity, can well describe the growth trend of the strength parameters of low-heat cement gallery concrete, under the curing temperature of 10 °C to 40 °C and humidity of 30% to 98%.

Table 6. Equivalent age of environmental chamber-cured concrete specimens.

Test Age (Days)	Equivalent Age (Days) (RH = 98%)			Equivalent Age (Days) (T = 20 °C)	
	T = 10 °C	T = 20 °C	T = 40 °C	RH = 60%	RH = 30%
7	3.48	7.00	16.85	4.72	3.00
14	6.96	14.00	33.71	9.15	5.51
28	13.92	28.00	67.42	17.79	10.13
60	29.82	60.00	144.46	37.26	20.20
90	44.73	90.00	216.70	55.40	29.46

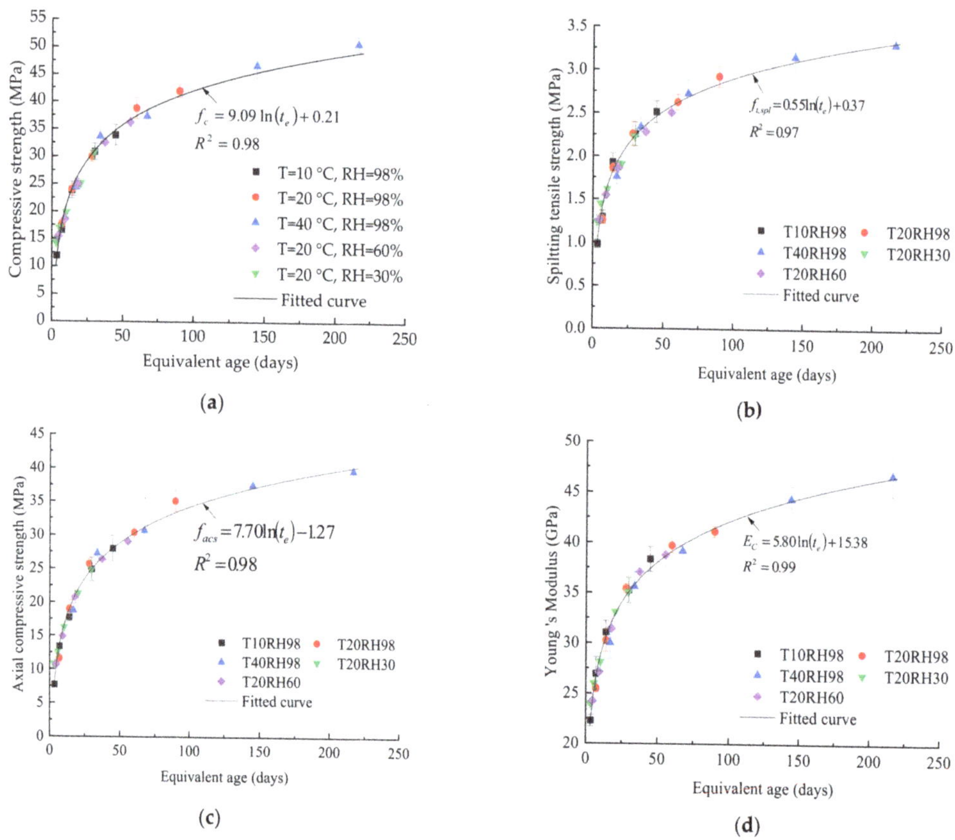

Figure 13. Relationship between strength parameters and equivalent age: (**a**) compressive strength; (**b**) splitting tensile strength; (**c**) axial compressive strength and (**d**) static compressive elastic modulus.

3.2.3. Fracture Parameters

Figure 14 shows the relationship between the equivalent fracture toughness and fracture energy of the gallery concrete, under different curing temperature and humidity conditions and the equivalent age. It can be seen from Figure 14 that the coefficients of determination, for the relationship equations between the fracture parameters and the equivalent age, are greater than 0.96, indicating that the temperature and humidity maturity method and the logarithmic function can well describe the development trend of the fracture parameters of low-heat cement gallery concrete.

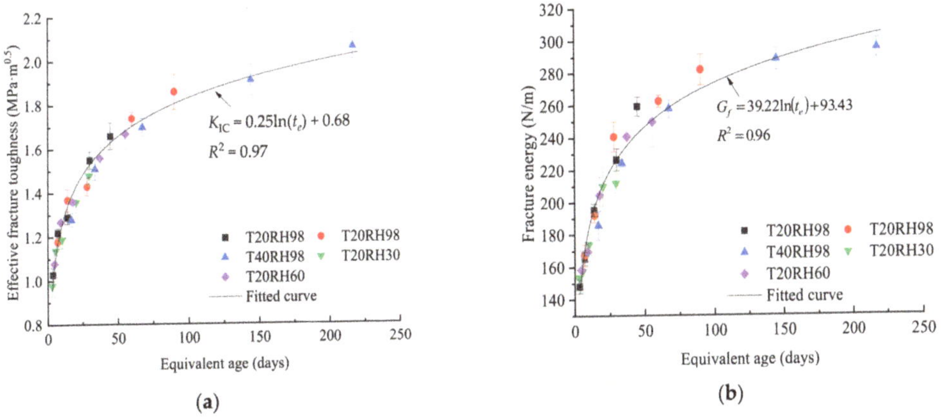

Figure 14. Relationship between fracture parameters and equivalent age: (**a**) equivalent fracture toughness; (**b**) fracture energy.

3.3. Determination of Real Strength and Fracture Parameters of Dam Gallery Concrete

3.3.1. Maturity Index Calculation

In engineering, the maturity of gallery concrete is constantly changing, due to the combined influence of environmental temperature and humidity, cement hydration heat release and water-cooling, which determine the development process of concrete strength and fracture parameters. To obtain the actual temperature and humidity history of the surface and interior concrete of the dam gallery, in this study, small weather stations were set up inside and outside the gallery to collect the internal and external environmental temperature and humidity, and thermometers were embedded between the top of the gallery and the cooling water pipe, to collect the internal temperature of the gallery concrete. Figure 15 shows the design layout and photograph of the thermometers, buried in gallery concrete. It is shown in Figure 15a that a total of six internal thermometers were embedded between the top gallery arch and the cooling water pipes, arranged perpendicular to the top of the gallery, and the distances from the top of the gallery were 2 cm, 5 cm, 10 cm, 20 cm, 35 cm and 50 cm; the serial numbers are from No. 1 to No. 6, respectively.

Figure 16 shows the actual hourly temperature history of the gallery concrete, measured by the above-mentioned, pre-embedded thermometers. It can be seen from the figure that the temperature of the gallery concretes, at different locations, all presented a trend of first increasing and then decreasing. During the first 2.0 days or 2.5 days, the concrete temperature rose rapidly, from a low temperature state to the peak temperature, around 24 °C, and then gradually decreased due to the combined effect of water cooling, hydration heat release and ambient temperature. Comparing the temperature data measured by thermometers at different positions, in the temperature rise stage, the temperature measured by thermometers No. 1 to 6 are basically the same; in the cooling stage, the temperature measured by thermometers No. 1 to 3 are similar, and the temperature measured by thermometers No. 4 to No. 6 are basically the same, which is lower than the temperature measured by No. 1 to No. 3. If it is assumed that the internal gallery concrete has sufficient moisture for hydration, the temperature maturity method can be used to calculate the equivalent age of the internal concrete, and the strength and fracture parameters can be predicted, based on the equivalent age. Evidently, the equivalent age and mechanical properties of the concrete close to the top of the gallery are greater than those near the cooling water pipe.

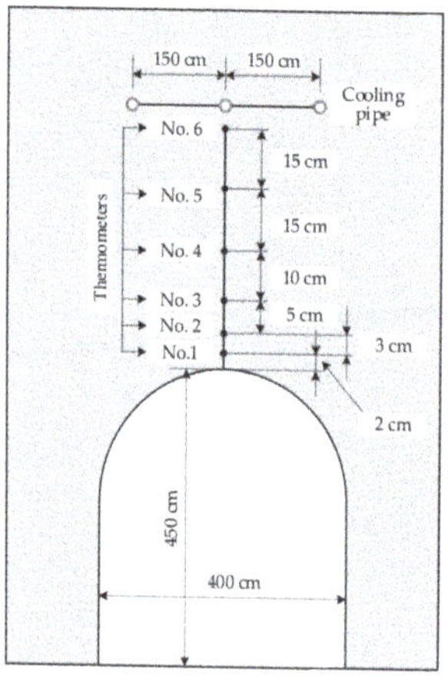

(a)

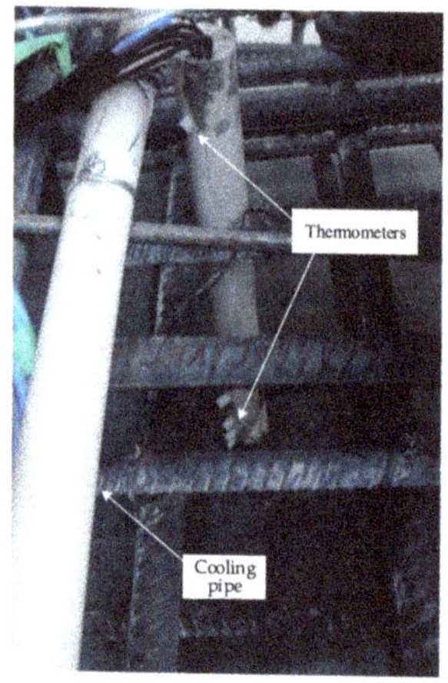

(b)

Figure 15. Arrangement of thermometers in gallery concrete: (**a**) design layout drawing; (**b**) actual buried thermometers.

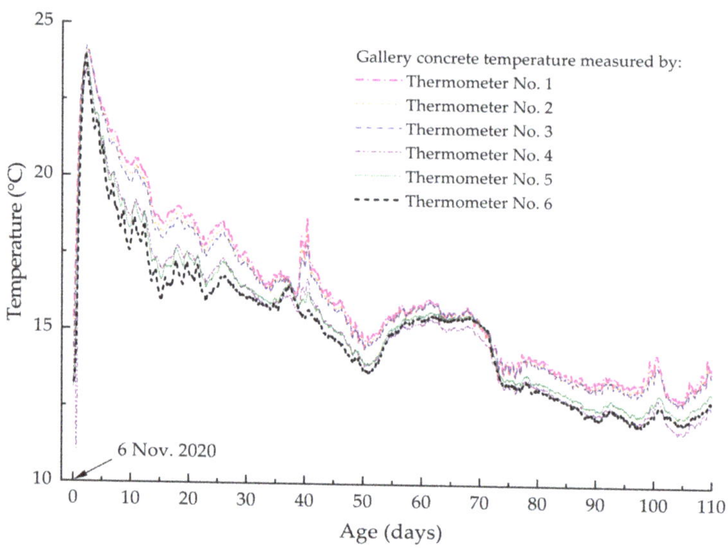

Figure 16. Hourly temperature history of the gallery concretes at different positions measured by the pre-embedded thermometer.

The concrete within 2 cm of the arch surface of the gallery is a shallow surface concrete, and the development of its material mechanical properties is greatly affected by the temperature and humidity of the environment in which it is located. During the construction process, it is generally necessary to install wind-proof and heat-insulating doors at the entrance of the gallery, to reduce the possible adverse effects of external wind speed, temperature and humidity changes on the concrete materials. To quantitatively analyze the influence of temperature and humidity changes, inside and outside the gallery, on the shallow surface concrete of the gallery, this study set up small weather stations, inside and outside the gallery, to record hourly temperature and humidity data. Figure 17 shows the temperature and humidity data, outside the gallery and inside the gallery, respectively. Looking at Figure 17a,b, in the 110 days starting from 6 November 2020, the fluctuation range of the temperature outside the gallery is significantly greater than the temperature fluctuation range inside the gallery. This shows that the temperature in the gallery is relatively stable, which can help in avoiding the possible cracking of the concrete, caused by the sudden temperature drop. In addition, from Figure 17c,d, the relative humidity inside the gallery is significantly greater than the humidity outside the gallery, and the relative humidity reaches 100% during certain periods. This is mainly due to the fact that, during the construction process, the construction unit used artificial humidification to increase the relative humidity in the gallery during certain periods. From the above analysis, it can be seen that the temperature and humidity values, and the fluctuation range inside and outside the gallery, are quite different. It is necessary to calculate the equivalent age of the gallery concrete by using the temperature and humidity history inside and outside the gallery, respectively, so as to determine the most unfavorable conditions for the development of the mechanical properties of shallow surface concrete of the gallery.

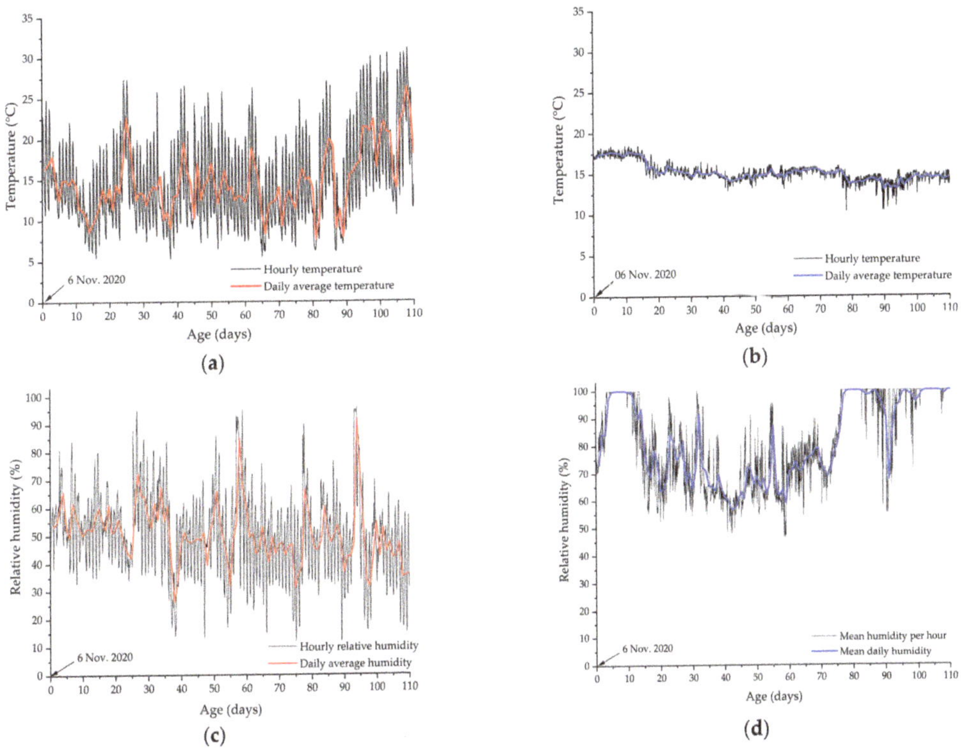

Figure 17. The variation of temperature and humidity inside and outside the gallery: (**a**) hourly and daily average temperature outside the gallery; (**b**) hourly and daily average temperature inside the gallery; (**c**) hourly and daily average humidity outside the gallery and (**d**) hourly and daily average humidity inside the gallery.

Table 7 shows the calculation results of equivalent ages for the shallow surface, surface and deep concrete of the top arch of the gallery. Among them, the equivalent age of the surface and deep concrete is calculated based on the temperature history, measured by the embedded thermometers, according to Equation (10). The equivalent age of the shallow surface concrete is calculated based on the hourly temperature and humidity data, recorded by small weather stations, located outside and inside the gallery, according to Equation (11). It can be seen from the calculation results in Table 7 that when the actual ages are the same, the equivalent age, calculated based on the temperature data of thermometers No. 1 to No. 3, is almost the same, and greater than the equivalent age calculated based on No. 4 to No. 6. In addition, the equivalent age of the shallow surface concrete is significantly lower than the equivalent age of the surface and deep concrete, and the equivalent age, considering the influence of temperature and humidity, is smaller than the equivalent age considering only the influence of temperature.

Table 7. Equivalent age calculated by hourly temperature and humidity history.

Actual Age (Days)	Equivalent Age (Days) Calculated by Temperature History							Equivalent Age (Days) Calculated by Temperature and Humidity History		
	Measured by Thermometer						Measured by Weather Station Installed		Measured by Weather Station Installed	
	No. 1	No. 2	No. 3	No. 4	No. 5	No. 6	Outside the Gallery	Inside the Gallery	Outside the Gallery	Inside the Gallery
3	3.25	3.23	3.22	2.99	3.08	3.01	2.53	2.58	2.16	2.42
7	7.69	7.64	7.62	7.20	7.28	7.12	5.38	6.11	4.61	5.94
14	14.79	14.65	14.56	13.80	13.80	13.44	9.68	12.27	8.24	12.04
28	27.72	27.43	27.24	25.85	25.75	25.04	19.63	23.28	16.69	22.11
60	53.60	53.08	52.81	50.36	50.38	49.28	41.53	46.91	35.07	43.21
90	75.28	74.65	74.40	71.02	71.43	70.07	61.48	68.89	51.51	64.15

3.3.2. Determination of Strength and Fracture Parameters

After the equivalent ages, shown in Table 7, are calculated, the strength and fracture parameter prediction equations, presented in Figures 13 and 14, can be substituted to obtain the compressive strength, splitting tensile strength, axial compressive strength, elastic modulus, equivalent fracture toughness and fracture energy of the gallery concrete, respectively. The calculation results are summarized in Table 8. In traditional dam structure analysis, the strength and fracture parameters of concrete, at a specific age, under standard curing conditions, are generally taken as the basic material parameters. Therefore, in order to compare the difference between the proposed method, presented in this study, and the traditional method, Table 8 shows the strength and fracture parameters of gallery concrete, under standard curing conditions, from 3 to 90 days. For the concrete within the range of 2 cm to 50 cm, for the arch of the gallery, in order to consider the most unfavorable working conditions, only the minimum equivalent age, calculated by the temperature data measured by thermometer No. 6, and its corresponding predicted strength and fracture parameters are presented. Similarly, for the shallow surface concrete, Table 8 gives the equivalent age and the prediction results of strength and fracture parameters, considering the effects of temperature and humidity. Since the temperature and humidity data for the equivalent age calculation in this study are all collected from the actual dam gallery structure, it can be considered that the predicted strength and fracture parameters calculated, based on these equivalent ages, are the real strength and fracture parameters of the gallery concrete. While the strength and fracture parameters of specific age concrete, under the standard curing conditions used in the traditional analysis process, are theoretical parameters, the deviation between the two is considered as the error between the theoretical parameter and the real parameter.

Table 8. Comparison between concrete strength and fracture parameters given by prediction model and traditional method.

Parameters	Standard Curing Conditions		Temperature Measured by Thermometer No. 6			Temperature and Humidity Inside the Gallery			Temperature and Humidity Outside the Gallery		
	Actual Age (Days)	Theoretical Values	t_e (Days)	Model Predictions	Error of Theoretical Value (%)	t_e (Days)	Model Predictions	Error of Theoretical Value (%)	t_e (Days)	Model Predictions	Error of Theoretical Value (%)
f_c (MPa)	3	10.20	3.01	10.23	−0.29%	2.42	8.22	24.09%	2.16	7.19	41.86%
	7	17.90	7.12	18.06	−0.89%	5.94	16.41	9.08%	4.61	14.11	26.86%
	14	24.20	13.44	23.83	1.55%	12.04	22.83	6.00%	8.24	19.38	24.87%
	28	30.50	25.04	29.49	3.42%	22.11	28.35	7.58%	16.69	25.80	18.22%
	60	37.43	49.28	35.64	5.02%	43.21	34.44	8.68%	35.07	32.55	14.99%
	90	41.11	70.07	38.84	5.84%	64.15	38.04	8.07%	51.51	36.04	14.07%
f_t (MPa)	3	0.97	3.01	0.98	−1.02%	2.42	0.85	14.12%	2.16	0.79	22.78%
	7	1.44	7.12	1.45	−0.69%	5.94	1.35	6.67%	4.61	1.21	19.01%
	14	1.82	13.44	1.80	1.11%	12.04	1.74	4.60%	8.24	1.53	18.95%
	28	2.20	25.04	2.14	2.80%	22.11	2.07	6.28%	16.69	1.92	14.58%
	60	2.62	49.28	2.51	4.38%	43.21	2.44	7.38%	35.07	2.33	12.45%
	90	2.84	70.07	2.71	4.80%	64.15	2.66	6.77%	51.51	2.54	11.81%
f_{acs} (MPa)	3	7.19	3.01	7.22	−0.42%	2.42	5.52	30.25%	2.16	4.65	54.62%
	7	13.71	7.12	13.85	−1.01%	5.94	12.46	10.03%	4.61	10.50	30.57%
	14	19.05	13.44	18.74	1.65%	12.04	17.89	6.48%	8.24	14.97	27.25%
	28	24.39	25.04	23.53	3.65%	22.11	22.57	8.06%	16.69	20.40	19.56%
	60	30.26	49.28	28.74	5.29%	43.21	27.73	9.12%	35.07	26.12	15.85%
	90	33.38	70.07	31.45	6.14%	64.15	30.77	8.48%	51.51	29.08	14.79%
E_c (GPa)	3	21.75	3.01	21.77	−0.09%	2.42	20.49	6.15%	2.16	19.84	9.63%
	7	26.67	7.12	26.77	−0.37%	5.94	25.72	3.69%	4.61	24.25	9.98%
	14	30.69	13.44	30.45	0.79%	12.04	29.81	2.95%	8.24	27.61	11.16%
	28	34.71	25.04	34.06	1.91%	22.11	33.34	4.11%	16.69	31.71	9.46%
	60	39.13	49.28	37.99	3.00%	43.21	37.22	5.13%	35.07	36.01	8.66%
	90	41.48	70.07	40.03	3.62%	64.15	39.51	4.99%	51.51	38.24	8.47%
K_{IC} (MPa·m$^{0.5}$)	3	0.95	3.01	0.96	−1.04%	2.42	0.90	5.56%	2.16	0.87	9.20%
	7	1.17	7.12	1.17	0.00%	5.94	1.13	3.54%	4.61	1.06	10.38%
	14	1.34	13.44	1.33	0.75%	12.04	1.30	3.08%	8.24	1.21	10.74%
	28	1.51	25.04	1.49	1.34%	22.11	1.45	4.14%	16.69	1.38	9.42%
	60	1.70	49.28	1.65	3.03%	43.21	1.62	4.94%	35.07	1.57	8.28%
	90	1.80	70.07	1.74	3.45%	64.15	1.72	4.65%	51.51	1.67	7.78%

Table 8. *Cont.*

Parameters	Standard Curing Conditions		Temperature Measured by Thermometer No. 6			Temperature and Humidity Inside the Gallery			Temperature and Humidity Outside the Gallery		
	Actual Age (Days)	Theoretical Values	t_e (Days)	Model Predictions	Error of Theoretical Value (%)	t_e (Days)	Model Predictions	Error of Theoretical Value (%)	t_e (Days)	Model Predictions	Error of Theoretical Value (%)
G_F (N/m)	3	136.52	3.01	136.66	−0.10%	2.42	128.01	6.65%	2.16	123.57	10.48%
	7	169.75	7.12	170.44	−0.40%	5.94	163.34	3.92%	4.61	153.38	10.67%
	14	196.93	13.44	195.34	0.81%	12.04	191.01	3.10%	8.24	176.16	11.79%
	28	224.12	25.04	219.74	1.99%	22.11	214.86	4.31%	16.69	203.82	9.96%
	60	254.01	49.28	246.29	3.13%	43.21	241.14	5.34%	35.07	232.95	9.04%
	90	269.91	70.07	260.10	3.77%	64.15	256.63	5.17%	51.51	248.03	8.82%

From the comparison results in Table 8, it can be seen that for the deep concrete of the gallery, its real strength and fracture performance were relatively close to the theoretical values used by traditional methods. This is mainly because the calculation result of the equivalent age of deep concrete is close to the actual age of concrete. Taking compressive strength as an example, when the actual age is 7, 28 and 90 days, the equivalent age calculated by the temperature history of thermometer No. 6 is 7.12 days, 25.04 days and 70.07 days, respectively. The deviations between the theoretical compressive strength and the real compressive strength, predicted by the proposed method, are all less than 10%, at −0.89%, 3.42%, and 5.84%, respectively. For the shallow surface concrete of the gallery, when the measured temperature and humidity history inside the gallery is used to calculate the equivalent age of the concrete, the actual ages of 3 days, 7 days, 28 days, and 90 days, correspond to equivalent ages of 2.42 days, 5.94 days, 22.11 days, and 64.15 days, respectively. This will result in a deviation of 24.09% between the theoretical compressive strength and the real value at 3 days, and the maximum deviations at 7, 28, and 90 days are reduced to 9.08%, 7.58%, and 8.07%, respectively. This means that when the actual age of concrete is used to determine the theoretical strength or fracture parameters of the shallow concrete of the gallery, it may lead to an overestimation of the theoretical value, resulting in insufficient structural design safety reserves. In addition, if the gallery entrance is not completely closed during the construction process, the temperature and humidity outside the gallery may affect the development of the concrete strength and fracture parameters. Table 8 also shows the equivalent ages calculated by using the external temperature and humidity data and the corresponding predicted values of the concrete mechanical parameters. Consequently, because the external temperature and humidity is lower than the internal temperature and humidity of the gallery, when the actual age is the same, the equivalent age calculated from the external temperature and humidity is lower, which leads to a greater difference between the theoretical strength and fracture parameter and their real values. For example, at 3 days, 7 days, 28 days, and 90 days, the deviations of compressive strength are 41.86%, 26.86%, 18.22% and 14.07% respectively. In actual engineering, when the strength and fracture parameters of the gallery concrete, determined by the method proposed in this article, are relatively small compared with the strength and fracture properties of the gallery concrete obtained by the traditional method, special attention should be paid to whether the strength and fracture parameters of the dam gallery concrete meet the requirements of the bearing capacity.

4. Conclusions

A determination method for the strength and fracture parameters of the dam gallery concrete, considering the environmental temperature and humidity, is proposed in this study. Based on the strength and fracture test results of gallery concrete, at different ages and under various curing conditions, the evolution equations of the mechanical parameters of gallery concrete are established, and then combined with the measured temperature and humidity data of the dam gallery structure, meaning the real strength and fracture parameters of gallery concrete at any age are determined. The main conclusions obtained in this study are as follows:

(1) In the range of 10 °C to 40 °C, the higher the curing temperature, the greater the strength and fracture parameters of the gallery concrete, and the faster the growth rate of the strength and fracture parameters at the same age. As the age increases, the strength and fracture parameters of concrete, under different curing temperatures, will gradually approach their final stable values.

(2) The curing humidity has a negligible effect on the strength and fracture parameters of gallery concrete at early ages, but as the age increases, the adverse effects of insufficient curing humidity on the strength and fracture parameters gradually appear, and the difference between the material parameters under different humidity conditions demonstrates a law that first expands and then reduces. The low-humidity curing condition may result in a reduction in the strength growth rate, up to 58.72%, and

strength up to 33.81%; hence, its effect on the mechanical parameters of gallery concrete cannot be neglected.

(3) The coefficients of determination for the relationship equations between strength and fracture parameters and equivalent ages are all greater than 0.96, indicating the temperature–humidity maturity function and logarithmic relationship between strength and maturity index can be used to quantitatively describe the growth law of strength and fracture parameters of low-heat cement gallery concrete, under conditions of curing temperatures of 10 °C to 40 °C, and relative humidity of 30% to 98%.
(4) The method of embedding thermometers in the gallery concrete and setting up weather stations inside and outside the gallery can determine the most unfavorable environmental conditions for the development of the gallery concrete's strength and fracture parameters, which can help to correct a maximum deviation of 54.62% on mechanical parameters, calculated using actual ages.
(5) A case study based on the real temperature and humidity data of a dam gallery shows that the development of the strength and fracture parameters in the shallow surface concrete, within 2 cm of the gallery top arch, is greatly affected by the ambient temperature and humidity.
(6) In engineering, when the strength and fracture parameters of the gallery concrete, determined by the method proposed in this study, are much lower compared with the values obtained by the traditional method, special attention should be paid to whether the strength and fracture parameters of the dam gallery concrete meet the requirements of bearing capacity.

Author Contributions: Conceptualization, X.G. and Y.H.; formal analysis, X.G., Z.L. and C.L.; funding acquisition, Y.H. and Q.L.; investigation, X.G. and Z.L.; methodology, X.G., Y.T., N.Y., C.L. and Y.L.; supervision, Y.H. and Q.L.; validation, X.G. and Y.T.; writing—original draft, X.G. and Q.L.; writing—review and editing, X.G., Z.L., J.Z. and K.W. All authors have read and agreed to the published version of the manuscript.

Funding: This research was funded by the National Natural Science Foundation of China (Nos. 52009123 and 51839007), Zhejiang Provincial Natural Science Foundation of China (No. LY22E090004), and Research Projects of China Three Gorges Corporation (No. BHT/0809).

Institutional Review Board Statement: Not applicable.

Informed Consent Statement: Not applicable.

Data Availability Statement: The data presented in this study are available on request from the corresponding author.

Conflicts of Interest: The authors declare no conflict of interest.

References

1. Li, Q.B.; Lin, P. Demonstration on intelligent dam. *J. Hydroelectr. Eng.* **2014**, *33*, 139–146. Available online: http://www.slfdxb.cn/CN/Y2014/V33/I1/139 (accessed on 4 January 2022). (In Chinese)
2. Zhong, D.H.; Shi, M.N.; Gui, B.; Wang, J.J.; Guan, T. Research progress of the intelligent construction of dams. *J. Hydraulic. Eng.* **2019**, *50*, 38–52+61. (In Chinese) [CrossRef]
3. Tan, Y.S.; Fan, Q.X.; Wang, Z.L.; Chen, W.F.; Guo, Z.G.; Lin, E.D.; Lin, P.; Zhou, T.G.; Zhou, M.X.; Liu, C.F.; et al. Intelligent construction methods for the Baihetan super high arch dam. *J. Tsinghua Univ. (Sci. Technol.)* **2021**, *61*, 694–704. (In Chinese) [CrossRef]
4. Lin, P.; Wei, P.C.; Wang, W.H.; Huang, H.F. Cracking Risk and Overall Stability Analysis of Xulong High Arch Dam: A Case Study. *Appl. Sci.* **2018**, *8*, 2555. [CrossRef]
5. Lin, P.; Guan, J.F.; Peng, H.Y.; Shi, J. Horizontal cracking and crack repair analysis of a super high arch dam based on fracture toughness. *Eng. Fail. Anal.* **2019**, *97*, 72–90. [CrossRef]
6. Lin, P.; Zhou, W.Y.; Liu, H.Y. Experimental study on cracking: Reinforcement and overall stability of the Xiaowan super-high arch dam. *Rock. Mech. Rock. Eng.* **2015**, *48*, 819–841. [CrossRef]
7. Wieland, M.; Kirchen, G.F. Long-term dam safety monitoring of Punt dal Gall arch dam in Switzerland. *Front. Struct. Civ. Eng.* **2012**, *6*, 76–83. [CrossRef]
8. Malla, S.; Wieland, M. Analysis of an arch-gravity dam with a horizontal crack. *Comput. Struct.* **1999**, *72*, 267–278. [CrossRef]

9. Ma, K.; Zhuang, D.Y.; Tang, J.A.; Jin, F.; Tang, S.B. Study on formation causes of gallery cracks in Dagangshan high arch dam based on microseismic monitoring. *Chin. J. Rock. Mech. Eng.* **2018**, *37*, 1608–1617. [CrossRef]
10. Mcdonald, J.E.; Curtis, N.F. *Repair and Rehabilitation of Dams: Case Studies*; U.S. Army Engineer Research and Development Center: Vicksburg, MS, USA, 1999.
11. Froehlich, D.; Kapoor, V.; Flint, E. India's large dams are showing their age. Part 1: Concrete and masonry dams-examples from drip. In Proceedings of the Third National Dam Safety Conference, Roorkee, India, 18–19 February 2017.
12. Mirzabozorg, H.; Ghaemian, M.; Roohezamin, A. The reason of cracking in bottom gallery of SefidRud Buttress Dam and earthquake and post earthquake performance. *Struct. Monit. Maint.* **2019**, *6*, 103–124. [CrossRef]
13. Wang, W.M.; Ding, J.X.; Wang, G.J.; Zou, L.C.; Chen, S.H. Stability analysis of the temperature cracks in Xiaowan arch dam. *Sci. China. Technol. Sc.* **2011**, *54*, 547–555. [CrossRef]
14. Lv, Y.N.; Li, T.; Lin, P.B.; Huang, J.W. Cause analysis and effect of gallery cracks on dam block 19 at Shaxikou dam. *Dam Saf.* **2007**, *4*, 36–38. Available online: http://magtech.dam.com.cn/CN/Y2007/V0/I4/36 (accessed on 4 January 2022). (In Chinese).
15. Lin, P.; Ning, Z.Y.; Shi, J.; Liu, C.; Chen, W.F.; Tan, Y.S. Study on the gallery structure cracking mechanisms and cracking control in dam construction site. *Eng. Fail. Anal.* **2021**, *121*, 105–135. [CrossRef]
16. Mi, Z.X.; Li, Q.B.; Hu, Y.; Liu, C.F.; Qiao, Y. Fracture Properties of Concrete in Dry Environments with Different Curing Temperatures. *Appl. Sci.* **2020**, *10*, 4734. [CrossRef]
17. Viviani, M.; Lanzoni, L.; Savino, V.; Tarantino, A.M. An Auto-Calibrating Semi-Adiabatic Calorimetric Methodology for Strength Prediction and Quality Control of Ordinary and Ultra-High-Performance Concretes. *Materials* **2022**, *15*, 96. [CrossRef] [PubMed]
18. Mi, Z.X.; Hu, Y.; Li, Q.B.; An, Z.Z. Effect of curing humidity on the fracture properties of concrete. *Constr. Build. Mater.* **2018**, *169*, 403–413. [CrossRef]
19. Brzozowski, P.; Strzałkowski, J.; Rychtowski, P.; Wróbel, R.; Tryba, B.; Horszczaruk, E. Effect of Nano-SiO2 on the microstructure and mechanical properties of concrete under high temperature conditions. *Materials* **2022**, *15*, 166. [CrossRef]
20. Mahmood, W.; Khan, A.-u.-R.; Ayub, T. Carbonation Resistance in Ordinary Portland Cement Concrete with and without Recycled Coarse Aggregate in Natural and Simulated Environment. *Sustainability* **2022**, *14*, 437. [CrossRef]
21. Fan, Q.X.; Duan, Y.H.; Wang, Y.Z.; Wang, X.H.; Yang, S.M.; Kang, X.S. Intelligent closed-loop control of concrete moisture levels. *J. Tsinghua Univ. (Sci. Tech.)* **2021**, *61*, 671–680. (In Chinese) [CrossRef]
22. Gao, X.F.; Hu, Y.; Yang, N.; Wu, K.; Zhong, Y.H.; Tan, Y.S.; Li, Q.B. Method for determining real strength of full-graded low-heat cement concrete. *J. Hydroelectr. Eng.* **2021**, *40*, 91–104. (In Chinese) [CrossRef]
23. Kim, J.; Han, S.H.; Song, Y.C. Effect of temperature and aging on the mechanical properties of concrete. Part 1: Experimental results. *Cem. Concr. Res.* **2002**, *32*, 1087–1094. [CrossRef]
24. Castellano, C.C.; Bonavetti, V.L.; Donza, H.A.; Irassar., E.F. The effect of w/b and temperature on the hydration and strength of blastfurnace slag cements. *Constr. Build. Mater.* **2016**, *111*, 679–688. [CrossRef]
25. Lee, C.; Lee, S.; Nguyen, N. Modeling of compressive strength development of high-early-strength-concrete at different curing temperatures. *Int. J. Concr. Struct. Mater.* **2016**, *10*, 205–219. [CrossRef]
26. Shoukry, S.N.; William, G.W.; Downie, B.; Riad, M.Y. Effect of moisture and temperature on the mechanical properties of concrete. *Constr. Build. Mater.* **2011**, *25*, 688–696. [CrossRef]
27. Li, Q.B.; Guan, J.F.; Wu, Z.M.; Dong, W.; Zhou, S.W. Equivalent maturity for ambient temperature effect on fracture parameters of site-casting dam concrete. *Constr. Build. Mater.* **2016**, *120*, 293–308. [CrossRef]
28. Yu, B.J.; Ansari, F. Method and theory for nondestructive determination of fracture energy in concrete structures. *Struct. J.* **1996**, *93*, 602–613. [CrossRef]
29. Mi, Z.X.; Hu, Y.; Li, Q.B.; Zhu, H. Elevated temperature inversion phenomenon in fracture properties of concrete and its application to maturity model. *Eng. Fract. Mech.* **2018**, *199*, 294–307. [CrossRef]
30. Mi, Z.X.; Hu, Y.; Li, Q.B.; Gao, X.F.; Yin, T. Maturity model for fracture properties of concrete considering coupling effect of curing temperature and humidity. *Constr. Build. Mater.* **2019**, *196*, 1–13. [CrossRef]
31. Powers, D.C. A discussion of cement hydration in relation to the curing of concrete. *Highw. Res. Board Proc.* **1948**, *27*, 178–188.
32. Liao, W.C.; Lee, B.J.; Kang, C.W. A humidity-adjusted maturity function for the early age strength prediction of concrete. *Cem. Concr. Comp.* **2008**, *30*, 515–523. [CrossRef]
33. Liang, J.L.; Wang, H.; Zhang, Q. Effect of curing temperature and humidity conditions on flexural tensile strength of pavement cement concrete. *J. Highw. Transp. Res. Dev.* **2011**, *28*, 32–38. (In Chinese) [CrossRef]
34. Cebeci, O.Z. Strength of concrete in warm and dry environment. *Mater. Struct.* **1987**, *20*, 270–272. [CrossRef]
35. Saengsoy, W.; Nawa, T.; Termkhajornkit, P. Influence of relative humidity on compressive strength of fly ash cement paste. *J. Struct. Constr. Eng.* **2008**, *73*, 1433–1441. [CrossRef]
36. Un, H.; Baradan, B. The effect of curing temperature and relative humidity on the strength development on the Portland cement mortar. *Sci. Res. Essays* **2011**, *6*, 2504–2511. Available online: https://academicjournals.org/journal/SRE/article-abstract/DE9732F23029 (accessed on 4 January 2022).
37. Bazant, Z.P.; Prat, P.C. Effect of temperature and humidity on the fracture energy of concrete. *Aci. Mater. J.* **1988**, *85*, 262–271. [CrossRef]
38. Lau, D.; Buyukozturk, O. Fracture characterization of concrete/epoxy interface affected by moisture. *Mech. Mater.* **2010**, *42*, 1031–1042. [CrossRef]

39. *DL/T 5055-2007*; Technical Specification of Fly Ash for Use in Hydraulic Concrete. National Development and Reform Commission: Beijing, China, 2007. (In Chinese)
40. *DL/T 5150-2017*; Test Code for Hydraulic Concrete. National Energy Administration: Beijing, China, 2017. (In Chinese)
41. Xu, S.L.; Reinhardt, H.W. Determination of Double-K criterion for crack propagation in quasi-brittle fracture. Part III: Compact tension specimens and wedge splitting specimens. *Int. J. Fract.* **1999**, *98*, 179–193. [CrossRef]
42. Nurse, R.W. Steam curing of concrete. *Mag. Concr. Res.* **1949**, *1*, 79–88. [CrossRef]
43. Saul, A.G.A. Principles underlying the steam curing of concrete at atmospheric pressure. *Mag. Concr. Res.* **1951**, *2*, 127–140. [CrossRef]
44. Rastrup, E. Heat of hydration in concrete. *Mag. Concr. Res.* **1954**, *6*, 79–92. [CrossRef]
45. Freiesleben Hansen, P.; Pedersen, E.J. Maturity computer for controlling curing and hardening of concrete. *Nord. Betong.* **1977**, *1*, 21–25.
46. Malhotra, V.M.; Carino, N.J. *Handbook on Nondestructive Testing of Concrete*; CRC Press: Boca Raton, FL, USA, 1991.
47. *ASTM C1074-11*; Standard Practice for Estimating Concrete Strength by the Maturity Method. ASTM International: West Conshohocken, PA, USA, 2011.
48. Plowman, J.M. Maturity and the strength of concrete. *Mag. Concr. Res.* **1956**, *8*, 13–22. [CrossRef]
49. Utepov, Y.; Aniskin, A.; Tulebekova, A.; Aldungarova, A.; Zharassov, S.; Sarsembayeva, A. complex maturity method for estimating the concrete strength based on curing temperature, ambient temperature and relative humidity. *Appl. Sci.* **2021**, *11*, 7712. [CrossRef]
50. Freiesleben Hansen, P.; Pedersen, E.J. Curing of concrete structures. *CEB Inf. Bull.* **1985**, 166.

MDPI
St. Alban-Anlage 66
4052 Basel
Switzerland
www.mdpi.com

Buildings Editorial Office
E-mail: buildings@mdpi.com
www.mdpi.com/journal/buildings

www.ingramcontent.com/pod-product-compliance
Lightning Source LLC
Chambersburg PA
CBHW040759250926
51850CB00007BA/1034
* 9 7 8 3 7 2 5 8 0 7 1 4 7 *